Cumulative Effects in Wildlife Management

Impact Mitigation

Cumulative Effects in Wildlife Management

Impact Mitigation

Edited by
Paul R. Krausman
Lisa K. Harris

CRC Press
Taylor & Francis Group
Boca Raton London New York

CRC Press is an imprint of the
Taylor & Francis Group, an **informa** business

CRC Press
Taylor & Francis Group
6000 Broken Sound Parkway NW, Suite 300
Boca Raton, FL 33487-2742

First issued in paperback 2019

ISBN-13: 978-1-4398-0916-7 (hbk)
ISBN-13: 978-0-367-38301-5 (pbk)

Library of Congress Cataloging-in-Publication Data

Cumulative effects in wildlife management : impact mitigation / editors: Paul R. Krausman and Lisa K. Harris. -- 1st ed.
p. cm.
Includes bibliographical references and index.
ISBN 978-1-4398-0916-7 (alk. paper)
1. Wildlife conservation. 2. Wildlife management. 3. Animals--Effect of human beings on. 4. Environmental impact analysis. I. Krausman, Paul R., 1946- II. Harris, Lisa K.

QL82.C86 2011
333.95'416--dc22 2010038462

Visit the Taylor & Francis Web site at
http://www.taylorandfrancis.com

and the CRC Press Web site at
http://www.crcpress.com

Contents

Preface

Cumulative effects (i.e., the influence on the environment resulting from activities when added to other past, present, and foreseeable future actions) and their influence on fish and wildlife populations have been discussed and incorporated into law and public policy for nearly three decades. Cumulative effects can have serious consequences on fish and wildlife populations, and they should be addressed in land management plans that influence natural resources. Unfortunately, the only attention they often receive is a check-off during an effects analysis indicating they have been considered but without much serious thought or mitigation for possible future altered landscapes. Furthermore, practitioners often address single issues only; after all, how can anyone be responsible for what others are doing, especially when their plans are not available to the public? It is an easy defense to make and much easier to simply address single issues without considering how they will fit into the broader picture. Cumulative effects have been given lip service only, in part, because there is no set guideline to follow in addressing them and it has been difficult to predict what will happen in the future. However, if the only actions that are addressed in relation to environmental influences are those that have direct or indirect effects, information about the most important influence (i.e., cumulative effects) will not be considered. Good examples are river sedimentation as a result of housing developments or forestry practices, and polar ice melting as a result of climate change. In both cases if only direct and indirect effects are considered, serious damage to the landscape would not be recognized initially. Cumulative effects are important and have to be considered seriously if society is to obtain a complete view of how anthropogenic influences affect natural resources.

Over 100 years ago, there were those who recognized that big game would not survive without serious changes to the way society viewed them and their use. The result was the Great American Experiment from which the North American Model of Conservation has evolved and become the envy of countries and conservation organizations worldwide. That period was marked with declining populations, populations that rested on the verge of extinction before our forefather's very eyes. We would argue that over 100 years later we are in the same situation with wildlife habitat. Management has restored big game populations across North America but as the human population increases, wildlife habitat is decreasing before our eyes, in quality and quantity. A housing development here, a shopping mall there, a few more trees cut here, another road put in there, another hectare plowed or grazed, another oil or gas well dug, with a scattering of resorts, ski lodges, off-highway vehicle trails, exurban developments, power lines, airports, and other associated "necessities" of successful societies. At the same time, each of these cuts into the habitat available for wildlife. Unless cumulative effects are recognized and incorporated at the beginning of project development, we will continue to see wildlife habitat disappear at unprecedented rates.

During the 14th Annual Meeting of The Wildlife Society in Tucson, Arizona, in 2007, Dr. Lisa K. Harris and Bruce Pavlick of Harris Environmental Group organized a symposium entitled "Cumulative Effects: Implications and Analysis on Sensitive Resources." Only six papers were presented at the time, and many dealt with endangered species. We realized that there was clearly much more that needed to be presented to push the importance of how and why cumulative effects should be addressed in wildlife management and conservation. Starting with the symposium as the genesis of this book, we expanded the depth and breadth of the topic. In particular, we wanted to bring the importance of cumulative effects to the forefront for managers and practitioners that deal with wildlife, their habitats, and changing landscapes. As humans continue to encroach on wildlands and wildlife habitat, we need to be aware of the actions of our particular projects and those of our neighbors on federal and private lands. Without a conscious knowledge of what is happening around us, we will not be able to incorporate an effective land ethic and natural resources will lose.

The book is divided into two parts. Section 1, Understanding Cumulative Effects, and Section 2, Case Studies. The chapters in Section 1 outline the differences between direct, indirect, and cumulative effects, and address the confusion that can be created by not considering them; the legal aspects of cumulative effects; and how cumulative effects are addressed in Canada. We initially planned a chapter to address cumulative effects in Europe and other countries but there was little to draw upon and many of the countries that did emphasize cumulative effects (i.e., Sweden, United Kingdom, Australia) had laws and policies similar to the United States and Canada. Section 1 also presents the standard means of quantifying cumulative effects as proposed by the Council on Environmental Quality and a final chapter addressing the economics of dealing with cumulative effects.

Section 2 is a series of case studies about border issues with Mexico, scenic resources, and how cumulative effects are dealt with in the Canadian Arctic. The final three chapters addressing the numerous issues that need to be considered when dealing with cumulative effects in suburban and exurban landscapes, freshwater fishes, and the cumulative impacts of energy development on sage-grouse. Each of these chapters is presented to give the reader an appreciation of how anthropogenic influences are interconnected and the importance of understanding how human actions influence our ability to make informed decisions. Many of these chapters point to new and innovative means of addressing cumulative effects in a comprehensive manner. While the state of the art is not yet developed to the degree where there is a standard way to measure cumulative effects, these examples certainly point the way to more efficient means including moving from a project point of view to a landscape approach.

There is a much more to be done and managers and practitioners are at the forefront to ensure that cumulative effects are seriously considered. We hope this text helps resource managers make informed decisions regarding the effects of any proposed action on fish and wildlife and its habitat.

Paul R. Krausman
Missoula, Montana

Lisa K. Harris
Tucson, Arizona

Acknowledgments

Randy Brehm, Patricia Roberson, and John Edwards of the Taylor & Francis Group were all instrumental in the production of this book. Randy is an editor at Taylor & Francis who attended the 14th Annual Wildlife Society meeting in Tucson, Arizona, in 2007; she encouraged us to consider publishing a book on cumulative effects and wildlife. Patricia and John were project coordinators who helped with the numerous details that go into the publication of any text. Their consideration and input were invaluable.

Organizational assistance was provided by Marisa Franz who spent numerous hours making sure all of the citations were included and the format was consistent. Other day-to-day operations were accomplished by Jeanne Franz, who keeps the Wildlife Biology Program at the University of Montana on an even keel.

The following reviewed the book chapters: Sherry Barrett, U.S. Fish and Wildlife Service (USFWS); Susan J.M. Brown, Western Environmental Law; Andrea W. Campbell, U.S. Forest Service (USFS); Ruth Doyle, USFS; Laura López-Hoffman, University of Arizona; Julie Jonsson, Harris Environmental Group, Inc.; Andrew Laughland, USFWS; John Loomis, Colorado State University; Jason P. Marshal, University of Witwatersrand; Jim Rorabaugh, USFWS; Susan Sferra, USFWS; William Shaw, University of Arizona; and Jeffery Wright, University of Montana.

Two chapters in this book are modifications of previously published materials. Permission to use materials in this volume was graciously provided by the Montana Fish, Wildlife & Parks for Chapter 9 and by Island Press for Chapter 11.

The authors of Chapter 8 are grateful to the many participants of a February 2007 workshop in Yellowknife, Northwest Territories, who provided feedback and support to developing a demonstration project that would focus on assessing human activities on the summer range of the Bathurst caribou herd, with use of Traditional Knowledge. They thank K. Clark and A. Legat (Wek'èezhìi Renewable Resources Board) for their ongoing support and participation. T. Antoniuk, S. Francis, and B. Stelfox (ALCES® Group) contributed conceptually to the integrated modeling approach. Throughout the project, D. Taylor conducted spatial analyses using Geographic Information Systems software. C.J. Johnson was supported by the Natural Sciences and Engineering Research Council of Canada and the University of Northern British Columbia. The authors thank the Circumpolar Arctic Rangifer Monitoring and Assessment network for support to A. Gunn and D. Russell. The Government of the Northwest Territories provided funding to conduct the work described here. Funding was also provided by Indian and Northern Affairs Canada through the Northwest Territories Cumulative Impact Monitoring Program.

The authors of Chapter 10 thank the U.S. Geological Survey Cooperative Fish and Wildlife Research Units program and the University of Arizona who provided support and information for this project. They also thank the staff of the USFWS National Conservation Training Center who provided information on structured decision making and how it can be used to address cumulative effects.

The book was supported by the Boone and Crockett Program in Wildlife Conservation, University of Montana, Missoula and the Harris Environmental Group Inc., Tucson, Arizona.

About the Editors

Paul R. Krausman, Ph.D., received his B.S. in zoology from the Ohio State University, his M.S. in wildlife science from New Mexico State University, and his Ph.D. from the University of Idaho. He has taught and conducted research at Auburn University, the University of Arizona, Wildlife Institute of India, and the University of Montana. He has concentrated his research and teaching on wildlife management, especially in arid areas in the Southwest, North Africa, and India, and in areas where there are significant anthropogenic influences on wildlife habitats. Dr. Krausman is a certified wildlife biologist, active with The Wildlife Society (TWS) (currently president-elect), and a TWS Fellow. He is a recipient of the Leopold Award and Medal. He has published hundreds of scientific articles, reports, and papers, and several books. He is currently the Boone and Crockett Professor of Wildlife Conservation at the University of Montana, Missoula.

Lisa K. Harris, Ph.D., received her degrees from the University of Chicago and the University of Arizona. She has developed a successful environmental consulting firm guiding government agencies through compliance with the National Environmental Protection Act, Endangered Species Act, National Historical Preservation Act, and other environmental regulations. Identifying and analyzing cumulative effects are constant issues in her work. A prolific writer with wide-ranging interests, Dr. Harris has published scientific articles on natural resource conservation, land-use modeling, effects of military impacts on endangered species, and cactus transplantation, popular press articles, and essays on adventure travel, parenting, health and fitness, cycling, and food. She lives in Tucson, Arizona, with her two daughters and a menagerie of four-footed friends. Whenever possible, she slips on a backpack and heads to road's end.

Contributors

Jan Z. Adamczewski
Wildlife Division
Environment and Natural Resources
Government of Northwest Territories
Yellowknife, Northwest Territories, Canada

Scott A. Bonar
USGS Arizona Cooperative Fish and Wildlife Research Unit
University of Arizona
Tucson, Arizona

Matt Carlson
ALCES® Landscape and Land-Use Ltd.
Ottawa, Ontario, Canada

Holly E. Copeland
The Nature Conservancy
Lander, Wyoming

Brian Czech
U. S. Fish and Wildlife Service
National Wildlife Refuge System
Arlington, Virginia

Colin J. Daniel
APEX Resource Management Solutions Inc.
Ottawa, Ontario, Canada

Jonathan Derbridge
Boone and Crockett Program in Wildlife Conservation
University of Montana
Missoula, Montana

Kevin E. Doherty
U.S. Fish and Wildlife Service
Bismarck, North Dakota

Anne Gunn
Saltspring Island
British Columbia, Canada

Lisa K. Harris
Harris Environmental Group
and University of Arizona
Tucson, Arizona

Matthew J. Holloran
Wyoming Wildlife Consultants, LLC
Laramie, Wyoming

Chris J. Johnson
Ecosystem Science and Management Program
University of Northern British Columbia
Prince George, British Columbia, Canada

Matt Kenna
Attorney in Private Practice and of Counsel
Western Environmental Law Center
Durango, Colorado

Paul R. Krausman
Wildlife Biology Program
University of Montana
Missoula, Montana

Debby Kriegel
Coronado National Forest
Tucson, Arizona

William J. Matter
School of Natural Resources and the Environment
University of Arizona
Tucson, Arizona

Jerod Merkle
Boone and Crockett Program in Wildlife Conservation
University of Montana
Missoula, Montana

David E. Naugle
Wildlife Biology Program
University of Montana
Missoula, Montana

John S. Nishi
ALCES® Landscape and Land-Use Ltd.
Calgary, Alberta, Canada

Robert B. Richardson
Department of Community, Agriculture, Recreation, and Resource Studies
Michigan State University
East Lansing, Michigan

Don E. Russell
Shadow Lake Environmental Consulting
Whitehorse, Yukon Territory, Canada

Sonja M. Smith
Boone and Crockett Program in Wildlife Conservation
University of Montana
Missoula, Montana

Jason D. Tack
Wildlife Biology Program
University of Montana
Missoula, Montana

Lirain F. Urreiztieta
Harris Environmental Group
Tucson, Arizona

Brett L. Walker
Avian Research Program
Colorado Division of Wildlife
Grand Junction, Colorado

Section 1

Understanding Cumulative Effects

1 Grappling with Cumulative Effects

Lisa K. Harris and Lirain F. Urreiztieta

CONTENTS

INTRODUCTION

Environmental regulations exist, in part, to guide resource managers in the decision-making process relative to impacts and effects of their actions. In the United States, the National Environmental Policy Act (NEPA) requires review of government actions that occur on public lands or are funded with public funds. National Environmental Policy Act review identifies and analyzes environmental effects through a process often referred to as an Environmental Assessment (EA) or Environmental Impact Statement (EIS). To ensure consistency across agencies and land managers, the Council on Environmental Quality (CEQ) assists environmental planners in the environmental review process by interpreting NEPA regulations. Other countries such as Canada, Sweden, the United Kingdom, and Australia have regulations similar to those in NEPA, and they also follow a similar process in the identification and subsequent analysis of environmental impacts (Canadian Environmental Assessment Agency 1998; Court et al. 1994; European Commission 1999; Office of the Deputy Prime Minister 2005; Scottish Executive 2005). The purpose of this chapter is to define and analyze impacts or effects that fall into three categories: direct, indirect, and cumulative.

IDENTIFYING EFFECTS RATIONALE

The purpose in identifying environmental impacts is to help resource managers, policy makers, and the public make informed decisions regarding the effects of a proposed action. While originally designed to be used when an "on-the-ground" project, such as a dam or new road, was in its planning stages, the process of identifying and evaluating effects is also used to determine the impacts of policy (e.g., hunting regulations, predator control programs; Bonnell and Storey 2000).

Practitioners undertake an impacts or effects analysis to understand the implications of a proposed action requiring NEPA compliance on the environment. They start by categorizing the impacts or effects of the proposed action as direct, indirect, and cumulative. Effects can be defined as one type or several (i.e., direct and indirect, or solely one or the other). While this book focuses primarily on cumulative effects, we will discuss direct and indirect effects because they are not easily parsed apart.

Underlying the identification of environmental impacts *before* a project or policy is implemented is the belief that if the consequences are uncovered before a proposed action occurs, policy makers can make decisions (including not taking action or the "no action alternative") with the understanding of what may happen in the future. By identifying the consequences, managers can also develop mitigation programs that help offset negative impacts before they are irreversible.

Evaluating environmental consequences also allows resource managers to allocate funding where it is needed most. Spending funds equally in areas that vary in habitat suitability for a particular species may not be as effective as directing funds where they are most needed. Funds should be allocated based on the identification of effects across the resource. For example, in an examination of multiple sites within the John Day River basin, Wu et al. (2000) classified some as pristine and others as degraded. If the bulk of the available mitigation funding is set aside for the reconstruction of the degraded sites, the results will yield a healthier riverine ecosystem with sustainable populations of native fishes (Wu et al. 2000).

Bigger environmental issues, such as global warming or the extinction of imperiled species, can be argued to be the product of cumulative effects of human impacts. Global issues such as these cannot be resolved unless policy makers identify the culpable actions and determine their impacts. Effects determinations are frequently made by assuming the future will unfold the way policy makers predict. However, there may be unconsidered consequences that were unknown at the time decision-making processes are undertaken. Identifying and analyzing effects is therefore not a straightforward process. Predicting the future through a simplified formula is rarely possible.

USING THE CASE STUDY APPROACH

To understand how the process of identifying and analyzing effects, particularly cumulative effects, is carried out, this book will examine theoretical and real world perspectives, providing academic and practitioner viewpoints. Topics will be presented with theoretical aspects and case studies.

This chapter is intended to help define impact-related effects and explain why confusion exists for practitioners when they evaluate a proposed action. The following case study is based on an actual project. The proposed action, setting, and the subsequent EA analysis conducted as part of the subject land use permit application are real. The proponent's name was changed to maintain anonymity.

This case study was selected, in part, because the effects analysis was complicated despite the seemingly straightforward nature of the proposed action. This contradiction of a simple action versus a convoluted effects analysis allows us to dissect the conundrum resource managers and planners face each time an action is planned and evaluated.

Case Study: Arizona Mining Project

DWG Inc. proposes to develop a quarry for decorative stone, landscape materials, and boulders on Bureau of Land Management (BLM) land in Florence, Arizona, a small town approximately 100 km east of Phoenix, Arizona. The proposed quarry site is located in an undisturbed area adjacent to Sandman Wash (Figure 1.1). The wash (i.e., an ephemeral stream) drains into the Gila River located approximately 2.4 km away from the proposed site. The surface quarry will encompass 9.7 ha and will require widening an existing unpaved road to accommodate heavy earth-moving equipment. DWG Inc. will limit quarrying to blocks ≤ 2.0 ha at any given time and will ultimately restore the quarry blocks by backfilling and revegetating.

FIGURE 1.1 Sandman Wash, near Florence, Arizona, a potential site for a gravel quarry.

Initial consultation with BLM biologists and land managers identified the agency's concern for project impacts on recreational users. Agency personnel wanted to minimize negative impacts to the viewshed by situating the quarry in as discreet a location as possible. As a result of these discussions, the quarry and material source pits were relocated.

Proposed Mine Site Description

The proposed quarry is located within the Sonoran desertscrub biotic community (Brown 1994). The relatively low density vegetative structure is composed of medium-sized trees and assorted cacti, including saguaros (*Carnegiea gigantea*), a columnar cactus that serves as habitat for cavity-nesting birds and is protected in Arizona. In addition to Sandman Wash, the site contains several small ephemeral drainages that sporadically contain surface flow after significant rain events. The vegetative density within and adjacent to these washes is higher than that found in the surrounding upland areas.

The project area is within suitable habitat for two federally listed species (lesser long-nosed bat [*Leptonycteris yerbabuenae*] and Acuña cactus [*Echinomastus erectocentrus* var. *acunensis*]), and seven BLM special status species including three bats (i.e., cave myotis [*Myotis velifer*], fringed myotis [*Myotis thysanodes*], and pocketed free-tailed bat [*Nyctinomops femorosaccus*]); three reptiles (Sonoran desert tortoise [*Gopherus agassizii*], common chuckwalla [*Sauromaulus ater*], and Tucson shovel-nosed snake [*Chionactis occipitalis klauberi*]); and one plant (Parish Indian mallow [*Abutilon parishii*]).

The Acuña cactus has been documented in four areas, the closest to the project area being approximately 3.2 km from the proposed area of disturbance. While no species-specific surveys were conducted for this plant on the proposed quarry site, local conditions are similar to those where the cactus exists. The lesser long-nosed bat, a nectar-feeding species, forages on plants such as agaves (*Agave* spp.) that were documented within the site and in the surrounding area. However, no roosting areas such as caves, mines, or adits were present on the proposed site. No surveys for bats were conducted, so it is unknown if the species actually uses the site.

The desert tortoise was the species land managers were most concerned about. Throughout its range, the Sonoran Desert population is declining and its legal status is under review by the U.S. Fish and Wildlife Service (USFWS). Classified as a Species of Concern under the Endangered Species Act, the population is afforded limited protection through its current status. However, if as a result of agency review, the protection status changes to Threatened or Endangered, the tortoise population in the Sonoran Desert would be afforded additional protection.

According to BLM Policy Goal (IM AZ-91-15), the agency approves activities which "conserve and improve where feasible, the distribution, quantity and quality of desert tortoise habitat on public lands with no net loss of quantity or quality of Category I and Category II habitat." Category I habitats support medium to high densities of tortoises and are essential to maintaining large to high densities of tortoise; Category II supports medium to high densities that may be essential to maintaining viable populations. The proposed quarry site is located within Class II and Class III (areas that support low to medium density populations and are not essential to maintaining viable populations) desert tortoise habitat. The BLM's category definitions are qualitative.

Proposed Mine Effects Analysis

As part of the DWG Inc. quarry permit application to the BLM, an EA was developed and the direct, indirect, and cumulative effects of the proposed action were identified and analyzed. An effect is often defined as a change measured "by comparing starting and ending points of reference for a specific indicator and relative to some benchmark of magnitude" (Dube et al. 2006, p. 94). The starting point is the state of the environment before the proposed action is implemented. The ending point varies, depending on whether the effect is direct, indirect, or cumulative. Determining the ending point so that the effect's magnitude can be ascertained is not always clear. This is one of the key issues environmental planners face when determining effects and appropriate mitigation measures for their impacts.

Proposed Mine Direct Effects

A direct effect (also referred to as a direct impact) is caused by the action that occurs at the same time and place (National Environmental Policy Act 1969). In our case study, the direct effect of constructing the quarry and widening the ingress and egress road would include the removal of natural vegetation, the loss of habitat for special status species and the death of any special status species from the actual construction and operation of the quarry.

Determining the beginning and end points of direct effects in the proposed DWG quarry project is relatively simple. It begins when the first construction equipment appears on the scene to move dirt and remove vegetation. The end point is when the quarry ceases operation and the last 2.4 ha quarry block is revegetated with native plants (estimated as 40 years into the future).

The direct effect of the proposed action would result in a net loss of 10.8 ha (quarry and road widening combined) of habitat for the Acuña cactus and desert tortoise. The site would ultimately be revegetated once the quarry materials are extracted (40 years from the start date), but site restoration and revegetation take time to complete, particularly in a desert environment where plants grow slowly, so the time frame of determining effects is more than four decades. Understandably, no Acuña cactus would be able to grow on the site during quarry operations and the desert tortoise's use of the project area would be limited or nonexistent.

Of all the species included in the initial evaluation, only the desert tortoise and Acuña cactus were of notable concern to the BLM land managers in their evaluation of DWG's permit application. The federally endangered lesser long-nosed bat was excluded from further evaluation and subsequent mitigation because it feeds on agaves that are prolific within the region. Thus, the removal of 10.8 ha of foraging habitat for the bat was not considered a significant effect in the project's environmental analysis. The same reasoning was applied to the other species of bats, the chuckwalla, shovel-nosed snake, and mallow. These species were not evaluated further because there was ample habitat elsewhere to support these species.

Other direct effects of the proposed action include changing the morphology of the terrain. This landscape alteration would impact the viewshed and would affect the stability and deterioration of slope and vegetative composition. Operating the quarry and hauling rock materials from the site also would increase traffic noise; decrease public safety because of potential accidents with construction vehicles on the area's narrow, winding, dirt roads; degrade the unpaved roads; and negatively affect air quality because of the increase in dust particulates from added vehicular traffic.

Indentifying these direct effects is typically straightforward, and environmental planners can quantify them (e.g., area destroyed, traffic noise increase based on known ambient noise levels of trucks used, additional dust particles in air) and mitigate them. Analyzing the proposed action's effects beyond those caused directly by quarry construction and operation is a more complicated matter, as resource managers try to define and quantify indirect and subsequently cumulative effects.

Proposed Mine Indirect Effects

An indirect effect (also known as an indirect impact or secondary effect) is caused by the action and occurs later in time or is farther removed in distance, but is still reasonably foreseeable (National Environmental Policy Act 1969). Indirect effects may include growth-inducing effects and other effects related to induced changes in the pattern of land use, population density, or growth rate, and related effects on air and water and other natural systems, including ecosystems.

Indirect effects from constructing and operating the quarry include many of the same effects considered to have a direct impact on the environment such as increased traffic along the road, alteration of the viewshed, slope destabilization, increased sedimentation in adjacent steams, degradation of water quality, barriers to wildlife movement, increased genetic isolation, and habitat degradation and loss.

Although the quarry's proposed location was in a remote area that had few recreational users at the time the permit was applied for, once the unpaved road is widened and regularly maintained, there would be a strong likelihood that more recreational users would use the road, particularly on weekends. Additional vehicular traffic, above that associated with quarry operations, also would increase noise levels and degrade air quality from use of the unpaved road.

As is the case for other examples, the alteration of the viewshed is both a direct and an indirect effect. It is a direct effect because the construction and presence of the quarry throughout its working life would be visible from afar and alter the view of the area for recreationalists, residents, and people traveling in the vicinity. Alteration of the viewshed is also an indirect effect because the time frame of impact exceeds indefinitely beyond the quarry's closure. After rock materials have been extracted from each of the 2.0-ha blocks, the pits would be filled and slopes landscaped. But regardless of the extent of the restoration efforts, the end result may not match the surrounding vegetation and the former quarry site most likely would be visible from afar for years, perhaps decades. Since the revegetation efforts require a significant amount of time to restore the land to its preconstruction state, landscape characteristics would be marginal habitat for certain species for years, thus decreasing the use of the area by these species.

The construction of the quarry would alter the slope of the land, and to prevent landslides during quarry operations, the proposed slope configuration of each pit would require the maintenance of a 2:1 ratio. However, over time, weather conditions could alter the slope and the sides could destabilize, becoming steeper. This situation could cause a hazard for wildlife, including the desert tortoise, which could become trapped and perish in a pit with destabilized slopes. Slope destabilization is therefore an indirect effect because it could increase wildlife mortality in the foreseeable future.

Soil erosion would occur on the exposed slopes while the pits were excavated and in operation, increasing the likelihood that more sediment than normal would be carried away by heavy rainfall into nearby Sandman Wash. The sediment would alter water flow within the wash and affect plant growth, water may pool in areas where it would not in the pre-quarry state, potentially creating denser vegetative growth, or no growth at all. Over time, this sediment would be carried to the Gila River, 2.4 km away, where it would degrade water quality and affect fish habitat, including several fish protected under the Endangered Species Act.

The quarry may limit wildlife movement and create genetic isolation by cutting off current movement corridors for species like the desert tortoise, possibly resulting in genetic isolation and a decrease in the viability of local tortoise populations. Habitat fragmentation from human development (i.e., roads, subdivisions) and genetic isolation is a common indirect (and cumulative) effect in other wildlife species (Turner et al. 2004).

Proposed Mine Cumulative Effects

Cumulative effects are defined as "the impact on the environment which results from the incremental impact of the action when added to other past, present, and reasonably foreseeable future actions regardless of the agency (federal or non-federal) or person undertaking such other actions. Cumulative impacts can result from individually minor but collectively significant actions taking place over a period of time" (40 C.F.R. 1508.7). Direct and indirect effects measure immediate influences but cumulative effects analysis examines changes that may occur well into the future and at a further distance away from the proposed action.

Identifying cumulative effects is important because the analysis examines the full picture, not just individual impacts, which would lead to a disjointed and incomplete evaluation. A cumulative effects analysis can be described as studying the outwardly growing ripples in a pond after a rock, in this case the proposed action, has been thrown in. The swell moves across the pond's surface, altering the landscape far from where the rock initially broke through the water's surface. A cumulative effects analysis "provides an on-going mechanism to evaluate if levels of development exceed the environment's assimilative capacity (i.e., its ability to sustain itself)" (Dubc 2003, p. 723).

Environmental issues most recognized by the general public (e.g., global warming, acid rain, loss of biodiversity, river sedimentation, genetic isolation, habitat fragmentation) are the result of cumulative effects (Kennett 1999). Therivel and Ross (2007, p. 371) describe cumulative effects as "death by 1,000 cuts; each individual effect is insignificant but the accumulation of the many insignificant effects causes a significant adverse effect." Examples of this accumulation of consequences includes habitat fragmentation because of increasing roadway construction, addition of non-point source pollution to nutrient loading in lakes and rivers, genetic isolation, and progressive filling in wetlands. According to the CEQ (1997), the most devastating environmental effects may be the result of these cumulative effects and not necessarily the direct effect of a particular action.

Unfortunately, cumulative effects analysis is not always intuitive. Cumulative effects may extend over a large area (e.g., global warming) and may appear only after extended periods (e.g., river sedimentation caused by dams). Planners may not understand the degree of change from a single action at the time it is proposed. But even after individual cumulative effects are identified, how they are analyzed to determine their impact to the environment has received considerable criticism (Johnson et al. 2005). A primary concern is that because cumulative effects are not immediately associated with the proposed action's time and place, it is difficult to predict the impacts and determine the extent of those impacts (Parry 1990). How far into the future or how distant from the location of the proposed action analysts look for impacts is unclear and is left to the analyst and reviewer to determine (Therivel and Ross 2007).

Another criticism is that there is no accepted single methodology for quantifying cumulative effects (Dube and Johnson 2006; MacDonald 2000; Piper 2001; Smit and Spaling 1995; Sonntag et al. 1987; Therivel and Ross 2007). MacDonald (2000) contends that assessing cumulative effects is problematic because of the number of resources that could be affected, how they could be affected (i.e., could be multiple paths), and the uncertainty of temporal and spatial scales.

The CEQ (1997) attempted to sort out cumulative effects and created a conceptual framework for identifying and analyzing them (Chapter 10). The guideline includes information on cumulative effects definitions, how to identify cumulative effects, how to describe the affected environment, and how to determine consequences of cumulative effects. But the CEQ guidelines are only that, a process without any hard and fast steps, not a recipe with a repeatable outcome.

MacDonald reported (2000, p. 302) that the CEQ's guide is incomplete, downplaying "the more difficult decisions and limitations." Specifically, CEQ's framework (1997) does not include recovery rates of impacted resources, call for an evaluation of the role of impacted natural processes, or determine a guide for dealing with the temporal and spatial parameters (MacDonald 2000). MacDonald (2000) describes the variation in the processes of determining cumulative effects as a continuum between qualitative checklists, which costs less but contains less explicit results, to quantitative models with more thorough results and higher price tags. An example of using a qualitative methodology in a cumulative effects analysis is soliciting opinions from experts (Smit and Spaling 2000), where knowledgeable individuals determine the extent of effects of a given proposed action through a qualitative evaluation lacking quantitative analysis. A more intricate approach is the use of quantitative methods, which typically require more time and money. Quantitative methods include spatial analysis, ecological modeling, interactive matrices, land suitability evaluations, biogeographic analysis, remote sensing, and succession models, among others (Quiñonez-Piñon et al. 2007; Smit and Spaling 2000).

Using quantitative models has been criticized, too. Suring et al. (1998) suggest that models are often difficult to develop and maintain, and require complex data analysis. The results from quantitative models are often only useful for the application they have been designed for and cannot readily be used outside of those contexts. Forest landscape disturbance and succession models have been used for large-scale analysis and planning, including determining cumulative effects. They provide information in a spatial context related to forest planning, wildlife habitat quality, timber harvesting, fire effects, and land use change. Shifley et al. (2007) contends that applications of succession models have proven successful within a narrow use range and widespread applications are limited by several barriers. Chief among them are the technical skill level required to run the models and interpret findings. The documentation of initial conditions requires a high level of detail. The calibration required for a new region demands a high level of effort where user support is minimal. In addition, choosing from numerous management alternatives with multiresource interactions that vary over space and time is complex. The old saying "garbage in, garbage out," which applies to the use of modeling in general, certainly applies to determining impacts of cumulative effects: the model's outcome is only as good as the person using it and the validity of the initial data.

Many cumulative effects models are developed for the management of target species in a specific region (Dube et al. 2006; Gaines et al. 2003; Sorensen et al. 2007; Suring et al. 1998; Wu et al. 2000). They take into account landscape characteristics, disturbances, habitat quality, and population figures, and their application often aids land use managers in avoiding future crisis management (Suring et al. 1998). However, their use may be limited because they target one species in one area and because they often leave little room for fluctuations in variables (Sorensen et al. 2007). Long term changes with uncertain and unknown outcomes, such as global warming, reduce the accuracy of the models.

In real-world scenarios, most effects analyses are limited by time and budget constraints. Proposed projects come with timelines and costs; running over often translates into incomplete projects. Studying the environmental effects of a major landscape change, such as constructing a hydroelectric dam or developing long-term regional plans, may take years and cost millions of dollars. It is these types of projects that receive a lot of media attention and often involve the development of quantitative models where effects can be intensively identified and studied. However. projects like a dam or forest plan occur infrequently. The majority of environmental assessments are undertaken in support of projects that receive little to no outside attention, with short timelines and small budgets, as was the case in our quarry example.

DWG's quarry was privately funded, and the time and money used to undertake an EA was limited by how much the investors were willing to contribute, compared to the estimated income the quarry would generate in the future. The BLM, the permitting agency for the quarry, had no requirements about how thorough the EA had to be, nor did it dictate the types of information that had to be included before the agency undertook its evaluation and decided whether to issue a permit. The agency only required that the EA follow its format guidelines (i.e., specific sections, such as cumulative effects, and presented in specific order).

The cumulative effects section, taken directly from the approved EA (italic added for emphasis), states:

> Farming, mining, recreation, and grazing on BLM lands result in the loss, modification, and fragmentation of habitat which continues to contribute to the cumulative effects on vegetation, wetland and riparian resources, special status species, cultural resources, and other resources of the natural desert environment. This project demonstrates a *low-level* of proposed development with a plan to mitigate for project effect. There are *limited numbers* of new mineral lease requests in this area, and it is not expected that this project will contribute any adverse project effects that *reach a significant threshold level* for cultural resources, vegetation, special status species (including the transplanted population of bighorn sheep north of the Gila River), wetland and riparian resources, and water quality....

Applying the cumulative effects analysis continuum MacDonald (2000) described, the analysis undertaken for the quarry pit is qualitative in nature. The analysis uses descriptive words such as "low-level," "limited numbers," and "significant threshold" without seeking to define them. It does not call out individual impacts that might be considered cumulative, and brushes aside any impacts that could be considered cumulative, stating "project effects [will not] reach a significant threshold level for [resources]" (Environmental Consultants 2009, p. 23).

The analysis for the quarry took the "straw that broke the camel's back" approach, by analyzing whether the proposed action would create a consequence that tips the scales toward a larger environmental problem—does it "break the camel's back"? Does adding one more materials quarry to southern Arizona BLM lands affect resources to such a degree that its development may lead to species extinction, eliminate recreational use of the area, or negatively affect any other considered resource? Would the quarry's development reach a significant threshold level?

Presenting answers to these questions is what the BLM reviewer looked for in his analysis of the project (B. Bellew, BLM, personal communication December 14, 2009). While the cumulative effects section presents no new information that had not been presented earlier in the document under the effects review for each of the resources present, the EA was complete according to agency reviewers. Once submitted, the agency did not require additional information or clarification to determine if a permit for development of the quarry site was warranted.

Referring to McDonald's continuum (2000), the cumulative effects analysis for the proposed rock quarry is as qualitative as it gets. There are no figures, models, or numbers, not even reference to discussions with resource specialists who might have been familiar with the impacts of the project to the desert tortoise, Acuña cactus, or recreationists. There is no discussion on habitat fragmentation for either species, or the possible impacts of the quarry on genetic isolation to either the desert tortoise or Acuña cactus. Furthermore, there was no other mention of the translocated population of bighorn sheep (*Ovis canadensis*) within the EA and the potential concerns associated with the crucial intermountain habitat required for genetic flow between populations.

In terms of discussing impacts from the incremental addition of developing the quarry when added to other actions (federal or nonfederal) that are proposed in the present and reasonably foreseeable future, the cumulative effects section states only "There are limited numbers of new mineral lease requests in this area. . . ." The analysis does not define "limited" or "in this area," and leaves it up to the reviewer to put meaning behind these words.

The project budget and the owner's request to submit a finished report as soon as possible prevented further information from being gathered for analysis. If this had not been the case, additional data could have been presented. Information that would have given the environmental analyst more insight into the identification and analysis of cumulative effects includes the number of recreational users affected by a viewshed alteration (based on current use patterns), Acuña cactus survey and results, desert tortoise survey and results, computer rendering of future viewshed with developed quarry, identification and quantification of other mining permit applications within the BLM district, estimation of increase of future road traffic based upon other existing nearby mining operations, modeling of increased dust and noise generated based on estimated traffic increase, amount of sedimentation in nearby Sandman Wash from proposed quarry erosion and how long and at what quantity the sediment would reach the Gila River, information on translocated desert bighorn sheep, and amount of carbon dioxide added to the atmosphere by quarry operations and increased vehicular traffic. Unfortunately, the cost to complete such studies would most likely have exceeded the project proponent's budget.

These additional data would have created an elaborate picture of possible cumulative effects in order of complexity, with the number of recreational users impacted by a viewshed change as the information that would have been the easiest to gather (perhaps by a phone call to the BLM district office recreational planner) as compared to information that is most complex in nature and more difficult (time consuming and costly) to obtain (i.e., determining the increased load of carbon dioxide and how that may impact global warming). So at what point is there enough information to determine the cumulative effects of an action? Using the pebble-cast-onto-the-pond metaphor, as the rings of impact expand after the stone hits the water, how long does the planner wait until the ripples bump against the shore? Can the planner even see the shore? Perhaps the banks are too far away in both time and space. Are the possibilities unlimited?

Some cumulative analysis supporters believe the planner should wait until the waters become still to determine effects. But if that were the case, then almost every project proposed would ultimately lead to significant environmental issues such as global warming and species extinction. This approach works theoretically, but it does not apply empirically and is ultimately unfeasible and impractical. The location of the shore (where planners find the end point in their analysis) may come for agency evaluators when regulators dictate the degree of information required before they are able to make a project determination. But in most cases, determining the depth of information and how it is analyzed is left to the environmental planner charged with writing the assessment.

Proposed Mine Effects Analysis

In the DWG quarry case study, the only direction provided by the BLM was their initial concern about the project's impact to desert tortoise, Acuña cactus, and recreational users. Thus, the site for the proposed quarry was positioned within an area that received little recreational use and minimized impacts to the Acuña cactus and desert tortoise. Mitigation measures included the revegetation of the site after the final extraction of materials, and a fee to the BLM was incurred (based upon the amount of land disturbed).

However, the BLM ultimately ruled against the project and denied the permit, stating that the proposed quarry would have a negative impact on resources discussed above (B. Bellew, BLM, personal communication). Although the location of the proposed quarry was relocated to a site that would minimize disturbance to the viewshed and reduce exposure to recreationalists, and the quarry owner agreed to pay the mitigation costs for the loss of tortoise habitat (based on a standard BLM approved formula), the deal breaker, "the straw that broke the camel's back," was the removal of potential Acuña cactus habitat. This decision was made because there are only four documented patches of this cactus known to exist, with the closest being approximately 3.2 km away. The BLM biologist was concerned that the quarry would destroy undocumented plants and potential seed bank (viable seeds dormant in soil). The quarry proponents did not conduct a species specific survey for the cactus and the BLM did not request one. The EA called for a survey prior to construction. The only mitigation mentioned was the revegetation of the site after the operation ceased (decades into the future).

The quarry proponents did not appeal the BLM's decision. If they had, the acquisition of additional information may have been necessary (e.g., how many recreationalists would be affected by the project), and a survey for Acuña cactus may have been required. Ultimately, the quarry proponents abandoned the project. They had spent more than 2 years and thousands of dollars on environmental studies. A profit analysis determined that further investment was not warranted.

Proposed Mine: Scale and Its Impact on Analyzing Effects

In this case study, scale mattered. Agency reviewers examined the broader picture and made landscape-scale assumptions to support their determination. Time also played a role. While the project proponents planned to revegetate the affected area to preconstruction conditions, removing desert tortoise habitat for a minimum of four decades was too long and outweighed the benefits from the anticipated restoration.

Scale is a common missing component in many cumulative effects analyses (Therivel and Ross 2007). Limitations include poor consideration of likely human activities beyond the project. This is true in our case study. As stated in the EA, "Farming, mining, recreation, and grazing on BLM lands result in the loss, modification, and fragmentation of habitat which continues to contribute to the cumulative effects on vegetation, wetland and riparian resources, special status species, cultural resources, and other resources of the natural desert environment." All future modifications are lumped together and dismissed. It is unclear how the agency would have responded had scale been taken into account, but additional cumulative effects would have been identified.

Our case study showed that identifying direct and indirect effects was easier than identifying cumulative effects, which are often overlooked in analyses. Scale from a time and distance perspective is often missing or downplayed. To be effective, the EA should allow resource managers and policy makers to decide the best course of action, which requires that all effects be identified and their consequences evaluated.

CONCLUSION

Regulations must be taken seriously. Ingelson et al. (2009, p. 16) discuss examples from the Philippines, where there are laws in place to evaluate environmental impacts, but the process is "designed to make it appear as though projects were subject to environmental scrutiny when in reality the system merely facilitates project approval." Analyzing the effects of a proposed action using a checklist or a model depends on how much time and money is invested. Politics also play an important role. Cooper and Sheate (2002) suggest that government agencies need to develop stronger guidelines for what should be included in a cumulative effects analysis and not leave it to the action's proponents to decipher the process.

Currently, the responsibility of determining effects rides on the shoulders of whoever develops the analysis and reviews it. As the DWG Inc. proposed quarry case study shows, identifying and analyzing cumulative effects is difficult. Dube (2003) suggests that cumulative effects analysis is a science and as such should not be undertaken on a project-by-project basis where the process is fragmented, but instead by an entity that examines an entire region. It is unrealistic to expect the environmental analyst conducting a cumulative effects analysis or the reviewing agency to know other projects that may be planned within the proposed action's immediate vicinity—particularly by nonfederal entities. How, then, can an analysis be realistically conducted? Because we do not know, the process of determining the impacts of effects becomes a guessing game. We now have 30 years experience (since NEPA was signed into law) at speculating. It may be time to throw out our paradigm of examining effects on a piece-meal basis and, instead, develop environmental policies that evaluate projects at regional, national, and international levels. This landscape level approach may be more effective at analyzing effects than guessing if the proposed project is the "straw that breaks the camel's back."

2 The NEPA Process
What the Law Says

Matt Kenna

CONTENTS

INTRODUCTION

While the desire to make environmentally enlightened decisions can lead to the preparation of environmental impact statements (EIS), legal requirements are often the main drivers of producing such documents, and certainly proscribe their scope. Accordingly, it is incumbent upon the drafter of an EIS or other environmental review documents to know at least the basics of what the law requires. While it is agency personnel who set the sideboards for such reviews, the preparers will do well to have some knowledge of their own.

Legal issues can seem daunting for the nonlawyer. However, just as a nonscientist should not shy away from a basic understanding of science, neither should the scientist avoid a basic understanding of the law applicable to his work. This chapter will provide a brief overview of the basic requirements of the National Environmental Policy Act (NEPA), followed by a discussion of what is required of a proper cumulative effects analysis, including the related but distinct question of the proper scope of the proposed action(s) to be analyzed.

NEPA OVERVIEW

NEPA is the overarching statute that seeks to protect the environment in the United States. It was passed by Congress in 1969 and signed into law by President Richard

Nixon on January 1, 1970, when the first Earth Day was recognized in the United States in response to growing environmental degradation. The "congressional declaration of purpose" for NEPA states:

> The purposes of this Act are: To declare a national policy which will encourage productive and enjoyable harmony between man and his environment; to promote efforts which will prevent or eliminate damage to the environment and biosphere and stimulate the health and welfare of man; to enrich the understanding of the ecological systems and natural resources important to the Nation; and to establish a Council on Environmental Quality ["CEQ"] (42 United States Code [U.S.C.] § 4321).

It is important to bear in mind that NEPA is a "procedural" statute, meaning that unlike other "substantive" environmental statutes such as the Clean Water Act, it does not impose substantive environmental standards or proscribe activities. Instead, it requires federal agencies to disclose and analyze the effects of proposed government projects and related activities, and to consider alternatives to such proposed projects. It does so primarily through the requirement that every federal agency must prepare an EIS for "proposals for legislation and other major Federal actions significantly affecting the quality of the human environment" (42 U.S.C. § 4322[C]).

While NEPA only imposes procedural safeguards, "... it is not better documents but better decisions that count. NEPA's purpose is not to generate paperwork—even excellent paperwork—but to foster excellent action. The NEPA process is intended to help public officials ... take actions that protect, restore, and enhance the environment" (40 Code of Federal Regulations [C.F.R.] § 1500.1[c]).

The preparation of an EIS promotes two interconnected goals of NEPA. First, it requires federal agencies to make fully informed and well-considered decisions. Compliance with NEPA occurs only when an agency first takes a hard look at the environmental consequences of its actions. The "hard look" requirement stems from NEPA's command that agencies comply with the statute "to the fullest extent possible" (42 U.S.C. § 4332), and it is improper for an agency to ignore the information it collects once it has taken a hard look. Second, an EIS requires an analysis of alternatives, which provides a mechanism for an agency to make better decisions once the impacts of each alternative are considered, which can include rejecting the project, choosing a wholly different alternative, or adding mitigation measures to the proposed action (42 U.S.C. § 4332[C][iii]; 40 C.F.R. § 1502.14). The consideration of alternatives to the proposed action is the "heart" of the EIS (40 C.F.R. § 1502.14).

The CEQ has issued regulations that fill out the requirements of NEPA, which provide the guiding standards for preparing EISs and other NEPA documents such as Environmental Assessments (EAs), which are like small-scale EISs for smaller projects (40 C.F.R. Part 1500). An EA has twin goals: to help an agency determine whether a full-scale EIS is necessary, and to aid an agency's compliance with NEPA when no EIS is necessary (40 C.F.R. §§ 1501.4[b], 1508.9). An EA results in either a Finding Of No Significant Impact (FONSI), in which case an EIS is not prepared, or a finding that the project will have a significant impact, in which case an EIS is required (40 C.F.R. §§ 1501.4[c], [e]).

As indicated, even if an EIS is not required, an EA plays an important role in agency decision making, and many of the requirements applicable to EISs apply to EAs, including the need to examine the impacts from the proposed action, and an examination of alternatives (40 C.F.R. § 1508.9[b]). This means that EAs need to consider cumulative effects and the related concepts discussed below, just like EISs do. In an average year, 100 times more EAs are prepared than EISs, so it is important that cumulative effects be fully considered in EAs (Council on Environmental Quality 1986).

Some proposed federal actions are so minor that they do not even need an EA, let alone an EIS. Such actions may be approved with a categorical exclusion (CE or CatEx), which are "categor[ies] of actions which do not individually or cumulatively have a significant effect on the human environment and which have been found to have no such effect in procedures adopted by a Federal agency . . . and for which, therefore, neither an environmental assessment nor an environmental impact statement is required" (40 C.F.R. § 1508.4). Each agency will have its own CE rules, and they include truly minor actions such as mowing the lawn at an agency office and more substantial actions such as timber sales under a certain size (see U.S. Forest Service CE rules at 36 C.F.R. § 200.6), although the propriety of these more substantial ones is the subject of ongoing debate and litigation.

It is important to note that NEPA only applies to federal actions, plus any nonfederal actions (e.g., state, tribal, private) that are implicated by federal action, such as private actions for which a federal permit (e.g., dredge-and-fill wetland permit under the Clean Water Act) is required. It is beyond the scope of this chapter to explore the topic of which actions contain a sufficient federal "trigger" to require NEPA compliance, and many states have "little NEPAs" that have similar requirements for state and nongovernmental actions. However, the CEQ regulations are clear that once it is determined that NEPA applies, cumulative impacts must be considered regardless of whether nonfederal entities carry out the other actions causing the cumulative impacts (40 C.F.R. § 1508.7).

With this background of NEPA law, the next topic explores ways to determine the scope of the action(s) to be analyzed (i.e., determining which projects, if any, in addition to the project directly at issue must be a subject of the environmental impact statement). This subject will be followed by a discussion of the scope of effects of those actions that need to be analyzed, including cumulative effects.

SCOPE OF ACTION(S) TO BE ANALYZED

Cumulative effects as understood by the nonlawyer (and even many lawyers and judges, for that matter) often includes several different concepts that are important to keep distinct for those who are investigating them in depth. These concepts include cumulative actions, connected actions, segmentation, similar actions, and indirect effects, and should be considered and kept somewhat distinct in one's mind. Most of these concepts actually define the "scope" of the action(s) to be analyzed (i.e., what actions are causing environmental impacts?), while cumulative effects and indirect effects go to the impacts resulting from these multiple causes once the scope of the actions causing impacts is determined (40 C.F.R. § 1508.25).

As an example, imagine you are tasked with analyzing a proposal to build a road though federal land, so that a private landowner can access an inholding of private land to build the first phase of a housing development in a national forest where similar developments are in early stages of planning, and where a nearby ski area has already been built. Must you analyze only the road, or do you have to analyze phase 1 of the development as a connected action? If you do, is phase 2 reasonably foreseeable such that you must include it in the scope of the action to be analyzed to avoid improperly segmenting the analysis? Is the existing ski area a similar action that you must consider? Each of these presents an important question and can drastically affect the size and validity of the environmental review document.

The CEQ regulations provide for a "scoping," whereby the NEPA process starts with the agency telling the public it is about to analyze the impacts of the project, and asking the public and government agencies to weigh in on what they believe should be included in the scope of the analysis (40 C.F.R. § 1501.7). By involving as many interested parties as possible, the action agency can effectively use the expertise of outside parties to help ensure it is not missing anything, and help make sure it is doing a proper NEPA analysis from the start.

Connected Actions

The CEQ regulations define the scope of actions to be considered to include connected action, cumulative actions, and similar actions (40 C.F.R. § 1508.25[b]). The definition of connected actions states:

> Connected actions . . . are closely related and therefore should be discussed in the same impact statement. Actions are connected if they:
>
> (i) Automatically trigger other actions which may require environmental impact statements.
> (ii) Cannot or will not proceed unless other actions are taken previously or simultaneously.
> (iii) Are interdependent parts of a larger action and depend on the larger action for their justification (40 C.F.R. § 1508.25[a][1]).

As stated by one court:

> The purpose of this requirement is "to prevent an agency from dividing a project into multiple 'actions', each of which individually has an insignificant environmental impact, but which collectively have a substantial impact." . . . We apply an "independent utility" test to determine whether multiple actions are so connected as to mandate consideration in a single EIS. The crux of the test is whether "each of two projects would have taken place with or without the other and thus had 'independent utility.'" When one of the projects might reasonably have been completed without the existence of the other, the two projects have independent utility and are not "connected" for NEPA's purposes (*Great Basin Mine Watch v. Hankins*, 456 F.3d 955, 969 [9th Circuit 2006]).

Applying the independent utility test to the hypothetical inholding development above, one could ask, could the development be built if the access road through federal land (the federal action providing the trigger for requiring review under NEPA) were not approved? If the road were simply one of several that the developer would like to provide, and there is legally sufficient access through nonfederal lands, then the road and the development would have independent utility, and the development would not be required to be included in the scope of the action to be analyzed. However, if it is the primary road or even an ancillary one required by the county or highway department as a condition of approval, where the development cannot go forward "but for" approval of the road, then the development is a connected action that must be considered in the scope of the EA or EIS.

Examples from past court cases include *Colorado Wild v. U.S. Forest Service*, 523 F. Supp. 2d 1213, 1225–26 (District of Colorado 2007), where a highway interchange for a proposed resort development, and the development itself, were connected actions to the Forest Service road connecting the interchange and development, even though they were not on federal land. Another is *Thomas v. Peterson*, 753 F.2d 754, 758 (9th Circuit 1985), where a proposed logging project and a road to facilitate the logging had to be considered in a single EIS because the timber sales could not proceed without the road, and the road would not be built but for the proposed timber sales.

It is important to note that, as discussed above, some states (e.g., California) have "little NEPA" that require environmental analysis documents for all major land development projects, including private land where most development occurs. In those instances these causal determinations need not be made, as the development itself would be the subject of analysis. But in other states, given that the scope of an EIS can be expanded orders of magnitude by finding a development to be a connected action to the federal action triggering the NEPA requirement, appeals and litigation often result between project proponents, the federal government, and project opponents. Other common federal triggers that can require a project on private land to be analyzed in a NEPA analysis include pollution discharge permits, particularly section 404 wetland fill permits under the Clean Water Act. In such cases, if the development on the uplands could or would not occur but for the development in the lowlands requiring the section 404 permit, the whole development must be analyzed in the EIS.

A sub-concept of the "connected action" issue is segmentation (i.e., splitting projects into smaller parts for purposes of environmental analysis, often so that a less elaborate EA can be completed instead of a complete EIS). Whether a complete EIS is required for a particular action depends on the context and intensity of the possible impacts from a project, and so in general the smaller the action the less likely it is that an EIS would be required (40 C.F.R. § 1508.27). Artificially dividing a project into smaller components to avoid the need to prepare an EIS is not permitted under NEPA regulations. Segmentation is part of the connected action issue although it often is used to describe situations where there is no doubt that each of the segments will need NEPA review at some point, and it is simply a matter of whether they need to be considered in a comprehensive EIS rather than separately. However, that is not always the case, and for instance a proponent of a state-federal project might try to avoid a comprehensive EIS by proceeding first with the state components of a project (in a state with no little NEPA).

In the instances where all segments will eventually receive NEPA review, one might wonder why segmentation is a problem. The reason is to avoid piecemeal approval of projects that might not be approved if their full scope is analyzed from the outset. It may be that each piece individually does not have a significantly adverse impact, and each additional segment might not appear to do so individually either, but if it were analyzed as a whole from the beginning, the decision makers might decide to forgo the project or choose a different, less environmentally damaging alternative.

The classic situation where segmentation arises is with highways, where one segment is being built and another is in some stage of planning. A project may be properly segmented if the segments (1) have logical termini; (2) have substantial independent utility; (3) do not foreclose the opportunity to consider alternatives; and (4) do not irretrievably commit federal funds for closely related projects (*Village of Los Ranchos de Albuquerque v. Barnhart*, 906 F.2d 1477, 1483 [10th Circuit 1990]). The first two factors are probably the most important; if a highway segment has no useful purpose if later segments are not built, then it cannot be properly segmented from the anticipated segments for purposes of NEPA review.

Cumulative Actions (as Distinct from Cumulative Impacts)

Cumulative actions, like connected actions, must be analyzed in the same EA or EIS (40 C.F.R. § 1508.25[a][2]). It is important to differentiate cumulative actions from cumulative impacts. As stated by Thatcher (1990:633):

> "The obligation to wrap several cumulative action proposals into one EIS for decision making purposes is separate and distinct from the requirement to consider in the environmental review of one particular proposal, the cumulative impact of that one proposal when taken together with other proposed or reasonably foreseeable actions." Section 1508.25(a)(2) requires the former, necessitating the coordinated analysis of proposals that "have cumulatively significant impacts." ... In contrast, section 1508.25(c)(3) requires the latter, namely, an analysis of the cumulative impact of [a single project] together with reasonably foreseeable future waivers (*Native Ecosystem Council v. Dombeck*, 304 F.3d 886, 896 n.2 (9th Circuit 2002).

Thus, cumulative actions go to the scope of what is being analyzed, while cumulative impacts address which impacts must be analyzed once the appropriate actions(s) to be analyzed are identified.

Cumulative actions are defined as those "which when viewed with other proposed actions have cumulatively significant impacts" (40 C.F.R. § 1508.25[a][2]) Unlike connected actions, cumulative actions need not depend on each other for their existence. And unlike similar actions they need not be of the same type of project. Rather, they merely need to be proposed actions that can have a cumulative effect with the primary action and a resource of concern. So, for instance, if bird habitat is an issue in a proposed housing development project area, and there are simultaneous proposals to build a pipeline in the same area unrelated to the housing project, they could be cumulative actions that would be required to be analyzed together in the same EA or EIS.

Similar Actions

The CEQ regulations state that similar actions are those "... which when viewed with other reasonably foreseeable or proposed agency actions, have similarities that provide a basis for evaluating their environmental consequences together, such as common timing or geography" (40 C.F.R. § 1508.25[a][3]). An example of similar actions would be timber sales in proximity to one another being approved around the same time. In contrast, the action of issuing agency regulations governing permit applications is not similar to issuing an individual permit under those regulations.

Unlike connected and cumulative actions that must be combined for analysis in the same EA or EIS, the CEQ regulations state that an agency merely "may wish to analyze these [similar] actions in the same impact statement" (40 C.F.R. § 1508.25[a][3]). Thus, it will really be up to the agency and its consultants to decide whether it wants to combine similar actions in the same EA or EIS, with quite a bit of latitude in making that determination. However, keep in mind that if a project were deemed both a similar action and a connected or cumulative action, then combining the projects for analysis would be required.

SCOPE OF EFFECTS TO BE CONSIDERED—DIRECT, INDIRECT, AND CUMULATIVE EFFECTS

Having defined the scope of the action(s) that need to be considered as discussed above, next the scope of the impacts must be considered. These include direct effects, indirect effects, and cumulative effects (40 C.F.R. §§ 1508.7, 1508.8). Note that the CEQ regulations clarify that the terms "effects" and "impacts" as used in the regulations "are synonymous" (40 C.F.R. § 1508.8).

Direct Effects

This term may seem fairly obvious, and its meaning usually is in referring to impacts to land, air, water, and other natural resources. However, NEPA requires an assessment of impacts to the "human environment," which raises the question of whether social impacts need to be considered. The CEQ regulations address this issue, stating that "... 'Human Environment' shall be interpreted comprehensively to include the natural and physical environment and the relationship of people with that environment. This means that economic or social effects are not intended by themselves to require preparation of an environmental impact statement. When an environmental impact statement is prepared and economic or social and natural or physical environmental effects are interrelated, then the environmental impact statement will discuss all of these effects on the human environment" (40 C.F.R. § 1508.14).

Accordingly, significant social effects will not themselves trigger the need for an EA or EIS, but once one is required to assess environmental impacts, social impacts must also be assessed. This includes addressing the economic assumptions underlying a proposed project, although economic and social impacts do not need to be as comprehensively analyzed as do ecological impacts.

Indirect Effects

The CEQ regulations state that "indirect effects" are those "caused by the action and are later in time or farther removed in distance, but are still reasonably foreseeable.

Indirect effects may include growth inducing effects and other effects related to induced changes in the pattern of land use, population density or growth rate, and related effects on air and water and other natural systems, including ecosystems" (36 C.F.R. § 1508.8[b]).

Thus, the Bureau of Land Management was required by a court to analyze the air pollution impacts from ore transportation when it approved expansion of a gold mine. The court held:

> The air quality impacts associated with transport and off-site processing of the five million tons of refractory ore are prime examples of indirect effects that NEPA requires be considered.... BLM is incorrect in asserting that these effects need not be considered simply because no change in the rate of shipping and processing is forecast. That may be so, but the mine expansion will create ten additional years of such transportation, that is, ten years of environmental impacts that would not be present in the no-action scenario (*South Fork Band Council of Western Shoshone of Nevada v. United States Department of the Interior*, 588 F.3d 718, 725-26 [9th Circuit 2009]).

This holding corresponds with the CEQ regulation's demand that it is not just the current intensity of impacts that must be considered, but also the long-term impacts (40 C.F.R. § 1508.27).

In contrast, the same court held that the government did not need to consider the environmental effects of a housing development when it granted loan guarantees to home buyers, finding: "The agencies' loan guarantees have such a remote and indirect relationship to the watershed problems allegedly stemming from the urban development that they cannot be held to be a legal cause of any effect on protected species for purposes of... NEPA. This case stands in contrast to those where the disputed agency action had a more direct, on-the-ground effect and where the environmental mandates thus had to be followed by the agencies" (*Center for Biological Diversity v. United States Department of Housing and Urban Development*, 2009 WL 4912592 [9th Circuit 2009]). Once again, the causal effect of an action is key: with the mine case, the air impacts would not have occurred but for the mine expansion, and so they needed to be considered. However, the housing development may very well have been built without the loan guarantees (although the case does not really indicate whether this is true).

Cumulative Effects

The CEQ regulations define a cumulative impact as:

> [T]he impact on the environment which results from the incremental impact of the action when added to other past, present, and reasonably foreseeable future actions regardless of what agency (Federal or non-Federal) or person undertakes such other actions. Cumulative impacts can result from individually minor but collectively significant actions taking place over a period of time (40 C.F.R. § 1508.7).

It is important to note that cumulative impacts do not refer to simply "'all [of a project's] expected impacts when added together.' [Rather,] cumulative impacts to which the regulation refers are those outside of the project in question; it is a measurement of the effect of the current project along with any other past, present, or likely future actions in the same geographic area" (*Taxpayers of Michigan Against Casinos v. Norton*, 433 F.3d 852, 864 [D.C. Circuit 2006]).

The purpose of considering cumulative impacts is to avoid the tyranny of small decisions; in isolation, the effects of a project might be minor, but once added to all the other actions affecting the resource, the project at issue might cause a tipping point at which, perhaps, the project should not go forward as planned. Once again, it is important to keep the issues of the scope of the actions and the scope of the effects distinct, and that even if an EIS is not required to include multiple projects as primary subjects to be analyzed in the EIS, the agency must still adequately analyze the cumulative effects of those other projects in the EIS.

Thus, suppose you are reviewing a housing development where the effects of sediment on a nearby fish spawning stream are at issue, and there is already an ongoing mining operation and a proposed timber sale in the area. The mining operation and timber sale may not be connected actions that must be included as the subject of the EIS (and therefore the EIS need not consider, for instance, alternatives to the mining operation and timber sale). However, for purpose of the effects analysis in the EIS, you still must not only analyze the housing development's contribution to sedimentation and the impact it causes, but also must consider that contribution combined with that of the mining operation and the timber sale, each of which might not cause significant problems for the fish species but when taken together would cause unacceptable impacts.

Other examples include the following: When approving a power line to connect Mexican power plants with southern California markets, the Department of Energy had to "expressly disclose the past [and] present levels of air emissions in the Salton Sea Air Basin, [and] consider the combined effects of the present actions when added to any unrelated, reasonably foreseeable future electricity generation projects in the air basin" (*Border Power Plant Working Group v. U.S. Department of Energy*, 260 F. Supp. 2nd 997, 1032 [Southern District of California 2003]). When approving timber sales, the Bureau of Land Management had to consider the cumulative effects of other timber sales on the spread of a harmful fungus infecting national forests (*Kern v. U.S. Bureau of Land Management,* 284 F.3d 1062, 1077–78 [9th Circuit 2002]). When approving an off-road vehicle project that was part of a larger system, the U.S. Forest Service had to analyze the impacts of the project along with the impacts of the interconnected off-road vehicle system as a whole (*Mountaineers v. U.S. Forest Service*, 445 F. Supp. 2nd 1235, 1246–49 [Western District of Washington, 2006]).

The CEQ regulations state that a cumulative impact is that which results from the "incremental impact" of an action, "when added to other past, present and reasonably foreseeable future actions." This means that an EIS should examine both the incremental and baseline impacts together, and a cumulative impacts analysis that "only fully accounts for the incremental environmental effect ... above its current use levels" is not permitted. Rather, "it is the additive effect of both agency and other actions taken together that constitutes the gravamen of appropriate cumulative impacts analysis under NEPA." (*Mountaineers v. U.S. Forest Service*, 445 F. Supp. 2nd 1235, at 1247–1248.)

It is important that a cumulative effects analysis must be considered with the direct effects, and may not be segregated into a separate section if that separate section merely contains a listing of other projects without the kind of analysis contained in the direct effects section. In analyzing the cumulative effects of a proposed action, an EA or EIS must do more than just catalogue relevant past projects in the area, it "must also include a 'useful analysis of the cumulative impacts of past, present and future projects,'" with sufficient detail to assist "the decisionmaker in deciding whether, or how, to alter the program to lessen cumulative impacts" (*City of Carmel-by-the-Sea v. U.S. Department of Transportation*, 123 F.3d 1142, 1160 [9th Cir. 1997]). While a separate cumulative effects section in an EA or EIS is not prohibited per se, it will have to repeat much of the analysis found in the direct effects section to be sufficient, and it probably makes sense to integrate cumulative effects into the direct effects section so as to avoid either duplication or an insufficient effects analysis.

As the courts have stated, a "calculation of the total number of acres to be [impacted by the other projects] is a necessary component of a cumulative effects analysis, but it is not a sufficient description of the actual environmental effects that can be expected from logging those areas" (*Klamath–Siskiyou Wildlands Center v. BLM*, supra, 587 F.3d at 995). "[I]n assessing cumulative effects, the Environmental Impact Statement must give a sufficiently detailed catalogue of past, present, and future projects, and provide adequate analysis about how these projects, and differences between the projects, are thought to have impacted the environment" (*Lands Council v. Powell*, 379 F.3d 738, 745 [9th Cir. 2004]). An EA or EIS "cannot simply offer conclusions. Rather, it must identify and discuss the impacts that will be caused by each successive [project], including how the combination of those various impacts is expected to affect the environment, so as to provide a reasonably thorough assessment of the project's cumulative impacts" (*Klamath–Siskiyou Wildlands Center v. U.S. Bureau of Land Management,* supra, 587 F. 3rd at 1001).

Sometimes the question arises of what actions and impacts are "reasonably foreseeable." This requirement has been described as those projects that are "sufficiently likely to occur that a person of ordinary prudence would take [them] into account in reaching a decision" (*Louisiana Crawfish Producers Association-West v. Rowan*, 463 F. 3rd 352, 358 [5th Circuit 2006]). If a Notice of Intent to prepare an EIS for a project or even an agency press release announcing a proposed project has been issued, that makes it reasonably foreseeable (*Northern Alaska Environmental Center v. Norton*, 361 F. Supp. 2nd 1069, 1081–82 [District of Alaska 2005]). In contrast, where a future project is too speculative, it need not be included in a cumulative effects analysis (Louisiana Crawfish Producers Association, supra, 463 F. 3rd 358). This determination requires a case-by-case examination for which generalizations may not be especially helpful.

Where some impact is expected from a reasonably foreseeable project, there is an information gap as to the extent of those impacts, which does not excuse a refusal to address the impacts. For example, the effects of future coal combustion spurred by a coal railway expansion were required to be considered, even if the contours of future power plants that would burn the coal were not yet defined (*Mid-States Coalition for*

Progress v. Surface Transportation Board, 345 F. 3rd 520, 549–50 [8th Cir. 2003]). In such a case, the EA or EIS should include:

> (1) A statement that such information is incomplete or unavailable; (2) a statement of the relevance of the incomplete or unavailable information to evaluating reasonably foreseeable significant adverse impacts on the human environment; (3) a summary of existing credible scientific evidence which is relevant to evaluating the reasonably foreseeable significant adverse impacts on the human environment, and (4) the agency's evaluation of such impacts based upon theoretical approaches or research methods generally accepted in the scientific community (40 C.F.R. § 1502.22).

Accordingly, when preparing an environmental document, the practitioner must assess the impacts of a project, even if not absolutely sure about underlying scientific issues. By disclosing any scientific uncertainties, you can appropriately reduce the risk of relying on bad science.

CONCLUSION

The cumulative effects requirement of NEPA and the related issues discussed in this chapter are intended to take a holistic approach to disclosing and assessing the true impact of a proposed development or other agency action. While the legal issues can sometimes be tricky, always keep in mind that the role of an EA or EIS is to help the government decide whether a project should be approved, and whether a better alternative exists that can achieve that project's goals but with less environmental impact. If you approach each issue with that in mind, it will serve the environmental review document and process well.

3 Regulating and Planning for Cumulative Effects

The Canadian Experience

Chris J. Johnson

CONTENTS

INTRODUCTION

Biologists, land-use planners, and resource managers working in Canada recognize the importance of the cumulative effects of human developments when managing and conserving wildlife (Schneider et al. 2003; Johnson et al. 2005; Aumann et al. 2007; Nitschke 2008). Although acute occurrences of disturbance and habitat change are of immediate concern during most regulatory decisions, it is the cumulative changes in natural systems resulting from incremental or continuous anthropogenic activities that are the most difficult to understand, manage, and, if necessary, mitigate. There are too many case studies to ignore or dismiss the long-term influences of humans on the distribution and abundance of wildlife (Mattson and Merrill 2002; Laliberte and Ripple 2004; Vors et al. 2007).

Canadian landscapes are still some of the most wild and undeveloped in the world. Compared to other countries with much longer histories of resource development and greater densities of people and associated disturbances, Canadians still have a wealth of natural capital and a range of options as to what resources they choose to develop, at what rate, and at what costs relative to other ecosystem services. However, rapid increases in the price of oil and natural gas, continuous expansion of the forest industry into boreal regions, recent discoveries of rare minerals such as diamonds, and even efforts to develop more sustainable power sources have increased pressures on these wild spaces (Houle et al. 2010; Schneider et al. 2003; Nitschke 2008). With the exception of a transformative shift in global economics, the cumulative effects of these activities will likely not diminish over the short or long term.

By area, Canada is the second largest nation in the world, and fronting three oceans has the longest coastline of any country. This huge area supports a population of only 33.6 million people. Ranked 2nd and 20th in proved oil and natural gas reserves, respectively, being the largest exporter of softwood lumber, and a top 10 producer of many rare minerals, gold, and other base metals, Canada is rich in natural resources (Central Intelligence Agency 2009). The global environmental sustainability index ranks Canada 4th out of 146 countries for Environmental Systems, but 104th for Reducing Environmental Stresses (Esty et al. 2005). Similarly, a poor track record of sustainable practices was reported by Boyd (2001): compared to the 29 other countries in the Organisation for Economic Cooperation and Development (OECD) Canada ranked 28th overall across 25 environmental indicators. Considering the management and conservation of wildlife, there are examples from across the country that illustrate declines in highly visible species resulting from the cumulative effects of development (Schaefer 2003; Nielsen et al. 2004; Courtois et al. 2007; Vors et al. 2007; Gibbs et al. 2009; Williams et al. 2009); the results are likely more striking for the less-known elements of Canadian biodiversity.

At a national scale, there is a disconnect between the wealth of natural capital and the effectiveness of natural resource management. Although Canada has many of the structural elements necessary for effective environmental policy and legislation, there is obvious room for improvement. A thorough review, analysis, and understanding of this problem would cover a wide range of topics from intergovernmental affairs to regional economic development. Focusing on the elements that directly influence the natural resource professions, environmental policy and legislation are relatively transparent and easily reviewed. There is a strong body of work critiquing these areas, and covering a range of ecosystem types and values. In particular, the concept of cumulative effects in both its formalized and conceptual form and cumulative effects analysis as practiced across Canada has received considerable attention (Ross 1998; Bonnell and Storey 2000; Baxter et al. 2001; Duinker and Greig 2006).

Cumulative effects can be defined as the synergistic, interactive, or unpredictable outcomes of multiple land-use practices or development that aggregate over time and space, and have significant impacts for valued components of the environment (Ross 1998; Harriman and Noble 2008). Given the temporal, spatial, and cross-sectoral breadth of potential effects, some argue that cumulative effects can only be dealt with through integrated approaches that consider the broader ideas of sustainability, including limits on regional growth or development activities (Duinker and Greig 2006). As currently represented in federal and some provincial legislation, cumulative effects are definable outcomes of industrial developments that are to be measured during approval or permitting processes. Although resource professionals working in Canada often consider the legislative elements of the environmental assessment process when assessing and mitigating cumulative effects, there are a number of alternative or complementary approaches that could result in the better management of Canadian landscapes for wildlife values.

In this chapter, I introduce the concept of cumulative effects from the perspective of regulation and land-use management in Canada. I do not focus explicitly on the characteristics of different effect types (Fox et al. 2006; Spies et al. 2007; Yamasaki et al. 2008) or methods for quantifying effects, but instead mechanisms that might be applied to the management of effects. I use the terms effect and impact, where effect is a change to the environment and impact represents the consequences of such changes (Wärnbäck and Hilding-Rydevik 2009). The supporting chapters are focused on cumulative effects in the context of wildlife management and conservation, which offers a somewhat limited scope when reviewing and assessing the regulatory elements of the topic. Cumulative effects will directly affect populations of animals; often this is a recognized focus of environmental assessment processes and includes the terrestrial or aquatic communities that constitute habitats. Thus, I present a perspective that includes a range of ecological elements not all directly linked to the distribution or abundance of individual wildlife populations.

From a regulatory perspective, cumulative effects are often considered within the context of supporting environmental assessment legislation. Although useful for assessing the impacts of acute events, I believe that the management of wildlife in the context of cumulative effects needs to move beyond the prescriptive environmental assessment process. Indeed, others have echoed this thought both within and beyond Canada (Dubé and Munkittrick 2001; Duinker and Greig 2006; Soderman 2006; Halpern et al. 2008; Samarakoon and Rowan 2008; Zhu and Ru 2008). Working outside the bounds of federal and provincial legislation that speaks directly to the inclusion of cumulative effects within the environmental assessment process, I broaden my focus to include other regulatory, policy, and planning mechanisms. Certain tools, such as federal environmental assessment legislation, require a national perspective. However, given the range of mechanisms applied at the provincial and territorial level, I will focus much of my attention on the experiences and approaches from British Columbia (B.C.). There are a number of reasons to focus on Canada's third largest province. From an ecological perspective, B.C. is the most diverse in terms of species richness and ecosystems (Austin et al. 2008). From an effect perspective, B.C. is the second largest producer of natural gas, exports the greatest amount of forest products, is Canada's third largest generator of hydroelectricity, and has significant deposits of coal, minerals, rare, and base metals. Also, B.C. has implemented regional land-use planning since the early 1990s; those lessons are useful when looking at alternative strategies for implementing cumulative effects management and monitoring.

The information presented in this chapter is retrospective and builds on the theses and research of others. Counterintuitive to past process, I make the argument that, in Canada, environmental assessment legislation is not the starting point or the most important tool for dealing with cumulative effects. Instead, we should focus on regionally scaled frameworks that encompass many of the elements of integrated land-use planning; monitoring is an integral component of that approach (Dubé et al. 2006). Traditional environmental assessment legislation continues to play a role, but only in regulating developments in the context of project-specific effects. Indeed, this form of regulation should be nested within broader planning initiatives (Gunn 2009).

I begin at the national level with a review and critique of federal environmental assessment legislation with an emphasis on the elements that apply to cumulative effects. Also, I discuss the interplay between the federal and provincial assessment processes. In the context of the federal review process, I then discuss the opportunities and successes of strategic environmental assessment (SEA). Applied in numerous countries, some have argued that SEA offers solutions to the limitations of project-specific cumulative effects analyses as practiced in Canada. From a review of cumulative effects as defined in legislation, I extend the scope of review to a broader set of potential tools. I discuss the application and utility of cumulative effects assessment and management frameworks. Finally, I move beyond cumulative effects as a focal point and discuss some of the successes, failures, and limitations of integrated land-use planning.

APPROACHES FOR MONITORING AND CONTROLLING CUMULATIVE EFFECTS

Environmental Assessment Legislation

Canada is a federation of ten provinces and three territories. Under the constitution, the federal and provincial governments have specific and sometimes overlapping jurisdiction over the various elements of policy, law, and taxation. Management of surface and subsurface natural resources, with the exception of the continental shelf, and the majority of environmental policy and law is governed by the provinces. In the case of industrial development and the environmental assessment process, however, there is an overlap between the federal and provincial legislation. The federal government administers and implements the Canadian Environmental Assessment Act across all Canadian lands and waters. Each province has developed parallel legislation and regulations that best meet the needs and issues of that jurisdiction (Noble 2004a). Although this approach has resulted in notable regulatory gaps in marine systems (McDaniels et al. 2005), the issue is less acrimonious for most developments that affect terrestrial wildlife. In the case of B.C., efforts have been made to find efficiencies in the assessment process when both sets of legislation apply. This includes a Cooperation Agreement between the Canadian Environmental Assessment Agency and the Environmental Assessment Office of B.C. promoting integrated delivery of cooperative environmental assessments that minimizes duplication, and equivalency agreements that recognize the authority or equivalence of the federal environmental assessment process.

At the federal and provincial levels, with a few exceptions, the environmental assessment legislation is focused at specific projects or activities, normally with acute and highly noticeable effects over a relatively small area. This is especially true for projects that fall within the provincial mandate. In both cases, there are no mechanisms to review developments that fall outside the legislation, but over the long term may have extremely large cumulative effects. The assessment process, however, is about more than simply providing government authorization to implement a publicly visible project such as a mine or gas pipeline. Through a series of steps, stakeholders and affected parties are informed of the scope, intent, benefits, and impacts of the project, the magnitude of the socioeconomic and environmental effects are determined, and mitigation strategies are developed (B.C. Environmental Assessment Office 2003).

The Federal Environmental Assessment Act applies to projects where the Government of Canada is the decision-making authority either as a proponent, being responsible for the land, as a source of funding, or as a regulator of a particular type of development. The federal process is hierarchical, with an incremental level of scrutiny and public involvement proportional to the expected magnitude of effects. At the lowest level, known as a screening, the environmental effects of a proposed project are documented, and mitigation strategies are defined that minimize or eliminate negative project effects. A comprehensive study is a more intensive process applied to projects that are thought to have a more significant level of environmental effect. When a project involves only a few interest groups, likely with a limited range of environmental effects, the Minister of Environment may appoint a mediator who can facilitate the mediation process. At the largest scale of effect, either involving a diverse range of values or large level of effect, a review panel is appointed. The panel will normally hear testimony from experts, interest groups, and the general public on the benefits and impacts of the project and then submit a report to the minister for decision.

At the provincial level, the B.C. Environmental Assessment Act comes into force after special request by the Minister of Environment, the proponent, or, as in most cases, the project exceeds a threshold set in regulation. Based on the category of development, thresholds are set according to project size, production capacity, or other criteria related to the presumed significance of effect (B.C. Environmental Assessment Office 2003). For example, a new coal or mineral mine will exceed the review threshold if it produces ≥250,000 tonnes of coal or ≥75,000 tonnes of mineral ore per year.

Working from a premise of development thresholds, the provincial environmental assessment legislation explicitly excludes smaller less-significant projects. Although this process is more streamlined and efficient from a proponent and governmental perspective, many smaller projects fall outside the scope of regulation, but could potentially result in a large magnitude of cumulative effects. Also, developments outside of the categories of industrial, mine, energy, water management, waste disposal, food processing, transportation, and tourist destination resort projects are not considered by the B.C. Environmental Assessment Office. Thinking about the area, and perhaps not the magnitude of effect, the most obvious omissions are forest harvesting and exploration activities for minerals, oil, and gas. These activities fall under separate permitting processes, and from a policy perspective integrated land-use planning (see the following text). Also, there is no explicit acknowledgement or requirement of cumulative effects in the act or supporting regulations. This is a notable omission given the recent implementation of the act and the formal recognition of cumulative effects in other jurisdictions including the federal legislation.

The Canadian Environmental Assessment Act is less exclusionary than the provincial counterpart. More project types across a broader inclusionary scope fall within the realm of this legislation. However, it is worth noting that lands under federal jurisdiction, including national parks, national defense installations, and Indian Reserves, cover approximately 4% of the B.C. provincial land base. Furthermore, many projects that receive a federally mandated environmental assessment are relegated to the lowest level of review, screening. More stringent assessments based on independent panels or comprehensive studies are less common and normally focused on large-scale projects such as nuclear or large hydroelectric facilities, pipelines, oil sands developments, or large mines. Once again, the legislation is not fine-tuned to smaller levels of developments that can have incremental cumulative effects over time. Indeed, many of the outcomes of these assessments are mitigation strategies that can limit or avoid the site-specific or immediate effects of a development but fail to mitigate impacts in the broader context of landscapes and mobile populations such as wildlife.

Although there are criticisms of the process and requirements under the Canadian Environmental Assessment Act, relative to some provincial legislation, including that of B.C., cumulative effects are legally recognized as a required element in all assessments. Looking to Subsection 16(1) of the act, each assessment must consider "any cumulative environmental effects that are likely to result from the project in combination with other projects or activities that have been or will be carried out." Under the act, cumulative effects are much broader than biophysical factors and include all elements of "environmental effects" including changes in health and socioeconomic conditions, physical or cultural heritage, or the use of lands and resources for traditional purposes by aboriginal persons. Given the complexity of the requirement, including proper scoping and analysis, the Canadian Environmental Assessment Agency worked with an independent multistakeholder committee to develop a Practitioners Guide (Hegmann et al. 1999) that followed up on earlier written guidance (Federal Environmental Assessment Review Office 1994).

One of the more difficult aspects of applying the cumulative effects requirement under the act is the consideration of the effects of the current proposal in the context of future projects that "will be carried out." In this case, responsible authorities should consider projects that are certain and reasonably foreseeable. The act does not require or limit consideration of hypothetical projects. Although not a legislative requirement, scenario planning and forecasting based on habitat supply or land-use models has been a recognized and somewhat fruitful area for research and development (Schneider et al. 2003; Peterson et al. 2003; Johnson and Boyce 2004). Unfortunately, many of these tools have seen little practical application (Duinker and Greig 2007).

Through federal legislation, the concept of cumulative effects has taken on a formal role in the environmental assessment process beyond just best practices or policy. There is considerable evidence, however, to suggest that cumulative effects are not adequately addressed during project evaluations (Burris and Canter 1997; Baxter et al. 2001). Many have argued that cumulative effects are inadequately represented in existing legislative frameworks or, at a more fundamental level, not served well by the structure and application of the environmental assessment process (Creasey 1998; Kennett 1999; Davey et al. 2001). Cumulative effects are not immediately associated with the time and place of a proposed development or activity and, therefore, it may be difficult to define the extent or magnitude of an impact (McCold and Saulsbury 1996). Also, under the current legislated process, there is often an inherent lack of a strategic vision for a process that is expected to encompass spatial and temporal domains that exceed the footprint of the proposal that triggered the assessment. Although cumulative effects is not a new idea, the environmental assessment process was developed and is structured to consider the effects of individual projects, not multiple projects that may span large areas, jurisdictional boundaries, and considerable time periods. Furthermore, environmental assessment in Canada and beyond (Dixon and Montz 1995) is a reactive proponent-driven process that in many cases occurs without guidance from longer-term geographically broader planning processes. This failing of the legislation and process has been recognized by the Canadian government. In 2003, the Canadian Environmental Assessment Act was amended to allow the use of studies that document cumulative effects across regional areas (Canadian Environmental Assessment Agency 2007).

Two recent works provide a critique of the role of cumulative effects in the environmental assessment process. Looking at fundamental principles, not individual case studies, Duinker and Greig (2006) make an impassioned argument for the failure of cumulative effects as recognized and practiced within Canada. Drawing on experience as practitioners within the federal assessment process, they outline six major conceptual problem areas and offer solutions for repositioning cumulative effects in a framework that meets broader goals of environmental sustainability. From a structural perspective, Duinker and Greig (2006) claim that cumulative effects are not well represented in project-specific approval processes. Economic pressures and the philosophy and practice of environmental assessment leave proponents focused on approval and regulators on minimizing impacts. Within the constructs of the review process, neither of these key players is fully engaged in quantifying the complete range of past, current, and future effects. Even where the proponent is interested in full accounting of effects, current or future development activities of other or similar industries may be proprietary or difficult to establish.

Duinker and Greig (2006) also make the case that the process is flawed and the practitioners are not well prepared for the task. They argue that cumulative effects should be the primary focus, not a secondary consideration once the project-specific effects have been presented. Also, practitioners and regulators require more training and a better understanding of systems approaches for representing human–environment interactions; such ideas were echoed by Bellamy et al. (2001). They advocate adaptive environmental assessment and management as a tool to facilitate such understanding; however, these approaches do not go without criticism (Simberloff 1998) and certainly violate the efforts of some jurisdictions, such as B.C., attempting to increase business competitiveness by streamlining the assessment process. Considering the time-horizons over which cumulative effects analyses are conducted, Duinker and Greig (2006) state that sustainability is a long-term concept. To be effective, practitioners must broaden the scope of analysis to include a wider range of potentially interacting effects over a longer time period.

The science of cumulative effects analysis does not escape Duinker and Greig's (2006) attention. They argue that, cumulative effects, once summarized, must be interpreted within the context of biophysical thresholds. This is a concept that has been the focus of ecologists (Sorensen et al. 2008) and regulators alike (AXYS 2001), and transcends the application of environmental assessment legislation. Although Duinker and Greig (2006) recognize the difficulty in identifying useful species-development thresholds, including the nonlinear dynamics of such effects (Crain et al. 2008), they offer no prescription for solving this vexing problem.

Baxter et al. (2001) were also critical of the cumulative effects process in Canada. Adopting an analytical approach, they used eight criteria centered on the three main steps of effects analysis—context and scoping, analysis, and management—to evaluate 12 Canadian cases. Reporting three key findings, Baxter et al. (2001) concluded that the process of evaluating cumulative effects was not distinct from project-specific effects. They argue that these sets of effects are distinct and demand unique methods and consideration. Second, most of the studies they reviewed had an insufficient level of scoping. Scoping exercises did not specifically address cumulative effects, failed to identify and validate all valued ecological components, and did not define boundaries for future impact scenarios. Third, many studies did not conduct analyses of the magnitude and range of cumulative effects for the majority of valued ecological components identified in the assessments. Often a component of the process that is challenged by a lack of empirical data, rigorous methods are available (Johnson and Boyce 2004; Shifley et al. 2008; Yamasaki et al. 2008). As noted by Baxter et al. (2001), these approaches are relatively well developed and are used when considering the cumulative impacts for wildlife species.

As currently practiced, environmental and cumulative effects assessment is likely to persist in Canada. The emphasis on resource development as a mechanism for economic growth, while maintaining other environmental values, requires a process with defined timelines and certainty. Baxter et al. (2001) provides recommendations for improvements that do not require a radical restructuring or scrapping of the current process. They suggest that cumulative effects analysis should be considered fully in the terms of reference for the underlying environmental assessment. This will explicitly identify and mandate the practice of cumulative effects analysis at the start of the process, provide a framework for critical feedback by the public and other participants, and serve as a benchmark against which the analysis can be judged. Second, Baxter et al. (2001) state that cumulative effects scoping is the most important part of the process; thus, they recommend the inclusion of methods or a process at the start of the assessment that identifies the range of potential cumulative effects, including their spatial and temporal boundaries. Finally, they advocate for more effective follow-up provisions including monitoring and environmental management plans. Without such systems, the effectiveness of the assessment process and mitigation strategies will remain unchecked.

Baxter et al. (2001) make a fourth recommendation that requires a more systemic change to how cumulative effects analyses are conducted across Canada. They argue for direct linkages between project-specific and strategic or regional-level cumulative effects assessments. This is consistent with the thoughts of Duinker and Greig (2006) and others (Conacher 1994; Creasey 1998; Kennett 1999; Davey et al. 2001; Gunn 2009) who argue for a strategic-tiered framework that sets land-use objectives across sectors and documents the full range of effects while supporting the decision-making process.

Strategic Environmental Assessment

Strategic environmental assessment is a process for considering the cumulative effects of development before project-specific assessments are required by federal or provincial regulatory agencies. Although SEA is practiced in a large number of countries (e.g., Retief et al. 2008; Zhu and Ru 2008; Sinclair et al. 2009), there is surprisingly little consensus on methodological approach or even criteria defining good practice. Broadly defined, SEA is a high-level comprehensive analysis of the environmental effects or impacts of policy, plan, or program (PPP) alternatives at an early stage in the decision-making process. Policy is a somewhat nondescript term that recognizes the intent, formal, legal, or otherwise, to follow some action; a plan is a set of coordinated and timed objectives and strategies for implementing a policy; and a program is a collection of projects, typically within one development sector, within a defined area that corresponds with guidance provided in a plan (Wood and Dejeddour 1992; Noble 2002). Strategic Environmental Assessment is intended to identify the environmental impacts of a range of strategies, choosing and then implementing a PPP that can direct future proposals for individual developments. Thus, SEA aids decision making at more detailed or site-specific levels of the regulatory process.

Although no consistent method or practice of SEA exists, most jurisdictions and practitioners agree that the approach is premised on meeting governmental or corporate objectives of sustainability (Brown and Therivel 2000; Nobel 2002). This is a structural deficiency in the process for project-specific cumulative effects analyses (Duinker and Greig 2006). Other advantages of SEA include a proactive as opposed to reactive framework; less emphasis on approval and mitigation of individual projects with an explicit focus on the acceptability of development objectives at the sectoral or regional scale; greater level of public involvement in scoping the range, magnitude, and acceptability of cumulative effects; establishment of impact thresholds; streamlining of project-specific EA process; and the collection of information necessary for the decision-making process including the identification of knowledge or data gaps and research/inventory priorities (Bonnell and Storey 2000; Davey et al. 2001; Harriman and Noble 2008).

In Canada, formal recognition of SEA is most well established at the federal level. Beginning in 1990 and then strengthened in 1999, the Canadian government introduced the Cabinet Directive on the Environmental Assessment of PPP Proposals (Canadian Environmental Assessment Agency 1999b). This directive requires that all federal government departments formally consider the environmental implications of PPP proposals submitted to the Cabinet.

Based on the theoretical strengths of SEA relative to the project-specific assessment process, one would think that the formal enactment of SEA through the Cabinet directive is a step forward for cumulative effects analysis and consideration. However, 10 years of implementation has identified some large weaknesses. From the perspective of implementing SEA in Canada, the triggering mechanisms are limited. At the federal level, the directive is enacted only if a proposal is submitted to an individual minister or the Cabinet for approval, and the implementation of the proposal may result in important environmental effects. There is likely much room for interpreting the importance of "environmental effects." This is compounded by the directive being a policy document, not a legislated requirement (Hazell and Benevides 2000). Thus, the requirements of the directive are less precise (i.e., legally defensible) and unchallengeable in the courts. To date, the majority of SEA initiated as a response to the directive have addressed plans or programs, not policy (Renton and Bailey 2002).

As discussed by Noble (2000a), policy formulation is an element of governance that is difficult to bound, perhaps leading to a conservative interpretation of what constitutes policy. However, policy, the highest level in the SEA hierarchy, can set in place government-sponsored actions, including international agreements that result in effects that have broad-scale and long-term impacts for the environment.

Most provincial jurisdictions lag behind the federal government in recognizing and practicing SEA (Noble 2004a). Within B.C., governmental policies and practices may be reviewable under the Environmental Assessment Act. As specified in the act (Sec 49), the Minister may direct the Environmental Assessment Office "(a) to undertake an assessment of any policy, enactment, plan, practice, or procedure of the government." However, such reviews are at the discretion of the government, a triggering mechanism more restrictive than the criticized federal directive. Based on a survey of the responsible provincial agencies found across Canada, respondents reported that implementation of the SEA was hindered by a lack of legislative requirements and a failure to understand the nature and benefits of higher-level assessments (Noble 2004a). Similar limitations were reported for other countries (Wärnbäck and Hilding-Rydevik 2009).

Working from a nonjurisdictional perspective, Noble (2009) assessed the state of SEA as implemented across Canada. Recognizing a lack of defined SEA practice, he assumed a liberal set of criteria for identifying ten SEA and SEA-type case studies. Reviewing applications of SEA that ranged from regional growth strategies to offshore petroleum development to aquaculture policy, Noble (2009) reported that the practice and achievements of SEA across Canada are variable. Successes included public participation and transparency; scoping that considered related strategic initiatives and identification and narrowing of valued ecological components; and a clear delineation of assessment roles and responsibilities. However, there were deficiencies in the SEA system, process, or results that spoke to failed opportunities for comprehensive reviews that, in some cases, consumed large amounts of human and financial capital and dealt with issues with significant environmental consequences. Most notably, the reviewed SEA failed to operate within a tiered system of assessment, planning, and decision making; provided few opportunities for participants to appeal the process or output; and did a poor job of considering cumulative effects. Noble (2009) demonstrated that SEA can and should be applied using context-specific frameworks and that examples of good practice can be found in review processes that were never branded or conceived as SEA.

Across Canada, SEA has become a much-championed approach for addressing the deficiencies of project-specific cumulative effects analyses (Davey et al. 2001). This set of ideas is touted as a cure-all for regulators and industries struggling to effectively represent and consider cumulative effects during the tenure issuance or licensing process (Bonnell and Storey 2000). With a broad temporal and often geographic focus encapsulating a range of options (e.g., regional suitability, location, intensity) for development before individual projects are proposed, SEA appears to be a near-perfect instrument for addressing cumulative effects. In particular, the advantages of SEA are well suited for nations such as Canada where strategic decisions on levels or types of development are still an option. There are still obvious challenges. As Noble (2009) discovered, some SEA processes have failed to include cumulative effects altogether.

Some have suggested that regional environmental assessment (REA), also referred to as regional cumulative effects analyses, is a more precise and effective form of SEA for many of the broad-scale environmental effects that apply to Canadian landscapes (Connacher 1994; Bonnell et al. 2001; Davey et al. 2001; Gunn 2009). Indeed, the importance of "regional studies" is now recognized in the Canadian Environmental Assessment Act. In the context of the environmental assessment process and SEA, region is a flexible concept with a range of applications in the context of cumulative effects analysis. For example, the World Bank (1999) defines REA as:

> An instrument that examines environmental issues and impacts associated with a particular strategy, policy, plan, or program, or with a series of projects for a particular region (e.g., an urban area, a watershed, or a coastal zone); evaluates and compares the impacts against those of alternative options; assesses legal and institutional aspects relevant to the issues and impacts; and recommends broad measures to strengthen environmental management in the region. Regional EA pays particular attention to potential cumulative impacts of multiple activities.

Here, the emphasis is still on PPP, but from a regional perspective. Harriman and Noble (2008) have added considerable precision to the term *regional*. Recognizing that conventional single-sector, proponent-driven assessments can encompass a regional area, they extend the definition delineating broad categories of analysis that focus on project-specific EA-type and strategic SEA-type approaches across large geographic areas. Thus, REA can serve to evaluate PPP or, at a less strategic level, after higher level direction is determined, the existing and projected cumulative effects of a range of developments across regions. Compared to project assessments that are inward focused on the specific impacts of one or more developments that may or may not have a regional scope, REA is outward focused on a range of interacting effects across a region (Harriman and Noble 2008). Failure to look outward beyond the development application limits our ability to address many of the deficiencies of project-specific approaches and develop decision-making frameworks that consider cumulative effects in all of their forms (Baxter et al. 2001; Duinker and Greig 2006).

The benefits of REA, in all of its forms, are numerous and include long-term development targets or plans in the context of sustainability, participation of all regulatory agencies and stakeholder groups, identification of a range of environmental effects early in the land-use decision-making process, assessment of baseline conditions and identification of data gaps, and development of monitoring and management frameworks that support the documentation of regional cumulative effects and the significance of project-specific impacts (Kennett 1999; Davey et al. 2001; Gunn 2009). The REA has been proposed or shown some success for a number of resource development sectors, and there are a range of technical approaches for understanding large-scale cumulative effects, especially from the perspective of animal and plant communities (Schneider et al. 2003; Johnson and Boyce 2004). However, a lack of multijurisdictional cooperation, no formal legislative or regulatory framework, difficulties in developing long-term involvement, and the need for considerable human and financial resources have hindered the application of REA across Canada (Barnes et al. 2001; Gunn 2009). Harriman and Noble (2008) argue that some of the perceived failures of REA are the result of an unrealized expectation for a one-size-fits-all model. They demonstrate that REA can take many forms and is often most successful when developed in the context of a particular region or set of land-use issues. As an example, assessment and management frameworks are a flexible and novel approach for monitoring and regulating cumulative effects across regional areas within the context of predefined objectives or targets for acceptable levels of effects.

Cumulative Effects Frameworks

In the absence of formal legislative or policy guidance, some jurisdictions in Canada have developed what are generically known as cumulative effects frameworks. An inherently flexible approach, frameworks have been defined as "an administrative structure that combines various initiatives that assist decision makers in assessing and managing the effects of human use on the land" (AXYS 2003a, *ii*). Examples include the Northwest Territories Cumulative Effects Assessment and Management Framework, Regional Cumulative Effects Management Framework for Cold Lake Alberta, and the Cumulative Effects and Management Framework for Northeastern B.C. Premised on many of the principles of REA, the frameworks are individual to a particular region and that region's environmental management, development sectors, and sources of cumulative effects. Founded not in provincial or federal legislation but in regional development challenges, these frameworks show some progress in addressing the diversity of cumulative effects found across Canada.

Gunn (2009) identified and reviewed ten Canadian cumulative effects frameworks. As with SEA, she concluded that the origin, goals, and development of each framework was unique, but there were some common themes including land-use planning and visioning, coordination among regulatory agencies, policy development, and monitoring of cumulative effects. In some cases, decision support tools were used to forecast cumulative effects over time and space that helped guide policy and planning decisions. The Crown of the Continent Managers Partnership (CCMP) is an excellent example of the objectives and the complexity of the issues confronted by these frameworks. Formed to address a number of cross-boundary challenges, including cumulative effects of human activities, sustainable use of wildlife populations, and data sharing and monitoring, the CCMP covers a mountainous region of 77,000 km^2 across southeastern B.C., southwestern Alberta, and northern Montana. The framework was designed to assist more than 20 government agencies responsible for jurisdictional and cross-jurisdictional issues including cumulative effects. Struggling with not only current-day administration of the Crown of the Continent area, partners were interested in developing a strategic vision based on an understanding of potential future cumulative effects. To assist this process, they used A Landscape Cumulative Effects Simulator (ALCES; Schneider et al. 2003) to model hypothetical futures in the context of natural and anthropogenic disturbance specific to the region. This modeling exercise was met with resistance, and the overall value for leading the CCMP was marginal (Gunn 2009).

The Northwest Territories Cumulative Effects Assessment and Management Framework (NWT CEAMF) further demonstrates the diversity of approaches and the flexibility of this particular model. Formed in the late 1990s as a response to the rapid development of the diamond mining industry, the NWT CEAMF is composed of a steering committee with representation from the territorial, federal, and First Nations governments and councils as well as nongovernmental and industry organizations. The committee is tasked with making recommendations or "refusable advice" to decision makers on a broad list of initiatives that encompasses ecological integrity, sustainable communities, and economic development. Although the NWT CEAMF implementation blueprint identifies baseline studies, research, and monitoring as necessary components of cumulative effects management, there has been little progress in this direction (NWT CEAMF Steering Committee 2007). This is despite a recognized concern by northern communities for the sustainability of wildlife populations, in particular, barren-ground caribou (*Rangifer tarandus*), and research on appropriate methods for conducting regional cumulative effects assessments (Johnson and Boyce 2004). Slow progress on such goals has influenced the overall legitimacy of the framework (Gunn 2009). Recognizing a broader mandate than cumulative effects, the Steering Committee recently changed the name of the Framework to the NWT Environmental Stewardship Framework.

Although the attributes of each framework are unique, Gunn's (2009) research has identified some common themes that reflect on the success and failures of this approach for managing or limiting cumulative effects. For the four frameworks she reviewed, a stakeholder-defined regional vision for future development was found to be very important. This is consistent with a broad spatiotemporal perspective on cumulative effects that engages a range of land-use sectors and their associated players. However, finding consensus across jurisdictions or representatives and implementing the vision is difficult and can hamper the overall effectiveness of the CEAMF.

Whereas predicting the impacts of development is important at the project level, from a more strategic perspective CEAMF committees were less dependent on modeling exercises that provided precise estimates of effects across complex regional environments. Instead, effects modeling (e.g., MARXAN, ALCES) provided an opportunity to identify valued ecosystem components and associated indicators, identify data gaps, and establish regional thresholds and targets. As a related problem, members of the frameworks often had difficulties linking the strategic nature of CEAMFs to regulatory decisions. Translating strategic visions into operational guidance and tracking progress toward meeting long-term goals was difficult, especially when participants were positioned in agencies tasked with project-level decisions.

Through Gunn's (2009) research, it was clear that members of CEAMFs were aware of the ever-increasing pace of development across many of the study regions and the need to act immediately. Limitations in knowledge or data were not reasonable excuses for failure to begin implementation of these strategic initiatives. And, when implementing frameworks, often with a large number of components or objectives (e.g., NWT CEAMF Steering Committee 2007), the primary outcome was to regulate the pace of development. This final point clearly differentiated frameworks that focused on strategic decisions about regional levels of effects over a particular time or area from project-specific assessments that resulted in a series of yes or no decisions with associated mitigation strategies.

Although CEAMFs have shown much potential for assessing and managing cross-sectoral cumulative effects at regional scales, effective implementation has not been without difficulties (Gunn 2009). One dramatic failure was the CEAMF for northeastern B.C. With a focus on oil and gas, the framework was initiated to address project-specific and regional cumulative effects, across an area with globally significant populations of ungulates and one of the largest remnants of undeveloped wilderness in North America, south of 60° latitude (Gustine et al. 2006). The stakeholder engagement and paper framework (AXYS 2003a, b) was completed in 2003 at a cost of more than $300,000 (Canadian). To date, there has been no implementation or formal achievements that draw from the outcomes of that work.

Land-Use Planning—An Alternative

When considering cumulative effects, our attention is first drawn to regulatory frameworks, such as the Canadian Environmental Assessment Act or the U.S. National Environmental Policy Act that approves or rejects project proposals based on the perceived significance of effects. Strategic EA and REA are more flexible, less-structured processes, but the emphasis is still on effects assessment, although at larger temporal, spatial, and sectoral scales. The one defining premise for this nested family of approaches is the occurrence of single or multiple developments and resulting effects. Integrated land-use planning is a more holistic approach for managing cumulative effects. Inherently proactive and cross-sectoral, planning processes are focused on developing goals and objectives for a range of land uses across a region, followed by strategies to achieve those aims, and finally monitoring to ensure that the goals and objectives are met. Acceptable types and levels of development are identified before cumulative effects assessment processes, at the project or regional level, are triggered. The linkages among land-use planning, environmental assessment, and cumulative effects analysis are well established, especially in the context of SEA and CEAMF (e.g., NWT CEAMF) (Noble 2009). As noted by Bardecki (1990, p. 322), “Assessing and managing cumulative impacts is planning.” However, integrated land-use planning is a complex process that might be initiated by cumulative effects concerns, but ultimately occurs outside the set of approaches found within the cumulative effects family.

In some Canadian jurisdictions, forestry or oil and gas activities that occur over large areas are deeply entrenched in local and provincial economies, and the effects of individual activities are often marginal and only significant from an incremental perspective (Schaefer 2003; Sorenson et al. 2008). In these cases, mechanisms for broad-scale public involvement and legislation that prescribe the magnitude and types of land use are essential for facilitating sustainable levels of growth. Such tools fall within the realm of integrated land-use planning. Of the state and provincial jurisdictions found across North America, B.C. is arguably the most experienced in implementing broad-scale and comprehensive natural resources-based participatory land-use planning (Frame et al. 2004). This is an excellent case study for understanding the role of planning in limiting cumulative effects.

During the 1980s and early 1990s, the forest industry in B.C. was under siege. The industry-government coalition faced increased scrutiny from British Columbians who demanded a larger say in the decision-making processes that affected public land. Led by well-organized grass roots environmental organizations, many citizens questioned forest practices and the economic and environmental sustainability of the business of forestry. Following highly publicized demonstrations, acts of civil disobedience, and efforts to have B.C. forest products boycotted by international markets, a new left-leaning government was elected in 1991 that promised "Peace in the Woods" (Jackson and Curry 2002). The primary tools for achieving that objective were consensus-based regional land-use planning and new, comprehensive, environment-focused forest practices legislation (Cashore et al. 2001). This was a radical swing in public policy that for the previous 40 years had supported the economic development of natural resources, namely, forest products, hydroelectricity, and oil and gas (Jackson and Curry 2002).

The first efforts at land-use planning in B.C. focused on four relatively large and contentious areas. Local stakeholder committees were directed by a nonpartisan Commission on Resources and the Environment to develop Regional Land Use plans. The ambitious geographic scale, poor guidance and coordination, and industry-led resistance forced the government to adopt a more streamlined flexible approach, Land and Resource Management Plans (LRMPs), for smaller subregional areas (15,000–25,000 km^2) (Jackson 2002; Jackson and Curry 2002). Initially, LRMPs were meant to provide higher-level guidance for environmentally and economically sustainable resource development. This broad goal was to be achieved through the resolution of land and resource conflicts, spatial delineation of suitable land uses, and provision of investment certainty for resource industries. Later, LRMPs became an instrument for engaging First Nations in government-to-government discussions on resource management. Each planning "round table" had different representation and a flexible process for reaching consensus; however, they all achieved their subregional goals by implementing a system of Resource Management Zones: General Resource Management, Special Management, Settlement/Agriculture, and Protected Areas.

The LRMP process in B.C. is ongoing with over 85% of the public land base (97 million ha) being covered by 26 land-use plans (Integrated Land Management Bureau 2006). Government, however, has recognized the costs of completing plans for outstanding subregions and, in particular, updating plans in the context of changing environments, such as a massive die-off of pine (*Pinus* spp.)-leading forests (Kurz et al. 2008), and new provincial priorities for land use and management (Integrated Land Management Bureau 2006). The process has cost from $50 to 100 million since inception, with costs per plan escalating in the last 5 years following meaningful engagement with First Nations (Integrated Land Management Bureau 2006). Given the large costs in budgetary and staff time, it is worth reflecting on the successes and failures of the regional and subregional land-use planning process as a model that might address cumulative effects in B.C. and beyond.

Following close to 17 years of practice, strategic land-use planning has received considerable review by academics and government. Many of the identified benefits of the LRMP process are related to involving local communities and stakeholder groups in developing a vision for the intensity, type, and location of land-use activities. Bringing the planning process to communities allowed the public to gain a much better understanding of the complexity of resource management; local-level relationships and trust also were developed between interest groups, industry, and government (Frame et al. 2004). The consensus-based decision-making process was successful in engaging a wide range of interest groups in dialogue and decision making that often involved very difficult and controversial trade-off decisions (Mascarenhas and Scarce 2004). Adoption of plans by the government also had a number of benefits for the economic objectives of the process including greater land-use certainty for the forest industry, improved clarity for approvals and tenure, and the cessation of market-based activities designed to boycott the sale of B.C. forest products. In addition, the planning process resulted in a considerable investment in resource inventory, mapping, and data management.

Based on the criteria of Noble (2009) and the research of Gunn (2009), many of the outcomes of integrated land-use planning in B.C. can serve as precursors or direct contributions to the development and implementation of CEAMF, SEA, or REA. This is confirmed by the work of Alex Grzybowski and Associates (2001) who reported strong correspondence between the B.C. LRMP process and outcomes and the characteristics of REA. In particular, both processes operate at large spatial and temporal scales; are comprehensive, considering a broad range of ecological and socioeconomic effects often across interacting resource sectors; assume a proactive and strategic perspective that considers present activities in the context of future developments; demand collaboration and the involvement of multiple agencies, stakeholder groups, and First Nations; are time consuming, taking several years to complete; and require effectiveness monitoring and adaptation to meet changing socioeconomic priorities and environmental conditions. They noted that the most important outcome of land-use plans for regional assessment was the articulation of a long-term strategic vision by stakeholders at the planning tables. When endorsed by government and local communities through a consensus-based process, this vision serves as a solid reference point for assessing individual projects within a regional context.

Although the outcomes of the strategic planning process for citizens, communities, and industry are largely positive, the process is not without faults. Each plan was formerly recognized by government; where consensus was not reached on a land-use or zoning decision, then government selected from a range of options. Following adoption, however, the plan became a policy document that was meant to inform and guide government and industry activities. Regulations existed, but they were not always used to elevate the plan or elements of the plan to legally binding objectives. Thus, some have questioned the desire of government to fully implement the guidance contained within each planning document (Jackson and Curry 2002). Others have criticized elements of the consensus-based decision-making process. Legitimacy was a key criterion identified by participants. Inequities in knowledge and resources among participants and structural limitations imposed by government were seen as limiting the overall value, acceptance, and success of the LRMP process (Mascarenhas and Scarce 2004).

Strategic land-use planning indirectly influenced or was coordinated with a number of other initiatives that addressed the amount or management of cumulative effects across B.C. Most notably, planning tables were given a percentage target and asked to nominate parcels of public land for inclusion within the provincial parks system. This was part of the province's strategy to comply with United Nations commitments and double the amount of park land from 6% to 12%. This initiative provided long-term protection for natural, recreational, and cultural values effectively eliminating the cumulative effects of industrial development across 14 million ha of B.C.

In association with the LRMP process, the government enacted parallel forest practices legislation that put into place a number of policy and planning processes to conserve biodiversity across managed landscapes. Initially, the government had adopted many of the core principles of conservation biology and outlined an ambitious process for an extensive network of linked old growth reserves where each landscape planning unit would have a target for percentage old forest based on the natural disturbance type of that ecosystem (B.C. Ministry of Forests 1995). The economic realities of allowing a large proportion of the managed forest to grow or remain old resulted in a lowering of targets and the abandonment of formal requirements for patch connectivity. However, old growth management areas remain a legislated requirement of forest planning guided by the principles of ecosystem-based management (B.C. Government 2004).

One might question why the cumulative effects of forest development are not an explicit consideration during the tenure and management process as is recognized in other jurisdictions (Noble 2004b; Shifley et al. 2008; Gustafson et al. 2007; Spies et al. 2007; Russell 2008). Regulation in the context of 5-year Forest Stewardship Plans addresses the acute environmental impacts of forest practices such as the amount of coarse woody debris remaining after cutting or impacts to soil or water quality. At a broader scale, the government is required by law to set a sustainable Allowable Annual Cut across large management units. The previously mentioned old growth management areas and other legislated conservation initiatives such as ungulate winter range and wildlife habitat areas provide a process for conserving high-risk or economically important species across managed landscapes.

Relative to forestry, there is much less regulation and fewer planning processes guiding the rate of exploration and development, location, or practices of the petroleum industry. Although seismic exploration, drilling, and pumping have created a large disturbance footprint across much of northeastern B.C., there is no formal process for recognizing or managing the cumulative effects or impacts (Nitschke 2008). With the exception of pipelines, these activities do not fall within the B.C. or federal environmental assessment legislation. Furthermore, the draft CEAMF for that area was never implemented. The one exception is the Muskwa–Kechika Management Area (MKMA), a serious of provincial parks and special management zones that cover 6.4 million ha in the northern portion of the province. This area is subject to separate legislation and more highly regulated resource development practices. Across the MKMA, exploration activities are expected to demonstrate best practices relative to wildlife habitat and disturbance, among other values (B.C. Oil and Gas Commission 2004). However, before the disposition of petroleum and natural gas rights, large planning units are subjected to a pretenure planning process. Similar in scope to the LRMP process, various stakeholder, community, and First Nations groups work to identify important ecological, cultural, and socioeconomic values across the planning area and then work collaboratively to define objectives, strategies, and indicators to ensure that oil and gas activities do not negatively impact those values. Aside from limited monitoring, there is no procedural or legislated recognition of the cumulative effects of those activities either at the exploration or development phase.

CONCLUSION

From a regulatory perspective, the cumulative effects landscape in Canada is not easily navigated. Environmental assessment legislation differs across provinces, and the process is complicated by the inter-provincial Canadian Environment Assessment Act. The issues confronting natural resource professionals are also highly variable and region specific. A long history of agricultural development in the central portions of the country is contrasted against oil and gas exploration and development in the west and mining for diamonds in the north.

Given Canada's reliance on natural resources to fuel the national economy, it is not surprising that there are numerous criticisms of current trends in development and the application of environmental assessment legislation (Kennett 1999; Baxter et al. 2001; Davey et al. 2001; Noble 2004a; Duinker and Greig 2006). Although not positive, such critical review offers hope for more effective practices in the future. Only through review of existing legislative frameworks and practices can we improve future regulatory processes and assessment methods. In response to current problems, there have been a number of innovative solutions. For example, Dubé and Munkittrick (2001) have proposed a systematic process for assessing cumulative effects that integrates both effects-based and stressor-based approaches. The development and application of ad hoc CEAMF is another example of Canadian innovation in the context of regional cumulative effects hotspots. There are other established approaches that only require coordination and adjustment to address some of the largest cumulative effects challenges facing the country. British Columbians, in particular, are facing unprecedented levels of development across a wide range of industrial sectors. If refocused on the assessment and regulation of regional cumulative effects, integrated land-use planning can help address many of those challenges. Although much work is needed, the lessons from B.C. (Mascarenhas and Scarce 2004) and the rest of Canada (Noble 2004a) can help guide regulatory and planning initiatives designed to address cumulative effects in other jurisdictions.

4 Quantifying Cumulative Effects

Paul R. Krausman

CONTENTS

INTRODUCTION

The Mitigation Symposium in 1979 (U.S. Department of Agriculture 1979) was an excellent collection of papers addressing mitigation for wildlife and fisheries and the habitats that support them. In a summary of the symposium, Larry John emphasized several points.

1. Anthropogenic influences would continue and, to minimize and reduce negative impacts to wildlife and its habitat, mitigation would be a way of life.
2. Mitigation should be essential to guide development and land-use conversion—not to stop development.
3. Mitigation is defined as a class of actions that have the purpose of counteracting effects of disruptions on the natural environment and on renewable resources associated with physical structures.

4. Mitigation is defined to:
    a. Avoid adverse impact altogether by not taking a certain action (i.e., preventative)
    b. Compensate for the impact by replacing or providing substitute resources or environments
    c. Minimize impacts by limiting the degree or magnitude of the action and its implementation
    d. Rectify the impact by repairing, rehabilitating, or restoring the effected environment; today known as habitat restoration
    e. Reduce or eliminate the impact over time by preservation and maintenance operations during the life of the action
5. To effectively incorporate mitigation, it should be part of the initial planning process, so projects can be designed to avoid degradation of natural systems. Mitigation should not be a "Whoops! We messed up and now need to fix the damage." Wildlife habitat will continue to be altered, and we need mitigation. Mitigation should be thought of as a positive measure in the planning process.
6. Mitigation varies from saving endangered species, to replacement of habitat, and preventing damage from occurring in the first place.
7. There are seven basic steps for mitigating where biological input is needed.
    a. Solid wildlife data base for the project and proposed mitigation site.
    b. Thorough and complete data analysis.
    c. Predictive models developed to create conceptual mitigation options.
    d. Design of required habitat modifications for mitigations.
    e. Designs must be implemented.
    f. Mitigation success must be monitored. Not just once but for the long term, the length of the project.
    g. Modifications to ongoing activities resulting from the monitoring program should be agreed to and be budgeted for by development agencies at the planning stages.

Cumulative effects of an action were not directly addressed at the symposium 30 years ago but, if this list were being derived today, an eighth category for consideration would certainly be cumulative effects. Several speakers came close to addressing cumulative effects by stating that mitigation should be done with an ecosystem approach because one action often led to others that had equally or stronger impacts on the resources being addressed. For example, oil and gas activity in Alberta being mitigated was a two-part process: mitigating the activity and mitigating the increases in access by the public as a result of mitigation. Several authors indicated that other times and places had to be considered, and true mitigation would not occur without a holistic approach.

Mitigation, when in the planning stage, is positive as it is not an afterthought, an add-on to planning (often unwelcome by developers and planners), or an attempt to compensate for errors; in part, a failure. With planning, these negative impacts of mitigation do not have to raise their ugly heads and mitigation should be in the planning stages; no one can effectively mitigate for lost habitat!

One of Leopold's (1949, p. 262) often used quotes is appropriate when thinking of changes being made on behalf of wildlife: "A thing is right when it tends to preserve the integrity, stability, and beauty of the biotic community. It is wrong when it tends otherwise." A slight modification in view of anthropogenic influences was made by Callicott (2002, p. 104): "A thing is right when it tends to disturb the biotic community only at normal spatial and temporal scales. It is wrong when it tends otherwise." This philosophy needs to be embraced. My objective in this chapter is to describe how cumulative effects are problematic, offer solutions to better understand how they influence wildlife resources, and discuss ways cumulative effects are measured.

THE BASICS OF CUMULATIVE EFFECTS

The use of the term cumulative effects has been common in the literature since the 1980s as biologists addressed how combined influences of variables (e.g., snowfall and temperature) influence an animal population. These, indeed, are cumulative effects but, when conducting environmental impact statements or similar documents, cumulative effects take on different meanings, including:

1. Increasing in force by successive additions
2. The specific considerations of effects due to other projects
3. Impacts that result from incremental changes caused by other past, present, or reasonably foreseeable actions, together with the project
4. Cumulative effects arise where several developments each have insignificant effects but together have significant effects; or where several individual effects have a significant combined effect

All of these points are summarized in the Council on Environmental Quality's (1997) regulations in accordance with the National Environmental Policy Act of 1969 as amended, which defines cumulative effects as ". . . the impact on the environment which results from the incremental impact of the action when added to other past, present, and reasonably foreseeable future actions regardless of what agency (Federal or non-Federal) or person undertakes such other action (40 CFR, 1508.7)." Cumulative effects defined in this manner have not been addressed much in the literature even though the definition touches on nearly every aspect of environmental impact evaluation. This definition could also be project based where a specific development may interact with other proposed or planned developments on a local basis. Planning-based cumulative effects could arise if a strategic development or plan interacts with other developments or plans on a regional or national basis.

There are numerous ways cumulative effects can operate.

1. Physicochemical transport; emissions interact with others at a distance.
2. Nibbling loss; gradual disturbance and loss of land and habitat by project proliferation.
3. Spatial and temporal crowding; too much happening in too small of a space or in too short of time.
4. Growth-inducing potential; one project induces others—particularly transport.

Even though cumulative effects have been poorly addressed or ignored, they need to be considered in planning, including:

1. A study of an area large enough to assess all important habitat components that may be affected by the project.
2. Other influences on the population or habitat in the past, present, or reasonably foreseeable future should be considered.
3. Incremental effects have to be assessed, so the entire habitat alteration can be evaluated.
4. Effects should be compared to thresholds, standards, or policies.
5. Mitigation should be proposed if the effects are negative.
6. The residual of significance effects should be established.
7. Initiation of mitigation and monitoring.

WAYS TO MEASURE CUMULATIVE EFFECTS

Analysis of cumulative effects is challenging because geographic and temporal boundaries are difficult to define. If defined too extensively, an analysis could be too large but, if too narrowly defined, some critical aspects may be missed. If the latter occurs, decision makers and managers will not have all the information necessary to understand how their project influences wildlife and the habitat it depends on.

Unfortunately, there is no single formula for determining the scope and extent of cumulative impacts. The practitioner has to determine the methods and necessary detail depending on the size and type of project, location, the potential to alter wildlife or its habitat (or other environmental resources), or the health of the potentially affected resource.

The Council on Environmental Quality (1997) reviews the literature on cumulative effects analysis and identifies cumulative effects analysis principles.

1. In scoping out the project, one should include past, present, and future actions.
2. Include all federal, nonfederal, and private actions.
3. Focus on each affected resource, ecosystem, and human community.
4. Focus on meaningful effects.

In describing the affected environment, practitioners should:

5. Focus on each affected resource, ecosystem, and human community
6. Use natural boundaries

Finally, in determining the environmental consequences of projects, managers should:

7. Address additive and synergistic effects
8. Look beyond the life of the project
9. Address the sustainability of resources, ecosystems, and human communities

However, in a review of nearly 100 methods to evaluate cumulative effects, none met all the criteria for cumulative effects analysis (Granholm et al. 1987). They described the problem but did not succeed in quantifying cumulative effects. Granholm et al. (1987) grouped the analysis methods into three categories.

1. Descriptions or models of cause-and-effect relationships via matrixes or flow diagrams.
2. Analysis of trends in effects or resource change over time.
3. Mapping techniques to overlay landscape features to identify areas of sensitivity, value, or past losses.

All three categories address important aspects of cumulative effects but do not provide a comprehensive approach to analysis. Several frameworks for analysis of cumulative effects have been developed (Council on Environmental Quality 1997) for the U.S. Army Corps of Engineers (Stakhiv 1991), U.S. Fish and Wildlife Service (Horak et al. 1983), Department of Energy (Stull et al. 1987), U.S. Environmental Protection Agency (EPA; Bedford and Preston 1988), and the Canadian Government (Lane and Wallace 1988). Also, the U.S. EPA and the National Oceanic and Atmospheric Administration developed two approaches that examined the influence of cumulative effects on wetland loss: an impact assessment and planning approach. The former analytically evaluated the cumulative effects of combined actions that influence the resources of concern and views cumulative effects analysis as an extension of environmental impact assessments. The latter optimized the distribution of cumulative stresses on resources regionally and is a correlation of regional or comprehensive planning. The impact assessment approach optimizes upon a community vision of future conditions in the absence of reliable data on the resources and the habitats they depend upon. However, the planning approach is gaining popularity as holistic management and sustainable development are considered. The methods are complementary because they combine environmental impact assessment with the planning process. Furthermore, Johnson (Chapter 3) makes a strong case for an analysis of cumulative effects to move beyond the prescription environmental assessment process.

Even though there is no single way to conduct cumulative effects analysis, those that do need to develop a study-specific method using the available methods and techniques to develop a conceptual framework from which to operate. The following discussion is the model proposed by the Council on Environmental Quality (1997), which presented the primary methods for developing the conceptual causal model for a cumulative effects study.

Conducting a Cumulative Effects Study

Understanding how cumulative effects influence wildlife and the habitat it depends on can be a complex process, but it has to be understandable to the public. If data used in cumulative effects analysis are ambiguous or do not have apparent relationships to the analysis because of intentional or unintentional manipulation, the analysis will not be user friendly and, thus, will be unacceptable to the public. The best efforts possible have to be made to present all adverse and beneficial effects to decision makers, so they can make informed decisions that are clear.

There are numerous tools available to assist in the development of cumulative effects analysis. These range from the simple, yet very useful, compilation of data into tables and matrices, to view the ranges of alterations available to Geographic Information System (GIS) technology that allows researchers to map out how alterations will change the shape of the landscape. Johnson et al. (2005) rated habitat for wolves (*Canis lupus*) across the Canadian central Arctic as poor, low, good, or high; calculated the size of the area occupied by each; and contrasted the modeled disturbance with the assumed disturbance. These types of displays clearly show how disturbance can alter habitats (Table 4.1). Other tools used to illustrate cumulative

TABLE 4.1
Variables Used to Model Resource Selection by Caribou, Wolves, Grizzly Bears, and Wolverines Monitored from May 1995 to January 2000 across the Canadian Central Arctic

Variable	Description
Resource Variables	
Esker density/patch[a]	Sparsely vegetated sand and gravel esker complexes
Forest density/patch	Continuous or discontinuous forested areas of dwarf white spruce, black spruce, and tamarack
Heath rock density/patch	Open mat heath tundra interspersed with bedrock and boulders
Heath tundra density/patch	Closed mat of heath found on moderate to well-drained soils on upland areas
Lichen veneer density/patch	Windswept, dry, flat topography covered with a continuous mat of lichen
Low shrub density/patch	Extensive areas of low birch and willow found on moist well-drained soils
Peat bog density/patch	Mosaic of uplands and lowlands with fens, bogs, mixed-wood forest, and peatlands
Riparian shrub density/patch	Active stream courses or areas of water seepage with a shrub layer of birch, willow, and alder
Rock association density/patch	Large areas of windswept bedrock or boulders with little vegetation cover
Sedge association density/patch	Wetland complexes of wet sedge meadow and drier hummock sites
Occurrence of caribou	Predicted likelihood of encountering caribou
Occurrence of grizzly bear	Predicted likelihood of encountering a grizzly bear
Occurrence of wolf	Predicted likelihood of encountering wolves
Human Disturbance Factors	
Major developments[b]	Operating mines, communities, winter road camps
Mineral explorations	Areas of mineral exploration activities
Outfitter camps	Seasonal guide-outfitter camps

[a] Vegetation was modeled as the percent area of land cover patches and the mean density of land cover types.
[b] Disturbance was modeled as the distance of animal and random locations from the nearest facility.
Source: From Johnson et al. (2005). Used with permission of The Wildlife Society.

effects include dose–response curves, cumulative frequency distributions, maps, videography, photography, and geographic information systems. For example, scientists examining habitat models generated GIS figures that illustrate important habitat for desert bighorn sheep (*Ovis canadensis*) in the Peninsular Ranges of California (Figure 4.1). Different areas are designated by the different models, allowing managers and decision makers to develop planning with adequate information (in relation to habitat use; Rubin et al. 2009). Different models (Figure 4.1A,B) depict sheep distribution related to important habitat features differently. This is a first step in deciding how to plan in a developing area, so humans and bighorn sheep can both continue to inhabit the landscape.

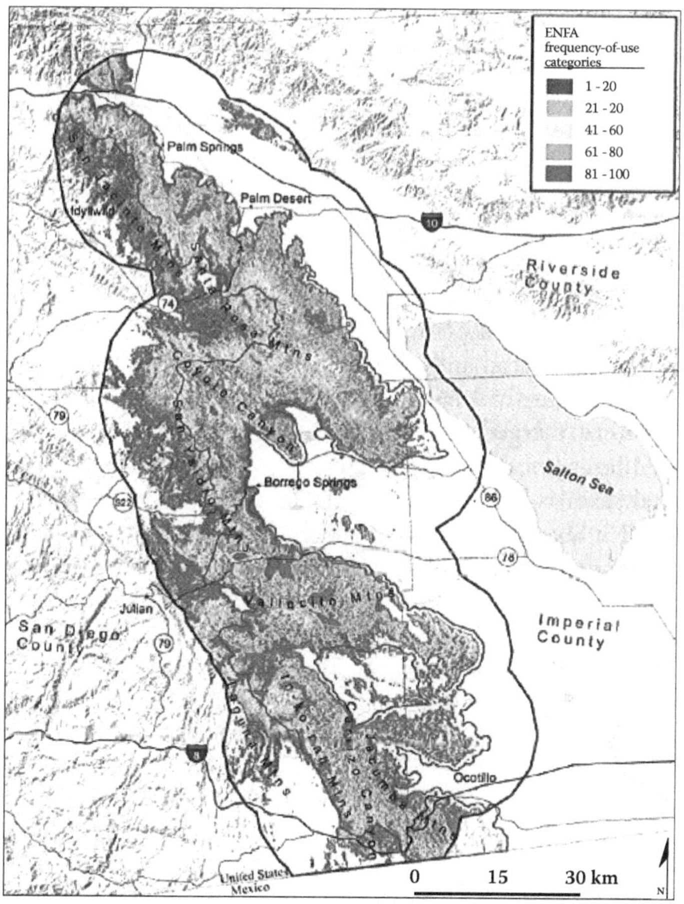

FIGURE 4.1A Habitat predicted by Ecological Niche Factor Analysis (ENFA) for male and female bighorn sheep (sexes combined) in the Peninsular Ranges, California, based on data collected during 2001–2003. Five frequency-of-use categories are shown. The study area is indicated with a black line, and the expert-based model boundary is shown as a lighter gray line. (From Rubin, E.S. et al. 2009. *Journal of Wildlife Management* 73: 859–869. Used with permission of The Wildlife Society.)

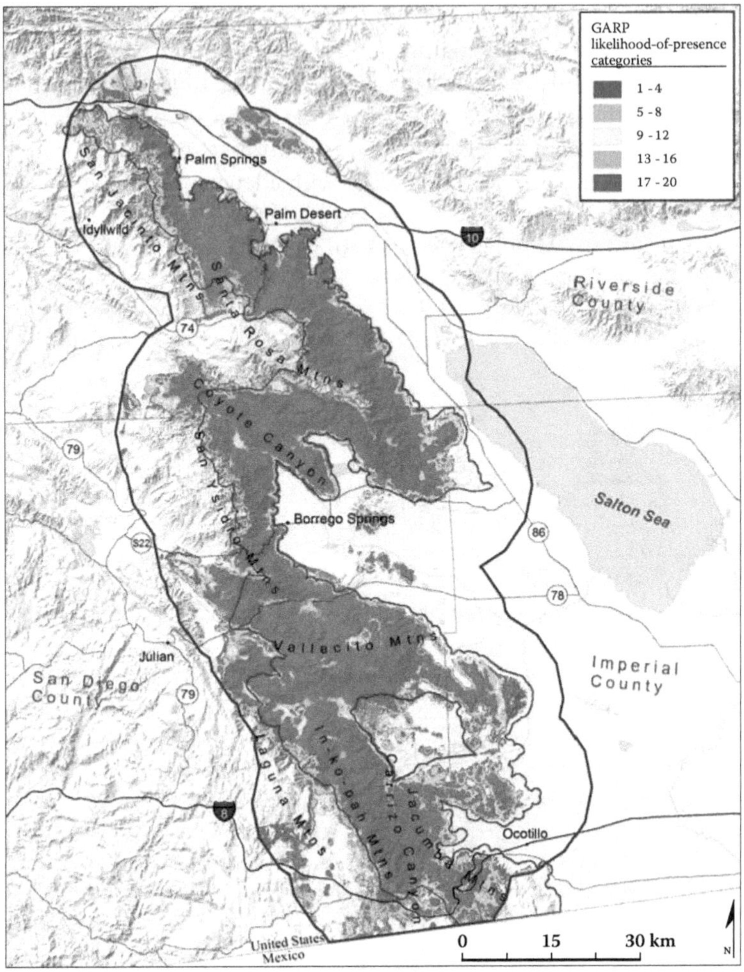

FIGURE 4.1B Habitat predicted by Genetic Algorithm for Rule-set Production for male and female bighorn sheep (sexes combined) in the Peninsular Ranges, California, based on data collected during 2001–2003. Five likelihood-of-presence categories are shown. The study is indicated with a black line, and the expert-based model boundary is shown as a lighter gray line. (From Rubin, E.S. et al. 2009. *Journal of Wildlife Management* 73: 859–869. Used with permission from The Wildlife Society.)

Development of Conceptual Models of Cumulative Effects

There are at least 10 methods used to understand the influences of cumulative effects: questionnaires, interviews, and panels; checklists; matrices; networks and system diagrams; modeling; trends analysis, overlay mapping, and GIS; carrying capacity analysis; ecosystem analysis; economic impact analysis; and social impact analysis. More than one of the techniques will be required in many cases.

Questionnaires, Interviews, and Panels

These methods are used to gather information about the wide range and actions needed for a cumulative effects analysis. These are the tools needed to obtain the wide range of influences that can influence the cumulative effects of the resources in question. Questionnaires can pool large audiences, interviews with knowledgeable people can refine the potential effects leading to cause-and-effect relationships, and expert panels can further refine the areas to be examined. These tools also engage the public, which is a critical part of the process. Questionnaires, interviews, and panels are a good place to start in acquiring knowledge. These information-gathering techniques are subjective but can be further strengthened by using the Delphi method (Linstone and Turoff 1975) or fuzzy model sets (Harris et al. 1994) to approach consensus among experts with differing opinions.

Planners will use these data in progressing toward meaningful analysis. For example, the National Park Service was concerned about noise from aircraft flights over Grand Canyon National Park, Arizona, and convened a technical team of experts to propose a framework to evaluate the impacts of aircraft noise. The framework was limited to four matrices with metrics or indicators and corresponding thresholds for impact intensities including visitor experience opportunities, soundscape, threatened and endangered wildlife, and ethnographic resources (Federal Aviation Association and National Park Service Technical Team Reports 2009). These thresholds represented degrees of increasing impact (e.g., negligible, minor, moderate, and major) that provide an overview of what is available in the scientific literature as a beginning point that could be used in evaluating cumulative effects. Acquiring the data is often a time-consuming and difficult task but one that is important if informed decisions are to be obtained.

Strengths of questionnaires, interviews, and panels are that they can be flexible and provide a mechanism to incorporate subjective information into the analysis. This disadvantage of using these tools is that they do not quantify values, and the comparison of alternatives can be subjective.

Checklists

Checklists are another tool useful in organizing the various environmental influences that are necessary to understand cumulate effects analysis. They provide a visible format to organize multiple actions and resources that can highlight cumulative effects. Checklists can be simple or complex, depending on the number of variables involved. If too simple, they may be incomplete and leave out important effects. If too complex, they could become irrelevant as they would not be useful. The strength lies in all components of a project that can contribute to cumulative effects of the project.

Two or more effects on a resource indicate a potential cumulative effect, which is determined by weighing the combined effects. The manager can easily view the systematic impacts and how they contribute to cumulative effects. The disadvantage of checklists is that they can be inflexible and do not necessarily address cause-and-effect relationships. The results from checklists can also be subjective.

Matrices

Matrices are two-dimensional checklists that quantify anthropogenic activities on environmental resources and can be used to assist in cumulative effect analysis, especially when there are multiple influences on the resources. Used alone, matrices do not quantify effects but are a useful way to present information that can be used in modeling and mapping. Matrices are well suited to combine the influences of different projects on resources and evaluate cumulative effects. They are mathematically straightforward, and managers can simply note presence or absence of an effect. The effects can also be qualified to a degree as high, medium, or low that reflects relative ranking, but these techniques tend to be subjective. However, they have been used to create stepped matrices that contrast resources against resources (Canter 1996) by addressing secondary and tertiary effects of actions and following them through the environment. As an example, the Federal Energy Regulatory Commission (1987) developed a relative index (total cumulative impact scores) in contrasting different resource components influenced by different projects.

Matrices can provide comprehensive presentations of the data with a comparison of alternatives that address multiple projects. However, they do not address space or time or cause-and-effect relationships and can be cumbersome.

Networks and System Diagrams

Any ecologist will be familiar with this technique as they are similar to food webs and relate the components of an environmental system and the factors that influence it in a series of potential links. Thus, the user can evaluate the various effects on resources, and networks and system designs are often the best method to identify cause-and-effect relationships that result in cumulative effects. They improve on the stepped matrix by sharing relationships of actions by using component boxes and linkage arrows to indicate interactions including secondary effects. These tools have been used for ecological and social systems but usually have to have separate diagrams for each. Weaver et al. (1987) used a system diagram to illustrate the management issues involved in assessing cumulative effect on grizzly bears (*Ursus arctos horribilis*; Figure 4.2). Networks and system diagrams facilitate conceptualization and can address cause-and-effect relationships.

Modeling

Modeling generally requires the acceptance of numerous assumptions; thus, public understanding and acceptance is low for many modeling exercises. However, modeling can be used as a powerful tool to quantify the cause-and-effect relationships of cumulative effects.

Unfortunately, project-specific models are costly in terms of time and money, so cumulative effects models generally use or modify existing models. There are usually only limited data available from which to determine cumulative effects, which also limits the use of sophisticated models. When an organization is interested in using models, they will have to develop a model or technique, or obtain baseline data for use in an existing model. Because baseline data can take years to collect, the investment in a specific model can yield significant rewards in the analysis of cumulative effects.

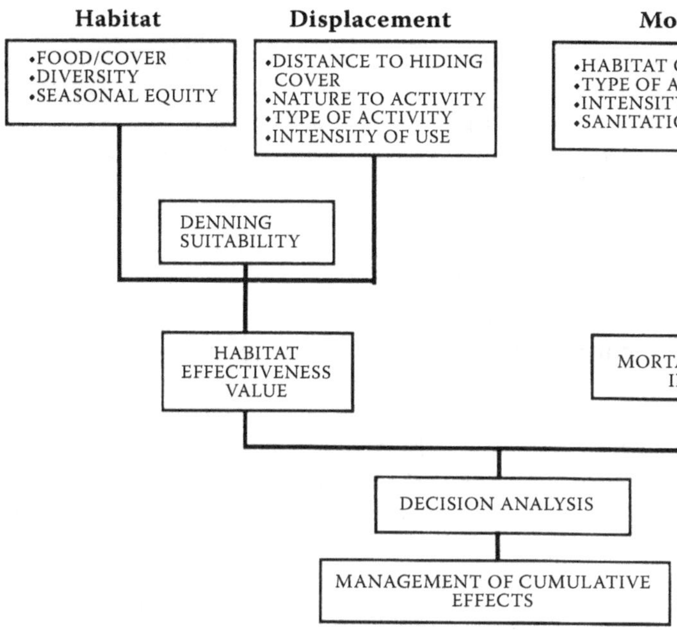

FIGURE 4.2 A framework for assessing cumulative effects on grizzly bears. (From Weaver, J.L. et al. 1987. A framework for assessing cumulative effects on grizzly bears. Transactions of the North American Wildlife and Natural Resources Conference 52:364–376. Used with permission from The Wildlife Management Institute.)

If a model is used, it should be justifiable and recognized in the scientific community. Common models have been designed to determine cumulative effects of soil erosion, sediment transport, and species habitat models (Polfus 2010).

Models, when properly developed with adequate data, can provide unequivocal results, address cause-and-effect relationships, are quantifiable, and can integrate time and space. Unfortunately, they require an abundance of data, can be expensive, and do not work with many interactions.

An appropriate approach to assess the cumulative effects on wildlife is to focus on those habitat features that influence the distribution and abundance of populations (Truett et al. 1994). The habitat-evaluation procedure (HEP) uses a series of habitat-suitability index (HSI) models to evaluate habitat, which can provide estimates of habitat quality (Anderson and Gutzwiller 1994). The Council on Environmental Quality (1997) then recommends the development of HSIs for each species in specific habitats that support the species. With these data, the cumulative effect on a species can be estimated by determining how each habitat for the animal will be altered. These models can use habitat suitability as the metric of interest in determining cumulative effects. Unfortunately, there are limited examples where these are put into practice by the federal government to assess cumulative impacts (Truett et al. 1994).

Trends Analysis

Trends analysis uses graphs or photographs to document how landscapes or resources can change over time. This analysis provides the historical context that is critical to understand the cumulative effects of proposed actions (Council on Environmental Quality 1997). There are at least three ways trends analysis can assist the cumulative effects analyst: identifying problems caused by cumulative effects, the establishment of baseline data, and projecting future cumulative effects.

When substantial losses to landscapes or resources can be documented and displayed, visualizing how cumulative effects problems can be magnified by future actions is straightforward. For example, the decline of a bighorn sheep population from greater than 200 animals to less than 10 easily suggests that the population may no longer be viable.

The establishment of baseline data is critical to management of natural resources. Any mechanism to demonstrate how landscapes or habitat components change is valuable in evaluating restoration efforts. Photographs of the same landscape over decades have been a very effective way of documenting change (Hastings and Turner 1965) and establishing a temporal baseline.

When similar environmental conditions exist, common cumulative effects can be used to predict future effects on resources and ecosystems from different stresses. Under same conditions, threshold points may be revealed where cumulate effects are significant. With increasing use of satellite imagery, population data, photo points, and video simulations, trend analysis has been used to document cumulative effects of declining animal populations, fragmentation of habitats, baseline establishment, resource degradation, and historical losses of wetlands.

The advantages of trend analysis include the ability to assist in problem identification, temporal baseline determination, and addressing cumulative effects over time. In many cases, abundant data over time will be necessary, and the extrapolation of the data can be subjective.

Overlay Mapping and GIS

Overlay mapping and GIS are very useful and common ways to provide visual representation of locational information for cumulative effects analysis. The technique is straightforward and uses simple mapping of the resources, ecosystems, or human communities of interest. The mapping can then be used to set boundaries on the area of interest, identify areas where the impacts are likely the greatest, and address concerns of interest such as the appropriate placement of biological linkages between different areas.

A common use of mapping is "impact oriented" where an impact, or combination of impacts, is placed on the landscape or could simplify outline areas stressed by different impacts or areas that are subjected to the greatest stresses. This is a resource-oriented map or resource capability analysis that delineates the site's natural and anthropogenic resources, and even indicates where development can and cannot occur. They are often used to visualize the responses of resources, ecosystems, and human communities when cause-and-effect models are not available. The advent of GIS-based computer overlays has replaced layering effects with transparencies. By using electronic GIS, overlay mapping rapidly produces maps depicting numerous layers of interest and, in some cases, it is possible to provide weights to each layer, so the map is less subjective. The advances made with GIS have provided benchmark technology in natural resource management by assisting with inventory and monitoring, management planning, policy setting, research, and consensual decision making (Franklin 1994).

Carrying Capacity Analysis

Carrying capacity is central to the management of wildlife populations (Krausman 2002) but has been used in so many ways the meaning is often obscured. For example, Macnab (1985) included carrying capacity with overpopulation, overharvesting, and overgrazing in his list of resource management shibboleths. Indeed, there are at least seven different meanings of carrying capacity (Miller and Wentworth 2000). Leopold (1933) was the first to define carrying capacity as the inherent threshold of a particular range is capable of supporting. When considering carrying capacity, analysis for cumulative effects is defined by the Council on Environmental Quality (1997, A-33) "... as the threshold of stress below which populations and ecosystem function can be sustained." In this case, carrying capacity applies to ecological functions that can continue with various levels of human use and is similar to optimum carrying capacity, which is the animal population density that best satisfies human expectations for it (Krausman 2002): or other ecosystem components.

In other words, it is the sum of human activities (e.g., hunting, livestock grazing, recreation) that can be accepted while still maintaining the desired outcomes by society. When cumulative effects exceed the carrying capacity of a resource or ecosystem, the results are significant. For example, impervious surface cover in urban landscapes of 10% can be tolerated by sensitive aquatic fauna but at 25%, fauna can be impacted and greater than 25% is often nonsupporting (Center for Watershed Protection 2003). In Columbus, Ohio, the health of streams declined when the amount of impervious cover exceeded 13.8%; declining biological integrity was noted in some streams at levels of urban land use as low as 4% (Miltner et al. 2004). In Wisconsin "... connected imperviousness levels between 8% and 12% represented a threshold region where minor changes in urbanization would result in major changes in stream conditions" (Wang et al. 2001, p. 255; 2003).

Carrying capacity analysis to examine cumulate effects can be useful if a threshold can be established and can be monitored effectively. The first and often most difficult step in using carrying capacity analysis is the identification of limiting factors (e.g., those factors that have the most significant impact on resource, ecosystem, or human community, e.g., predation, lack of food, and accidents). Mathematical equations can then be developed to describe the capacity of the resource in terms of numerical limits or thresholds that are imposed by each limiting factor. In this way, projects can be evaluated to determine the effects on the various limiting factors. For example, Mazaika et al. (1992) questioned if forage availability was adequate for bighorn sheep in Pusch Ridge Wilderness, Arizona, and based seasonal estimates of carrying capacity on the summation of half the total consumable forage divided by the mean daily intake of forage, and multiplied it by season length. Mazaika et al. (1992) were able to determine that the forage abundance was adequate for the population and searched for other limiting factors. Carrying capacity analysis in studies of cumulative effects on wildlife has been useful when examining wildlife populations, recreational use of natural areas, and land-use planning. The carrying capacity of public facilities (e.g., water supply, sewage treatment, traffic systems) is fairly straightforward and can be quantified. However, biological systems and attitudes of humans are more difficult. To assess human attitudes toward some form of recreation and to understand human values will require survey methodology. In some cases, carrying capacity for animals can be determined (Mazaika et al. 1992) but is usually for a single species and is limited when trying to understand ecosystems.

Carrying capacity analysis has the potential to identify thresholds and provides mechanisms to monitor the incremental use of unused capacity if true measures of cumulative effects are established against thresholds. This analysis can also address effects in a system context and address them temporally. However, it is rare that managers can directly measure carrying capacity and even when they can it is for a single species, and the requisite regional or local data are often lacking.

Ecosystem Analysis

It would be ideal if ecosystem analysis could be incorporated into cumulative impact assessment, but it is too broad as the analysis has to consider all the ecological resources and their interaction with the environment. Consideration of ecosystem analysis does force the analyst to focus on the resource or ecosystem, use natural boundaries instead of those created by man, and to address the issue of sustainability.

Ecosystem analysis addresses biodiversity and measures it from the genetic to species level at the local to regional ecosystem. The analysis includes watersheds or ecoregions as study sites incorporating large landscapes. As such, this analysis addresses sustainability of the structure and function of the system.

Cumulative effects analysis usually only considers a single entity (e.g., species, water quality, air, habitat alteration). However, with societies interest in biodiversity, entire ecosystems are receiving more attention; thus by their very nature, the maintenance of biodiversity is a cumulative effects topic. An ecosystem approach has to be considered whenever biodiversity is an issue.

The ecosystem approach involves three basic concepts: landscape approaches, using a suite of metrics that incorporate community and ecosystem levels, and address the array of interactions that allow healthy ecosystems to function (e.g., positive aspects of soil, earthworms, pollination, birds, mammals; Council on Environmental Quality 1997). Guidelines for an ecosystem analysis contain at least 11 points:

1. Consider a "big picture" or ecosystem view.
2. Ensure the integrity of communities and ecosystems.
3. Minimize fragmentation and promote natural biological connections.
4. Do not introduce exotic species and enhance survival of native species.
5. Protect rare species.
6. Protect unique and sensitive environments.
7. Maintain or minimize naturally occurring structural diversity.
8. Maintain or mimic ecosystem processes.
9. Protect genetic diversity.
10. Use restoration techniques to restore ecosystems, communities, and species.
11. Monitor, monitor, monitor.

By following these guidelines, one has to acknowledge uncertainty and keep an open mind. Clearly, it would be impossible to accomplish all of the goals during the life of a project (or a lifetime) but keeping them in the forefront of deliberations will enhance the usefulness and value of any analysis.

Ecosystem analysis is important because it considers large landscapes and all related components and interactions, addresses space and time, and considers sustainability of ecosystem functions. However, the analysis is limited to natural systems, often requires the suite of species surrogates for the system, and requires data collected over long periods, and having access to complete watersheds or ecoregions is often problematic.

Economic Impact Analysis

Economics is the study of the way society tackles limited resources; there are always more desires than can be accommodated. Economics is a system of allocating limited resources.

The economic system that has been derived in the United States and much of the world is a form of capitalist democracy in which people, markets, and government interact to allocate resources (e.g., all things physical including wildlife and their habitats). Economics is also an important aspect of cumulative effects analysis because the well-being of communities depends on it. At a minimum, any economic impact analysis should determine any changes in business activity, employment, income, or population.

There are three major steps in conducting an economic impact analysis. First, the region of influence has to be established, then the economic effects should be modeled, followed by a determination of the significance of the effects. Establishing the region of influence can be difficult but should include areas where economic linkages between the residential population and the businesses in the geographic area exist. After the region of influence is established, economic models can be used to analyze cumulative effects.

These models can be used to project effects associated with every alternative considered, and they can be simple or complex and can be combined with other tools to provide timely and cost-effective evaluations of the significance of effects. A complete description of these models is beyond the scope of this chapter. However, when considering wildlife issues in an ecological impact analysis, it is not simple, and neoclassical economics may need to be replaced with ecological economics. Neoclassical economy, using supply and demand curves, prices and quantities, and market equilibrium, is the common economic theory used. "According to the theory, as producers and consumers meet in the marketplace, voluntary exchange prices are established to set production and consumption at optimal equilibrium levels. Furthermore, increasing prices spur discovery to new technologies and new materials, and they encourage efficiency of production and consumption. Any departure from the market system, according to the neoclassicists, results in a less than optimal allocation of resources, and likewise, lower human satisfaction" (DeSteiguer 1995, p. 553–554).

When considering economics, most people are exposed to and take a neoclassical approach. However, this approach may not be applicable when considering wildlife for several reasons (Erickson 2000; Gowdy 2000; Hall et al. 2000).

1. Economic development often exploits resources in a nonrenewable manner.
2. "Unfortunately, major decisions that affect millions of people and vast areas of wildlife habitat are often based on economic growth models that, although elegant and widely accepted, are not validated" (Hall et al. 2000, p. 20).
3. The basic model of neoclassical economics views the economy as if it has no limits and often does not consider biology in the production of the raw materials required for growth. Neoclassical economists argue that should one resource decline, alternate technologies will develop alternatives.
4. The gross domestic product (GDP) is used by economists as a quality of life for humans. However, the GDP does not include costs for wildlife conservation and ecosystem functions. Consequently, the GDP underestimates the influence of human activities on wildlife population and vice versa.
5. Market decisions are often detrimental to wildlife, and using market economics is not the correct way to make large-scale decisions for wildlife (Geist 1988).
6. Clearly a more "... holistic approach is required for the purposes of wildlife conservation" (Hall et al. 2000, p. 23). The alternatives to neoclassical economics are sustainable development (i.e., development without alteration or destruction of wildlife and its habitat), biophysical economics (i.e., understanding economics from a biophysical perspective that focuses on the land and its resources rather than a social perspective, and ecological economics. Ecological economics embraces a broad interdisciplinary array of professionals that includes the natural and social sciences (Czech 2000a, b). Those who practice ecological economics consider it the science and management of sustainability (Costanza 1991).

The strengths of the economic impact analysis are that they do address economic issues and can provide quantified results. However, the utility of the models independent of the data and the use of neoclassical economics do not address nonmarket values such as wildfire.

Social Impact Analysis

The art and science of wildlife management includes the triad of wild animals, their habitat, and their interactions with humans. Thus, it is important to consider the human dimensions when considering cumulative impacts. There are more than five variables that need to be considered in the human dimensions arena (Interorganizational Committee on Guidelines and Principles 1994).

1. Human population characteristics including size, growth, ethnic and racial diversity, and type of residents (e.g., seasonal, permanent, leisure, business)
2. Community and institutional structure (e.g., size, structure, local government, historical and present employment, industrial diversification,

voluntary associations and related characteristics, religious organizations, and interest groups)
3. Political and social resources including the distribution of power and authority
4. Human attitudes and values related to the subject at hand
5. Community resources and use pattern of landscapes for recreation and housing, and community services (e.g., health, fire protection)

The consideration of social impacts in cumulative impact assessment is part of wildlife planning, which is an integrated system of management that incorporates everything that can influence the goals to be obtained through the mechanisms to reach the goals (Crowe 1983). All of the actions that influence wildlife and its habitats have to be considered in an analysis of cumulative effects as influenced by social condition. The following general categories describe the range of methods used to predict future social effects: linear trend projections (identifying, taking an existing trend, and projecting the same rate of change into the future); population multiplier methods (a specified increase in population implies designated multiples of some other variable); scenarios (characterization of hypothetical futures through a process of mathematically or schematically modeling the assumptions about the variables in question); expert testimony (experts can be asked to develop scenarios and assess their implications); and simulation modeling (mathematical formulation of premises and a process of quantitatively weighing variables)" (Council on Environmental Quality 1997-A-46).

This analysis differs from the other analysis of cumulative effects because it deals with subjective perception of effects. However, Decker et al. (2001) detail ways the human dimension of wildlife should be approached. The model obviously addresses social issues, and the models provide definitive and quantified results. However, the results and utility of the models are dependent on the quality of the data and assumptions made in developing the model. Also, social values are subjective and highly variable.

Environmental legislation has dictated that cumulative effects be taken into consideration when evaluating projects. Unfortunately, cumulative effects are rarely mentioned or analyzed adequately, and the results of projects that do consider them are often subjective. Johnson (Chapter 3) discusses opportunities and successes of Strategic Environmental Assessments that have the potential to overcome these shortfalls.

CONCLUSION

In the 1970s and 1980s, mitigation for altered wildlife habitat incorporated seven basic steps: a solid wildlife base for the project and mitigation site, thorough and complete data analysis, predictive models developed to create conceptual mitigation options, design of required habitat modifications for mitigations, implemented designs, monitoring of mitigation, and modifications to ongoing activities resulting from the monitoring project are agreed to and budgeted for by developing agencies at the planning stages. Cumulative effects are another step that would certainly be included in compiling a similar list in 2011.

Cumulative effects are the combined influences of variables on animal populations and the habitats that support them, and the effects have a range of meanings depending on how they are used. Similarly, there are numerous ways cumulative effects can operate (e.g., from physicochemical transport to where one project influences others). Cumulative effects need to be considered in all aspects of planning. This concept of quantifying cumulative effects was in its infancy in the 1970s but is becoming a common practice.

There are at least 10 methods we can use to understand the influences of cumulative effects: questionnaires, interviews, and panels; checklists; matrices; networks and system diagrams; modeling; trends analysis, overlay mapping and GIS; carrying capacity analysis; ecosystem analysis; economic impact analysis; and social impact analysis. More than one of the techniques will be required in many cases. The best possible efforts have to be made to present all adverse and beneficial effects to decision makers, so they can make informed decisions that are clear to them and the public.

5 The Economics of Cumulative Effects
Ecological and Macro by Nature

Brian Czech and Robert B. Richardson

CONTENTS

INTRODUCTION

As with economics in general, the economics of cumulative effects may be classified in various ways. Two such ways are especially relevant to the topic of cumulative effects: (1) macroeconomics relative to microeconomics and (2) ecological economics relative to environmental economics. Furthermore, these modes of classification are closely related.

Macroeconomics pertains to economic activities and processes in the aggregate. It covers such topics as economic growth, unemployment, and inflation. Microeconomics deals with nonaggregated or partially aggregated economic phenomena and the processes of production and consumption. It covers topics such as supply and demand, market equilibrium, and the derivation of prices.

There are shades of gray on the spectrum from microeconomics to macroeconomics. For example, the topic of "energy prices" encompasses numerous types or sources of energy, such as fossil fuels, nuclear power, and solar energy. Therefore, when speaking broadly of energy prices, some type of aggregation or synthesis must occur. We might think of such a topic as metaeconomics, a topic clearly broader than the price of gasoline and narrower than the rate of economic growth.

By definition, the topic of cumulative effects is not very microeconomic in nature, although it often is not completely macroeconomic, either. Rather, cumulative effects become more relevant as one proceeds analytically toward the macroeconomic end of the spectrum, beginning with the metaeconomics in the intermediate portion.

The distinction between macroeconomics and microeconomics is closely related to the distinction between ecological economics and environmental economics. Microeconomics came to dominate the study and practice of economics as the classical era of the 19th century gave way to the "neoclassical" economics of the 20th century (Heilbroner 1992). After Keynes (1936) wrote the *General Theory of Employment, Interest, and Money*, in the context of the Great Depression, macroeconomics came back into vogue in academia and public policy. However, many economists still equate economics with microeconomics, allowing little role for aggregate theory and (especially) policy. This is a long story with deep historical and ideological roots (Warsh 2006), but it suffices here to recognize that macroeconomics is not what most professional economists deal with. Furthermore, there are many other professionals with neoclassical training who use microeconomic philosophies and methods in their practices (e.g., financial managers, marketers, purchasing agents).

Meanwhile, the term environmental economics refers to the application of conventional or neoclassical economics to environmental issues. Given that neoclassical economics is concerned primarily with microeconomic efficiency (over secondary priorities such as equity or sustainable scale), environmental economics is likewise microeconomic in nature. It focuses on getting the prices right which, it is hoped, leads to the most efficient allocation of environmental resources among competing end uses (and users).

Ecological economics has more in common with macroeconomics, because the highest priority in ecological economics is the issue of "scale," or the size of the human economy relative to its containing, sustaining ecosystem. Indeed, ecological economics arose largely as a response to a lack of big-picture thinking in conventional economics. Neoclassical economists tend to resist the notion that limits to economic growth exist, which renders the scale issue moot. Ecological economists tend to be more versed in the natural sciences, especially the laws of thermodynamics and principles of ecology, and as such, they recognize limits to growth, and are highly critical of the conceptual shortcomings and risky policies associated with the assumptions of neoclassical economics. For over a decade, neoclassical economics has been deemed by a vocal and growing number of heterodox economists and social critics as a discipline in crisis for not adapting to environmental and related momentous challenges (Ormerod 1997).

A central theme of this chapter is that the economic implications of cumulative effects are, by their nature, ecological and macroeconomic. Economic growth itself contributes to the problem of cumulative effects. We must grapple today with the legacy of previous waste production and pollution, and examine how present activities will affect the future environment, which greatly complicates any economic analysis to support pollution control policy (Kolstad 2010). Furthermore, the ecological, macroeconomic nature of cumulative effects becomes clearer as the scale of the human economy approaches and exceeds the ecological capacity of the Earth and its ecosystems. This is not an economics that calls loudly for increasing private property rights and fine-tuning prices by estimating the social costs and benefits of private economic activity. Rather, it is an economics that calls for setting macroeconomic policy goals in a manner that will protect the environment from the deterioration that jeopardizes the economic prospects of the next generation.

Taken as a whole, wildlife is an emblematic example of this ecological, macroeconomic nature of cumulative effects. One economic activity, whether it is agricultural, extractive, manufacturing, or servicing, will affect numerous species (Czech et al. 2000). The growth of this single activity or sector will affect increased numbers of those species, plus other species as the sector's growth invariably interacts with other sectors that are required for its own growth (e.g., transportation, marketing, and waste disposal sectors). One of the most important points of this chapter, although it only bears explicating once, is that the economy grows as an integrated whole. Certain economic activities may wax and wane, and some may go "extinct," but there is a trophic structure that grows or shrinks as a whole (Czech 2008). At its most basic, this structure is comprised of producers and consumers, as in the economy of nature. The producers are agricultural and extractive agents such as miners, loggers, and fishermen. Manufacturing runs from the heaviest (e.g., mineral refining and timber milling) to the lightest (i.e., computer chip manufacturing). The cumulative effect of these activities is the competitive exclusion of nonhuman species in the aggregate.

If the breadth of the human niche were narrow, then the list of nonhuman species affected by human economic activity would be relatively short. Indeed, this was the case during the earliest stages of hominid development, when *Homo* species tended to specialize in the harvesting of select suites of species, depending on the ecosystem. For example, the Beringian immigrants that populated North America beginning with present-day Alaska specialized in the harvest of large mammals, eventually leading or contributing to extirpations and extinctions of many species, including the wooly mammoth (*Mammuthus primigenius*), mastodons (*Mammut* spp.), and the woolly rhinoceros (*Coelodonta antiquitatis*; Martin 1967).

Compared with extinct hominids, the history of *Homo sapiens* is one of technological progress whereby each invention or innovation that turned out to be useful in meeting material exigencies had the concurrent effect of broadening the human niche (Kingdon 1993). From a perspective focused on wildlife conservation, the natural and evolutionary history of *H. sapiens* has been one of continuously expanding niche breadth and therefore overlapping of niches previously occupied by other species. In other words, due to the tremendous breadth of the human niche, which expands via technological progress, the human economy grows at the competitive exclusion of nonhuman species in the aggregate (Czech et al. 2000). Failure to recognize this historical and ongoing phenomenon is tantamount to violating the spirit of cumulative effects as an academic concept and as a matter of law and policy. It may also be the case that a failure to recognize and address this phenomenon is what has led to the daunting, backlogged scenarios in which most cumulative effects research and decision-making efforts operate. Without a serious, substantial, explicit approach to macroeconomic policy, the study of cumulative effects and the application of findings will fall short of protecting the environment, conserving wildlife, and sustaining economic prospects.

At the same time, microeconomic tools and approaches remain necessary for addressing cumulative effects. To address any environmental and economic effects of management activities, including cumulative effects, there must be knowledge of how an individual action affects the environment and how the environmental effect affects economic values. Then, to work toward an assessment of cumulative effects, care must be taken to assess the potential for additive and emergent effects (i.e., effects that may emerge as more than a simple summation of individual effects). Perhaps the best approach to analyzing the economics of cumulative effects is a fusion of microeconomics and macroeconomics, as described below.

GOVERNMENTAL DECISION-MAKING AND THE FUSION OF MICROECONOMIC AND MACROECONOMIC APPROACHES

Government decision makers weigh many factors to select a management alternative. Environmental effects, politics, departmental strategy, stakeholders' concerns, and economic costs and benefits all play into the decision process. A government decision entails the allocation of resources by changing the regulatory framework through which society allocates its resources or by redirecting government resources to particular purposes. Microeconomics is the study of how individual producers and consumers allocate resources. Government decisions are microeconomic in two ways. First, the decision maker is an individual allocating resources. Second, the decision maker's actions affect how individuals in society will behave in allocating their resources. Both roles are suitable for microeconomic analysis.

Typically, the National Environmental Policy Act (NEPA) socioeconomic analyses consider how government actions will affect local and regional consumers and producers. The analysis shows how the government action will change their individual and collective decisions. Economic analysis of cumulative effects has the same inherent challenges as the analysis of cumulative effects themselves, because of the difficulty in defining the geographic (spatial) and time (temporal) boundaries (Council on Environmental Quality 1997). One can also distinguish different categories of cumulative effects, based on the sources of environmental change (single or multiple) or the process of accumulation. For example, there may be numerous disparate impacts on environmental or economic systems (i.e., accumulation of impacts) or an accumulated impact on a single environmental or economic element that is derived from several sources (Cocklin 1993). Like other environmental impact assessments, the selection of actions to include in an economic analysis of cumulative effects depends upon how the actions affect human uses and values of environmental resources. The process for conducting an economic analysis of cumulative effects follows the same pattern as economic impact analyses:

1. Define a region of influence.
2. Model the economic relationships within the region.

3. Estimate the effects the decision will have on the region.
4. Determine the significance of the effects.

Cumulative effects analysis has a wider scope than simple single-action analyses. Typically the economic model is expanded to include a wider geographic region and deeper interconnections of economic agents. Also, the model must be appropriate to address multiple actions. Many economic models are valid only for small (marginal) changes in prices and quantities, and assume there is no change in the social milieu (as indicated with the commonly used phrase *ceteris paribus*). If multiple actions go beyond marginal changes or the regulatory framework is shifted, the model may no longer be valid.

Economic analysis of cumulative effects embraces the same principles as other cumulative effects analysis. Effects have extensive and additive dimensions. The economics of cumulative effects deals with how impacts of individual decisions ripple out through the economy and how repeated decisions ultimately affect the economy as a whole. Cumulative environmental impacts may interact to create economic effects that also interact, and the cumulative economic impacts may be greater than their individual ones. Although we usually think of economics as dealing with smooth linear functions, economic tools can be adapted to include trigger or threshold effects also.

As we consider the cumulative effects of any decision or action, the layers of interaction and extension inevitably lead to national and global effects for good or ill. As the scale of the human economy approaches, and exceeds, the capacity of the Earth's ecosystems, recognizing the overall limits of the entire system becomes more relevant.

In the following sections, we introduce input–output analysis as a useful tool to assess cumulative effects at different geographical and economic scales. We then recommend enhancements to input–output analysis and the application of an ecological macroeconomic approach to analyzing cumulative effects. Finally, we discuss the microeconomic implications of cumulative effects for the analysis of net social benefits of environmental protection, degradation, or restoration.

INPUT–OUTPUT ANALYSIS

There is an intricate web of materials and economic values flowing among industries and households within the economy. Changes in spending within an economy (e.g., from economic expansion or environmental degradation) have indirect or multiplier effects on the regional economy through the increased demand of these businesses for the goods and services purchased from other local firms (a kind of "rippling" effect throughout the local economy). Leontief (1949) developed the input–output (I-O) model to illustrate the flows. Input–output analysis entails constructing a social accounting matrix representing the flows of spending between industries in an economy. The values in any single industry's column represent inflows to the industry, while values in the industry's row represent outflows to other industries. The social accounting matrix shows the interconnection of intermediate industries; each cell shows the value of an exchange from the row industry to the column industry. Rows are added to the bottom of the matrix to represent inputs from nonindustry sources such as payments for labor, taxes, and imports. Columns are added to the right side of the inter-industry matrix to show nonindustry final demand such as purchases by households or governments, and exports. The national I-O table is an essential element in balancing the national accounts to measure GDP and includes approximately 500 aggregated industries (BEA 2009).

After balancing the matrix to ensure that each industry's outputs equal its inputs, each column can be divided by its total to convert the values to proportions. Once divided, each cell in a column shows the value needed from the row industry to produce $1 of output in the column industry. The usefulness of I-O in forecasting the impacts of new spending comes from this quantification of the interactions among industries.

If the government is going to buy more tanks, for example, multiplying the spending by the coefficients in the tank industry column shows what other industries will benefit from the purchase, such as steel makers and engine manufacturers. The dollar increase in tank industry output also adds to steel makers' and engine manufacturers' output. Each time the dollar changes hands it is counted again in the I-O framework, so $1 in spending generates more than $1 in new output. This is called the multiplier effect. Conceptually, we can pick up the tank inputs column of spending apportioned to other industries, turn it on its side, and multiply the matrix of coefficients by it to show how the spending by secondary industries, like the steel makers and engine manufacturers, will in turn stimulate new economic activity in other industries, perhaps coal mining and scrap metal recycling. We can imagine repeating this operation many times to see where the tank spending ultimately flows through the economy. Thankfully, matrix algebra makes this unnecessary, and we can simply invert the inputs/dollar output matrix to calculate the results of all successive waves of spending and the resulting multiplier effect.

Input–output models were conceived and developed for conventional macroeconomic modeling, but they bridge a gap between microeconomic benefit–cost decision-making and impacts on the whole economy. The I-O begins where microeconomics ends. Microeconomics can help the individual determine his or her optimal behavior. Once that is determined, I-O can show the government decision maker how that action will impact the region and nation.

Most I-O models have been adapted to model regions as small as counties and zip code areas, which may be particularly useful for assessing the impact of cumulative effects. For such small geographies, most inputs come from outside the region, and many outputs flow out of the region, so industry interactions within the region are very limited. Regional impact modeling tools such as Impact Analysis for Planning, the Minnesota IMPLAN Group (IMPLAN), Regional Economic Models, Inc. (REMI), and Regional Input–Output Modeling System (RIMS11) use multipliers from the U.S. Bureau of Economic Analysis and rely on I-O tables for the estimation of regional economic impacts. These tools are frequently used in regional economic impact analysis for NEPA documentation (Weiss and Figura 2003).

Simple economic impact analyses of a single action define a region of interest and construct an I-O model based on the industries in that region. The model includes all of the interactions within the region, but interactions with industries outside of the region are considered exports and lost to the analysis. Cumulative effects may occur due to these exports outside the region. One response is to enlarge the geographic scope of the model. This helps, but the criticism remains that cumulative effects may be reverberating elsewhere in the economy. A solution would be to consider regional and national effects. Using an explicitly macroeconomic model ensures that all of the industries affected are represented and shows the overall affect of the action on the whole economy. Given the size of the national economy, any single project is an infinitesimal effect in the extensive dimension.

Cumulative effects of programs, however, can significantly affect national markets and output. Both the regional and national impacts are necessary to fully appreciate the additive interactions and emergent effects as changes flow through the economy.

Example: Banking on Nature

The U.S. Fish and Wildlife Service has issued a series of reports under the title "Banking on Nature" that use I-O analysis to examine the impact of national wildlife refuges on their local communities (Carver and Caudill 2007). Carver and Caudill (2007) studied 80 sample refuges in detail. They knew the number of refuge visitors and the activities they engaged in on the refuge from refuge management estimates. They extracted spending information from the National Survey of Fishing, Hunting, and Wildlife-Associated Recreation (conducted by the U.S. Fish and Wildlife Service) categorized by industry (U.S. Department of Fish and Wildlife Service 2006). They then estimated total spending by visitors to each refuge annually in each industry. The authors created regional I-O models covering nearby counties using the IMPLAN tool (MIG 2009). The total spending estimates could then be applied to the regional model to estimate the total impacts in the nearby communities.

For example, Chincoteague National Wildlife Refuge, a coastal barrier island on the Delmarva Peninsula, had 7.5 million visitors in 2006. Almost all of them enjoyed the beach and other nonconsumptive activities. Altogether, they spent $238.7 million during their visits. Applying this spending to the industries they patronized, primarily food, lodging, groceries, and sporting goods, the I-O model suggests that they added $315.4 million to the local output and 3,766 new local jobs (Carver and Caudill 2007, p. 210–213). The refuge has a substantial effect on the local economy.

Carver and Caudill (2007) extended their study to estimate the cumulative effect of all refuge visits on the national economy. They expanded the use estimates and the scope of the model. Using what they learned in the 80 sample studies, they developed estimates of aggregate visitation to all national wildlife refuges and total spending categorized by industry. Applying this vector of spending to a national I-O model indicated that refuge visits account for $1.7 billion of economic activity and 26,800 jobs nationwide.

Critiques and Extensions of I-O Analysis

Input–output analysis has limitations that reduce its credibility for long-term cumulative effects analysis. First, the I-O matrix represents economic flows within a single period of time. The flows of value reflect the technology, industrial linkages, and factor prices at the time they were measured. Prices and technology can change radically in the time frames considered in NEPA EIA analyses, rendering the I-O estimates inaccurate. Some regional economic models adapt the industry relationships dynamically (e.g., where prices are adjusted to reflect changes in demand), which may be more useful for long-term analysis of cumulative effects. Computable general equilibrium (CGE) modeling overcomes the fixed price and quantities limitation, but has much more demanding data requirements, including regional social accounts and parameters required for incorporating economic relationships among industries and institutions (Vargas et al. 2001).

A frequent criticism of national income accounting and traditional I-O has been that they fail to incorporate nonmarket effects (e.g., the values of household labor, ecosystem services) within the economy. Progressive macroeconomists are working to improve the accounting for environmental effects in national income measures and to create better indicators of national well-being (United Nations 2010). Alternative measures of socioeconomic well being such as the Genuine Progress Indicator (GPI) correct for some of these problems by taking into account the distribution of income, the value of nonmarket activities, and depletion of social and natural capital. In fact, the GPI includes the annualized value of the cumulative losses (i.e., reductions in stock) of farmland, wetlands, forests, and minerals, since the economy no longer receives the stream of benefits these resources once conferred. The indicator also incorporates the cumulative damages from activities such as emissions of greenhouse gases (which slowly but irrevocably affect the Earth's climate), the accumulation of radioactive wastes, and the accumulation of chlorofluorocarbon gases in the upper atmosphere that allow increased deadly ultraviolet radiation to reach the Earth (Talberth et al. 2007).

The I-O and CGE models have also been criticized because they are deterministic in nature and yield only single-point estimates. While superficially useful, such determinism is unrealistic. Regional impact estimates have an underlying stochastic element, which can be derived from the underlying statistical model or sample averages used to compute the inputs from which the output is derived. Weiler et al. (2003) combined statistical and deterministic approaches to generate confidence intervals for deterministic I-O models, which may be useful in assessing the relative significance of different economic drivers.

Another shortcoming of standard I-O analysis is that it ignores important inputs to production processes from outside its industrial framework or economic paradigm. Agricultural outputs, for example, are assumed to rely to a substantial degree on inputs from the manufacturing sector, and models disregard the support and function of ecosystem processes such as nutrient cycling and pollination, and the effects of exogenous variables such as weather. This approach may be reasonable for estimating regional economic impacts, but it fails to account for changes in ecological variables and their economic effects.

MICROECONOMIC ANALYSIS OF CUMULATIVE EFFECTS

Traditional economic approaches to environmental policy have been based on command-and-control regulation and have commonly focused on the measurement of the costs of pollution reduction (also known as abatement costs) for the emission of particular pollutants (Baumol and Oates 1988). For a given pollutant, the efficient standard (i.e., the level of pollution that maximizes net social benefits) would be set where the marginal benefits of pollution abatement are equal to the marginal costs of abatement. However, emission control may be of limited use in dealing with the more insidious problem of cumulative effects. The combined effects of previous activities on present environmental conditions and the effect of present activities on future conditions greatly complicates any economic assessment of the benefits and costs of environmental impacts.

Because cumulative environmental effects may have economic consequences for regions or groups of people, decisions about use or protection are ultimately social decisions. However, any action is likely to involve benefits for some people and losses to others. Translating individual preferences regarding the environment to social choices is a fundamental challenge to economic analysis. Individual preferences about the environment may be influenced by particular philosophical perspectives or worldviews, such as biocentrism, anthropocentrism, or sustainability. Economists use the utility function to represent individual preferences in consumption. An individual *i*'s utility is represented as U_i in the following utility function:

$$U_i = U_i\,(X_1, X_2, \ldots, X_n, e) \qquad (5.1)$$

where $X_1, X_2, \ldots, X_n$ represents the quantities of market goods consumed by individual *i*, and *e* represents the level of environmental quality (such as emissions of a particular pollutant). It follows then that U_i represents the utility or satisfaction that *i* derives from the consumption of market goods and the level of environmental quality. A basic microeconomic view of the environment would use indifference curves to depict the combinations of market goods (material consumption) and environmental quality that individuals would consume to maximize their utility (Kolstad 2010). The economic effect of a change in environmental quality would be represented by the quantity in material consumption that would be necessary to compensate for the losses associated with the environmental damages. For example, indifference curves (Figure 5.1) demonstrate the willingness of individuals to trade off levels of consumption and environmental quality. The preferences represented by the indifference curves (Figure 5.1) indicate greater values for environmental quality with higher levels of consumption. That is, at higher levels of *X*, the individual requires higher levels of material goods to offset a loss in environmental quality (*e*).

Social choices are represented by a social welfare function, which is comprised of the utility functions of individuals in a society. The social welfare function may simply be the sum of individual utilities (that assumes equal marginal utility of consumption), or individual functions may be weighted to incorporate equity (variable marginal utility of consumption), human rights (protection from unsolicited damages), or sustainability (effects on the utility of future generations). For a society with *m* individuals, social welfare is represented as *W* in the social welfare function:

$$\mathrm{W}\,(U_1, U_2, \ldots, U_m) = \Sigma_j \theta_j U_j, \theta \geq 0 \qquad (5.2)$$

where θ_j represents a weighting of the utility of individual *j* (θ may be equal, or it may vary among individuals in a society, perhaps on the basis of income). The benefits (of

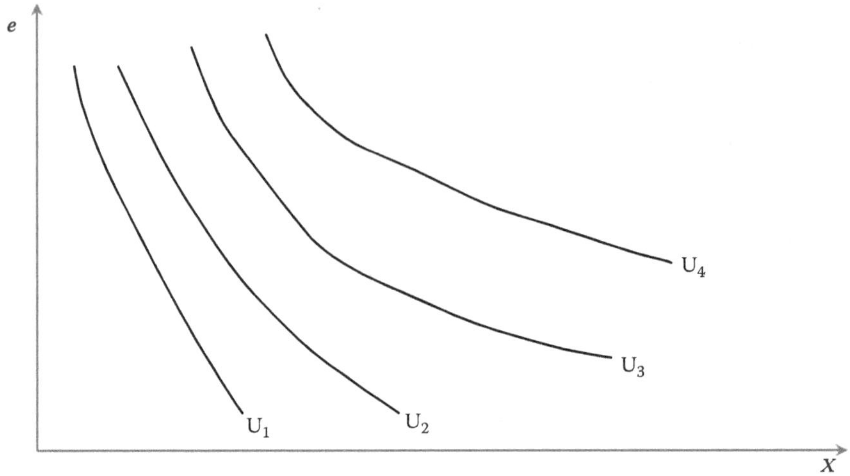

FIGURE 5.1 Indifference curves for the consumption of market goods (*X*) and environmental quality (*e*). (Adapted from Kolstad, C. D. 2010. *Environmental Economics*. Oxford University Press, New York.)

consumption) are weighed against the costs (of pollution and resource degradation), and economic efficiency is achieved when net benefits are maximized (i.e., where no other preferable allocation exists). Critics of the utilitarian approach emphasize that individual utility functions are not fixed, and preferences about consumption or demand for environmental quality may be influenced by advertising, technological developments, or knowledge.

The assumption that individuals will maximize utility subject to a budget constraint ultimately generates a demand curve for the consumption of goods and services. Measures of the gain (or loss) in welfare associated with consumption can be obtained by comparing what a consumer would be willing to pay for a quantity of goods with the market price; the difference is known as *consumer surplus*, which is interpreted as the extra value consumers get over and above the price paid. The concept of consumer surplus is easily illustrated with an example of recreational trips to a protected area or park. The relationship between the price of a trip and the number of trips taken is represented by the demand curve, which is downward sloping (Figure 5.2). As the price of a trip increases (perhaps because of higher travel costs), the number of trips is expected to decline. An individual will take seven trips/year when the price per trip is $30 (Figure 5.2). The consumer surplus (or net benefits) from the "consumption" of trips is represented by the gray triangle; the value of these benefits is equal to the area of the triangle (or $\frac{1}{2} \times 7 \times 60 = \$210$). Changes in environmental quality (e.g., impacts to wildlife population, water quality, or air quality) would be expected to affect the number of trips taken, and consequently, the consumer surplus or net benefits derived from the recreational trips (Loomis and Walsh 1997).

The demand for environmental quality is unique because most environmental "goods" are not valued in the market, so there are no observations for how much of an environmental good would be consumed at various prices. In most cases, this is

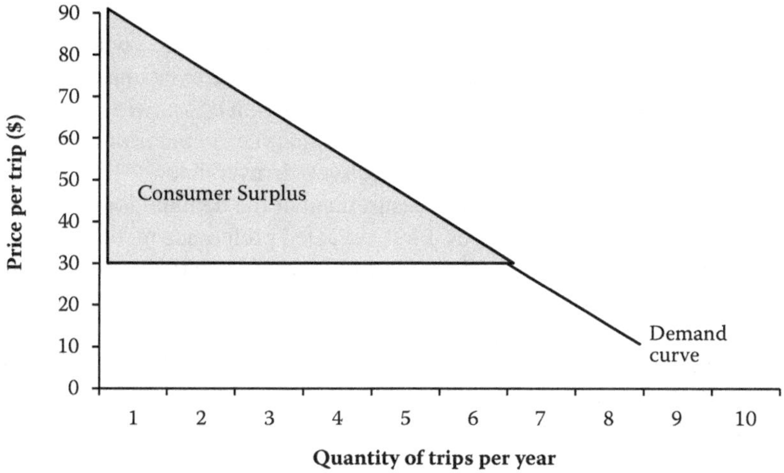

FIGURE 5.2 Consumer surplus for an individual's trips to a protected area.

because of environmental externalities or demand for public goods. Externalities (such as pollution) exist when some consumption or production choice impacts another entity's utility or production function without permission or compensation. The value of the externality may not be known because of the absence of compensation to victims. Public goods such as environmental protection are considered to be nonrival and nonexcludable in nature, and are frequently associated with the free-rider problem and high transactions costs from organizing. Examples of pure public goods include rainforests, wilderness areas, and many ecosystem services (e.g., climate regulation). Public goods are often unpriced and therefore are undersupplied by markets (Kolstad 2010).

Because the demand for environmental quality is represented in terms of the tradeoffs between using resources for material consumption and environmental protection, economists use the demand function to estimate how much an individual is willing to give up (or pay) for particular levels of an environmental good. Without a market, there is no price, so economists rely on the concept of marginal willingness to pay as a proxy for price, and net willingness to pay is analogous to consumer surplus. Eliciting true values is not easy, and the method of valuation depends upon the use or benefit associated with the environmental good.

Economists typically categorize environmental goods based on how individuals perceive environmental quality or realize the value of environmental amenities. This relates to whether consumers obtain utility through the use of environmental resources (e.g., the benefits of recreation such as hiking or fishing) or through nonuse values (e.g., the benefits of simply knowing that species exist and habitats are protected; Kolstad 2010). Use value is the gain in utility associated with the consumption of some good. In the case of environmental goods, use values may be affected by changes in environmental quality (e.g., health effects of breathing polluted air), damages to particular resources (e.g., recreational effects from a polluted lake), and damages to ecosystems (e.g., agricultural impacts of water degradation). Nonuse value is the gain in utility from the passive or indirect benefits of environmental resources. Nonuse values typically include existence value (i.e., benefits obtained from the satisfaction of knowing that species or ecosystems exist and are protected), bequest value (i.e., benefits obtained from the satisfaction of knowing that resources are protected for future generations), and option value (i.e., benefits of possible future uses of a resource for which depletion is effectively irreversible).

Two basic approaches exist for the measurement of the demand for environmental goods, in the absence of market prices. First, revealed preference methods involve the inference of value from the tradeoffs made in real choices. For example, the differential market prices for houses in regions with differing levels of air pollution could be used to infer the value of air quality, after controlling for other variables such as home size, neighborhood safety, and other influences. Revealed preference methods include hedonic regression (for the estimation of demand through real estate prices) and travel cost models (for the estimation of recreation benefits). Second, stated preference methods involve asking people questions about their individual willingness to pay (or accept) money for improvement (or degradation) to environmental quality. These methods rely on polls or surveys that use hypothetical scenarios to elicit tradeoffs for changes in the level of environmental benefits. Stated preference methods include contingent valuation (for the estimation of nonuse values) and contingent behavior analysis (for the estimation of hypothetical changes in some activity). Stated preference approaches have been criticized for their hypothetical nature, which may be associated with hypothetical bias, strategic bias, or embedding bias.

Other economic benefits of environmental resources may exist but may not be easily captured through the measurement of use or nonuse values using these approaches. Examples include scientific and educational benefits. Environmental resources have scientific value and their effects are frequently the subject of scientific research and publications. In some cases, environmental resources are used to understand the impact of human activities, and in others, to understand the influence of natural conditions on flora, fauna, and physical environments. Scientific research contributes to economic progress and understanding, and often to greater efficiency in policymaking, productivity improvements, and overall increased human well-being. The value of new discoveries and the knowledge that arises from the opportunity to study ecological processes has the potential to prevent costly natural resource management mistakes that lead to expensive endangered species' recovery efforts and environmental restoration activities. Natural environments also offer a living classroom for education, and there are a number of human development programs that use natural areas to train managers, promote teamwork, teach coping skills, and provide various forms of emotional and physical therapy. Wilderness expedition programs have also been used with troubled youths to foster better emotional development and adaptation skills. Cumulative environmental effects of human activities may impact, degrade, or disrupt the flow of scientific and educational benefits, particularly if the effects are irreversible.

Among the most important environmental benefits are the values of ecosystem services, which are the conditions and processes through which natural ecosystems sustain and fulfill life on Earth. Examples of ecosystem services include clean air and water (for health and economic security), essential support in producing renewable resources (e.g., agriculture and forest products), and the absorption and treatment of waste matter. Costanza et al. (1997) emphasize the interdependent nature of many ecosystem functions, which underscores their relevance in understanding the economic implications of cumulative environmental effects. For example, consumption of some of the net primary production in an ecosystem generates respiratory products necessary for additional primary production. These functions and services are interdependent, but in many cases, they represent joint products of the ecosystem and support human life and welfare. Because of this interdependence, the issue of cumulative environmental effects poses particular challenges for economic measurement and valuation (e.g., when particular functions of the same ecosystem service are affected differently). Ecosystem services may be valued in a number of ways. Some services are associated with market goods that carry a price. Indirect valuations may be used to measure society's willingness to pay for other services for which there is no market. Values of other services may be conceptualized as the costs avoided by protecting the ecosystem's ability to continue providing the services. Environmental impacts to ecosystem services and functions may interact, and the cumulative effect may affect human welfare and generate economic effects that ripple across regions, industries, and households.

Example: Economics and Ecosystem Services in the Desert

Deserts are frequently overlooked in studies of the economic value of ecosystem services (Sutton and Costanza 2002; Costanza et al. 1997), and hence, little is known about the economic implications of cumulative effects of human activities on desert ecosystems. Research on the global values of ecosystem services has not identified any valuation studies for desert biomes. Still, desert environments provide numerous services and functions that contribute to human utility, sustain and fulfill human life, and have measurable economic values.

Deserts serve as habitat for a diverse array of flora and fauna, including species of commercial importance such as the endangered desert pupfish (*Cyprinodon diabolis*) and several varieties of cactus. Desert bats, hummingbirds, and bees provide important pollination services for plants, including some commercial crops such as dates, almonds, alfalfa, olives, and sunflowers (Kremen et al. 2007). Deserts also provide abundant recreation opportunities for wilderness visitors, rock climbers, and motorized recreation enthusiasts. Natural vegetation and soil crusts in deserts protect the important function of erosion control. Desert vegetation and soils also support climate regulation services by sequestering carbon that would otherwise contribute to climate change (Luo et al. 2007; Housman et al. 2006; Jasoni et al. 2005), and by mitigating heat island effects through shading and evapotransporation (Stabler et al. 2005; Grimmond et al. 1996). Ants in deserts have been reported to provide waste treatment and nutrient cycling services, particularly in areas where grazing occurs (Wagner and Jones 2006). The economic effects of disruptions to the flow of these ecosystem services would interact to generate cumulative economic effects. The effects of air quality are used as an example for illustration.

Deserts of the southwestern United States in particular are threatened by pressures of urban expansion, population growth, motorized transportation, and industrial development. Population growth in desert-proximate counties of the Southwest has exceeded that of the United States, which has led to increases in construction, transportation, and recreation activities. Emissions of fine dust powder known as particulate matter (PM) are related to many of these activities, and they have contributed to a major source of air pollution associated with poor visibility and severe risks to human health. These tiny particles penetrate deep into human lungs and increase the risk of asthma and other health problems. Federal studies have reported that high levels of particulate matter are associated with premature death, aggravated asthma, childhood respiratory problems, chronic bronchitis, haze, and visibility in national parks and wilderness areas (Environmental Protection Act 1999). Woodruff et al. (1997) reported a statistically significant relationship between particulate air pollution in the United States and postneonatal infant mortality. A study of the effects of air pollution on children in southern California found that PM can also retard the growth of children's lungs (Gauderman et al. 2000). According to a report by the Natural Resources Defense Council, particulate pollution claims the lives of over 64,000 Americans every year (Natural Resources Defense Council 1996). Air quality levels in the California desert are particularly poor; the air basins in the region have the highest annual values of particulate concentration in the state, and the highest 24-hour concentrations occur where the problem of windblown dust is widespread.

Automobile emissions, cement production, military bases, and road dust all contribute to excessive concentrations of PM in the area. However, the dominant source is fugitive dust from roads (Alexis and Cox 2005). Gravel and unpaved roads, in particular, are sources of long-term soil loss and erosion, even in the absence of vehicular use (Havlick 2002). Road use exacerbates surface erosion, particularly on unpaved road surfaces. Construction, particularly in rapidly growing urban and suburban areas, contributes to emissions of PM through physical site disturbance and repeated use of machinery.

Over 9,000 deaths in California can be attributed to particulate pollution each year (more than motor vehicle accidents, accidental poisonings, and homicides combined; Sharp and Walker 2002). A study of the effects of PM emissions on hospital admissions in the San Joaquin Valley reported that higher levels of PM were associated with increased chronic and acute respiratory hospitalizations and emergency room visits (Hall et al. 2006). Visibility throughout the southern region of the state is frequently obscured by the effects of particle emissions, smog, and dust from wind erosion. Currently, over 99% of Californians breathe air that violates state PM10 standards at least part of the year (Alexis and Cox 2005). Although the Environmental Protection Agency (EPA) has mandated that air quality around national parks be subject to the most stringent level of protection, Joshua Tree National Park consistently exceeds the ozone and particulate matter concentration levels set by the EPA for human health (Joshua Tree National Park 2002).

In addition to the public health risks of particulate matter, emissions from vehicles and industrial activities cause pollution that can worsen visibility in parks and desert wildlands, harm vegetation, and increase the risk of wildfires. Visibility at Joshua Tree National Park is often obscured by haze caused by high concentrations of particulate matter, and cumulative concentrations of ozone in the park exceed levels known to cause injury to vegetation. Nitrogen pollution from vehicle exhaust, industrial emissions, and agricultural sources promote the growth of nonnative species, which have been shown to increase the risk of wildfires (Harrod and Reichard 2001). The sparse vegetation that is endemic at Joshua Tree protects against lightning-sparked fires; nonnative plants provide fuel for fires and can quickly consume thousands of hectares of slow-growing Joshua trees (*Yucca brevifolia*), juniper (*Juniperus* spp.), and pinyon pines (*Pinus cembroides*) (Joshua Tree National Park 2002).

In addition to emissions of particulate matter, over 17,000 tonnes of carbon monoxide and over 3,000 tonnes of nitrogen dioxide are emitted each day in California (28% of which is emitted from off-road recreation vehicles alone); both of these pollutants have been associated with increased risk of congestive heart failure, respiratory illnesses, birth defects, and in some cases, death (Alexis and Cox 2005; Sharp and Walker 2002).

Together, these environmental effects have economic implications for individual utility and overall social welfare. Generally, the economic effect of erosion in deserts can be estimated as either the willingness to pay (WTP) for better air quality (through reductions in emissions of particulate matter), or as the costs avoided by greater protection of the flows of the erosion control service. However, the economic effect is complicated by the various ways in which humans perceive environmental quality in deserts and use environmental amenities. Environmental impacts of degraded air quality in deserts may interact to affect human health, recreation benefits, property values, and nonuse benefits. As such, the example provides a useful illustration of how environmental effects interact to create cumulative economic effects. The economic effects related to public health would be based on the costs associated with hospital admissions, emergency room visits, and lost work days related to poor air quality. Sharp and Walker (2002) estimated that particulate pollution is responsible for public health losses to the state's economy of more $1 billion/year—not including the costs of thousands of less severe illnesses that result from excessive levels of PM.

The economic effects related to recreation could be modeled as the economic benefits (or net willingness to pay) of preserving visibility for recreation. Schulze et al. (1983) used the contingent valuation method to measure the economic benefits of preserving visibility in national parks in the southwestern United States (i.e., Arizona, California, and Colorado), and reported that average household willingness to pay ranged from $14.14 to $20.60 (2009 dollars per month) for the preservation of air quality and visibility in three national parks in the Southwest.

Other economic effects could stem from the benefits of biological diversity, including recreational wildlife viewing or existence value for particular species. The region provides habitat for roughly 360 nonfish vertebrate species, and there are 131 known fish taxa from the region. There are an unusually large number of threatened and endangered species in the region, which highlights the biodiversity of the desert and the increasing threats from the scale of human activities.

CONCLUSION

We have considered the economic implications of cumulative environmental effects and how such implications would be analyzed by ecological and neoclassical economists. Cumulative environmental effects are typically linked in part to socioeconomic activities, and as such they are associated with economic impacts that interact and accumulate within an economy, which may have substantial effects beyond the local level. Cumulative effects of environmental change may have economic consequences for the production of market goods (e.g., agriculture), the use of natural resources (e.g., recreation), or the function of ecosystem services (e.g., water regulation and supply).

We first considered the problem of cumulative effects through the lenses of neoclassical and ecological economics, and in terms of their relevance for macroeconomic and microeconomic analysis. We concluded that cumulative effects are fundamentally ecological and macroeconomic in nature, given their implications for economic growth. Growth itself contributes to the problem of cumulative effects, since it relates in part to what remains of previous waste and degradation and how present activities affect the future environment. The issue of scale, or the size of the human economy relative to the ecosystem that sustains it, becomes more important as the cumulative effect of economic growth approaches the ecological capacity of the Earth and its ecosystems.

We introduced I-O analysis as a tool for assessing cumulative effects of environmental change at different geographical and economic scales. The I-O matrix is a flexible tool for representing flows among multiple entities in an economy, and may be useful for estimating the regional economic impact of cumulative effects. Traditional I-O analysis shows exchanges of value measured in dollars, but there is no reason the same type of matrix could not express other exchanges using other units of measure. Ecologists, for example, have built I-O-style matrix models of energy exchange among species measured in kilojoules. We can imagine matrices for different materials. How many kilograms of iron are transferred from mining to smelters to auto manufacturers to auto dealers? Tracing those paths and relating them to dollars of output could elucidate connections from the ecological system to the human economy. Such a matrix could include both nonrenewable and renewable iron sources, and connect with models of iron extraction to illustrate the natural limits to growth. Care would be required to ensure that the units of measure remained consistent through each layer of processing. The United Nations Statistics Division is working on standards for models of water, energy, fisheries, and material flows (United Nations 2010). However, such models would still be bound by the limitations of I-O in that prices reflect a particular period and interactions reflect a frozen image of technology and social framework. We recommended enhancements to I-O analysis and application of an ecological macroeconomic approach that is more appropriate for analyzing cumulative effects.

Finally, we considered the microeconomic implications of cumulative effects for the analysis of net social benefits of environmental protection, degradation, or restoration. Decisions about environmental exploitation or protection are ultimately social decisions; however, any action is likely to involve benefits for some people and losses to others. We examined how individual preferences regarding the environment may be translated to social choices using the utility and social welfare functions. We noted that the demand for environmental quality is unique in that most environmental "goods" are not valued in the market, so there is little information about the relative value of environmental quality, usually because of environmental externalities or public goods. We discussed the various use and nonuse values of environmental resources and the methods for estimating the value of nonmarket benefits. Then, using the example of ecosystem services in deserts, we considered the cumulative effects of change in air quality on public health, visibility in public lands, and recreation uses.

In general, we conclude that economic analysis of cumulative effects embraces the same principles as other cumulative effects analysis. Both have extensive and additive dimensions, and the same inherent analytical challenges because of the difficult in defining the geographic (spatial) and time (temporal) boundaries. Economic effects may be distinguished by their sources of environmental change or by the processes of accumulation. The selection of actions to include in an economic analysis of cumulative effects depends upon how the actions affect human uses and values of environmental resources. However, many economic models are valid only for small (marginal) changes in prices and quantities, and assume that all other variables are unchanged. Environmental shocks through multiple actions or nonmarginal changes would weaken the ability of economic models to estimate valid effects.

Section 2

Case Studies

6 U.S. Department of Homeland Security, Environmental Waivers and Cumulative Effects

Lirain F. Urreiztieta and Lisa K. Harris

CONTENTS

INTRODUCTION

The environmental review process is designed to evaluate all effects (i.e., direct, indirect, and cumulative) and to develop mitigation strategies that eliminate, alleviate, or offset impacts. The evaluation process, however, can be short circuited. Projects can be implemented after an incomplete analysis, or in some cases the process can be waived altogether. In the latter, effects may never be evaluated, resulting in possible damage to the environment and possible future emergency mitigation methods that may be costly and inadequate.

Over a 3-year period (2005–2008) the Secretary of the U.S. Department of Homeland Security (DHS) exercised the authority granted by Congress to waive environmental laws to expedite the construction of physical barriers along the U.S.–Mexico border. This waiver represents a unique policy used to expedite a federal project. Decades before, the environmental review process had been truncated by petitioning the Endangered Species Act's (ESA) "God Squad," a seven-member Cabinet-level committee with the authority to exempt federal agencies from the ESA's Section 7 (i.e., consultation with the U.S. Fish and Wildlife Service [USFWS] so that a proposed action does not jeopardize listed species or adversely modify critical habitat) or through legislation such as the Arizona–Idaho Conservation Act of 1988. This act incorporated the terms of a Biological Opinion into law, thereby allowing the University of Arizona to build telescopes within ESA-listed Mount Graham red squirrel (*Tamiasciurus hudsonicus grahamensis*) habitat on Forest Service lands. The authorized use of the waiver by DHS to expedite the construction of border barriers enabled the agency to avoid the sometimes lengthy environmental review process to ensure the expeditious implementation of border infrastructure.

This chapter focuses on the San Pedro Riparian National Conservation Area (SPRNCA) in southern Arizona and the construction of vehicular and pedestrian barriers under environmental waivers. The waivers, which came under intense scrutiny, nearly eliminated the accepted methods for evaluating the action's effects and transformed the established timeframes set aside for reviews, from affected agencies and the public, ultimately lessening the scope and evaluation of cumulative effects.

SAN PEDRO RIPARIAN NATIONAL CONSERVATION AREA

The SPRNCA is a unique area located on Bureau of Land Management (BLM) lands in Cochise County, Arizona. The SPRNCA was designated in 1988 and contains nearly 23,068 ha between the international border and St. David, Arizona, including 64 km of the meandering upper San Pedro River. The primary purpose for special designation was to protect the desert riparian ecosystem, a rare remnant of what was once an extensive perennial riverine network throughout the American Southwest. The San Pedro River marks a transition area between the Sonoran and Chihuahuan deserts and is internationally known for its biodiversity (Steinitz et al. 2005). Originating from the Sierra La Mariquita, Sierra San Jose, and Sierra Los Ajos in north-central Sonora, Mexico, and the southern slopes of the Huachuca Mountains in Cochise County, the river flows north approximately 240 km to its confluence with the Gila River near Winkelman, Arizona (Makings 2005). The stretch of river within the SPRNCA is lush with vegetation, a sharp contrast to adjacent desert and grassland communities, and is often described as a ribbon of green in an otherwise barren landscape (Figure 6.1). It is one of the most important migratory bird corridors in the western hemisphere and a crucial movement corridor for wildlife (The Nature Conservancy 2010).

The San Pedro riparian corridor is habitat for towering cottonwoods *(Populus fremontii),* a diverse understory of plants, and a diversity of wildlife species including 84 mammals, 14 fish, 41 reptiles and amphibians, and more than 100 breeding bird species, and the area provides habitat for 250 additional species of migrant and wintering birds (Bureau of Land Management 2010). From the river's banks, visitors can spot beavers *(Castor canadensis)*, yellow-billed cuckoos *(Coccyzus americanus)*, gray hawks (*Buteo nitidus*), vermilion flycatchers (*Pyrocephalus rubinus*), yellow warblers (*Dendroica petechia*), summer tanagers (*Piranga rubra*), and hooded orioles (*Icterus cucullatus*).

People have lived near the San Pedro for eons; researchers discovered archaeological sites representing the remains of human occupation dating back 13,000 years. In recent times, the area near the SPRNCA has been used for rural agricultural purposes. Today, the area's population is sparse, with 19 inhabitants per 1.6 km within

FIGURE 6.1 San Pedro River within the San Pedro River National Conservation Area. Lush vegetation with cottonwood–willow overstory provides habitat for numerous species not found within the arid surrounding area. (Used with permission from Lyda Suzanne Harris.)

Cochise County (U.S. Census 2000). The dominant industry for the area comes from the military installation Fort Huachuca in nearby Sierra Vista, Arizona, from recreation, and to a lesser degree, cattle ranching.

Border Issues

The section of the U.S.–Mexico international border in Cochise County is 136 km long. Because of its remote location, this section became a hot spot for illegal activity, including drug smuggling and illegal entry by undocumented persons. Between October 2004 and February 2005, the U.S. Border Patrol (USBP) intercepted over 71,000 undocumented migrants (Thompson 2005). New trails and roads were created, "destroying cactus and other sensitive vegetation that can take decades to recover" (*Defenders of Wildlife and Sierra Club v. Chertoff* 2007). Ranchers complained of vandalism and theft: fences cut, gates left open for cattle to stray, cattle killed, watchdogs poisoned, water tanks drained, buildings broken into, and property stolen (Custred 2000). As DHS increased surveillance along other portions of the border, illegal activity near SPRNCA increased. In 2009, the U.S. Customs and Border Protection at the Tucson Sector of the border, which includes SPRNCA, accounted for almost half of all marijuana seized and illegal immigrants apprehended in the entire United States (U.S. Senate Committee 2009). Increased illegal activity brought with it more vandalism, robberies, abandoned cars, and damage to the environment. The land was littered by tonnes of trash and human waste. According to one estimate, each immigrant discards at least 3.6 kg of trash (Bureau of Land Management 2006). In addition, uncontrolled campfires left by smugglers caused wildfires that damaged habitat and property.

Proposed Border Barrier

The DHS proposed pedestrian and vehicle barriers within, and next to, SPRNCA in an attempt to control illegal immigration and secure the nation's borders. The DHS proposed building 3.7- to 4.3-m-high pedestrian barriers, constructed of vertical steel bollards or metal frames and sheets within SPRNCA (Figure 6.2). The proposed vehicle barriers would be 1.2–1.8 m high, constructed primarily of metal post on rail, and positioned within the San Pedro riverbed and nearby washes within SPRNCA.

Congress mandated that the barriers be constructed by the end of 2008. From an environmental review process perspective, this deadline was unrealistic. If the project were to undergo an environmental analysis following the National Environmental Policy Act (NEPA), DHS would not meet the 2008 deadline requirement. To fulfill requirements and construct the barriers within the mandated timeframe, DHS Secretary Chertoff issued the environmental waiver.

Environmental Waiver

Authorization for the environmental waiver was based upon four laws. The Illegal Immigration Reform and Immigrant Responsibility Act (IIRIRA) of 1996 directed the Attorney General to install barriers and roads at the border to deter illegal entry into the United States. The IIRIRA also authorized the Attorney General to waive applicable environmental laws when it was determined necessary to expedite barrier construction.

The Homeland Security Act of 2002 transferred responsibility for border security construction from the Attorney General to the Secretary of DHS. In 2005, Congress significantly expanded the authority of the Secretary of DHS with passage of the REAL ID Act. The Secretary of DHS was granted power to use sole discretion to waive all legal requirements that impeded construction of roads and barriers. A fourth law, the Secure Fence Act of 2006, directed the Secretary to achieve operational control over the border through all actions necessary and appropriate.

Under typical circumstances, a proposal from a federal agency to build barriers on federal land would have gone through the environmental review process outlined in NEPA. An Environmental Assessment (EA) or an Environmental Impact Statement (EIS) would have been developed and finalized, and the associated public

FIGURE 6.2 A section of the pedestrian fence constructed by the U.S. Department of Homeland Security along the border between Arizona and Mexico. (Used with permission from Lyda Suzanne Harris.)

review process incorporated. The assessment would have evaluated the effects of the barrier (i.e., the proposed action), and evaluated the impacts of other reasonable alternatives including a "No Action Alternative." However, Section 102 of the Real ID Act authorized DHS to issue an environmental waiver, bypassing NEPA and 18 other federal laws including the Endangered Species Act (Table 6.1).

In total, DHS issued five waivers to build border barriers: three were area-specific (San Diego [September 2005], Barry M. Goldwater Range [January 2007], and SPRNCA [October 2007]), and two were borderwide (U.S. Department of Homeland Security 2008). Secretary Chertoff's authorization of using waivers did not go unnoticed.

Reaction to Waiver

Government officials, including Representative Raul Grijalva (D-Arizona), Chairman of the National Parks, Forests, and Public Lands Subcommittee, expressed opposition to the waiver. "It broadsided us," said Natalie Rose, Grijalva's Press Secretary at the time (N. Rose, personal communication, January 26, 2010). In a letter to the

TABLE 6.1
Laws Waived to Build Barriers along the International Border between the United States and Mexico

Environmental Laws	
Administrative Procedure Act	Multiple Use and Sustained Yield Act
American Indian Religious Freedom Act	National Environmental Policy Act
Antiquities Act	National Forest Management Act
Archaeological and Historic Preservation Act	National Historic Preservation Act
Archeological Resources Protection Act	National Park Service General Authorities Act
Arizona Desert Wilderness Act	National Park Service Organic Act
Clean Air Act	National Parks and Recreation Act
Coastal Zone Management Act	National Wildlife Refuge System Administration Act
Comprehensive Environmental Response, Compensation, and Liability Act	Native American Graves Protection and Repatriation Act
Eagle Protection Act	Noise Control Act
Endangered Species Act	Otay Mountain Wilderness Act
Farmland Protection Policy Act	Religious Freedom Restoration Act
Federal Land Policy and Management Act	Rivers and Harbors Act
Federal Water Pollution Control Act/Clean Water Act	Safe Drinking Water Act
Fish and Wildlife Act	Solid Waste Disposal Act
Fish and Wildlife Coordination Act	Title I of the California Desert Protection Act
Historic Sites, Buildings, and Antiquities Act	Wild and Scenic Rivers Act
Migratory Bird Treaty Act	Wilderness Act

DHS Secretary, Grijalva (2007) wrote that the SPRNCA was "an extremely fragile ecosystem," providing "critical and irreplaceable habitat for thousands of migratory and endemic bird species, mammals and other living things." Grijalva believed the waiver "violated environmental laws," Rose said. Because the environmental review process was never completed, local communities, including Native American Indian tribes, were never consulted. According to Grijalva, the Bush Administration used antiterrorism and antiimmigrant sentiment as justification for the waiver. "They needed to show the American people that they (the Bush Administration) were doing something to fight the war on terror and undocumented immigrants," Rose said.

The administration "steam-rolled the environmental process," according Rose, "in order to shove their agenda through before his term ended." The environmental process, if allowed to occur, would have taken additional time, placing project construction under the next administration's watch.

In an interview, Ira Mehlman, Media Director for the Federation for American Immigration Reform, stated that "The drug cartels were taking advantage of our porous borders, wreaking havoc, and people who were seeking to do damage to the United States could enter freely." Mehlman believed that the environmental review process would have taken years to work through before a barrier was ultimately constructed. "Thousands of people were traipsing across our borders." The waiver speeded up the process of building a fence, and furthering DHS's mandate to secure the borders (I. Mehlman, Federation for American Immigration Reform, personal communication, January 29, 2010).

The BLM had developed an EA for constructing physical barriers within the SPNRCA in 2007. However, it was completed in-house and within approximately 3 months (typically an assessment of this nature would take more than 6 months to complete). The document was, according to one environmental watchdog group, not very thorough, and the cumulative effects section was lacking (M. Clark, Defenders of Wildlife, personal communication). In addition, there was no comment period when the "Finding of No Significant Impact (FONSI)" was released. As a result, the environmental groups, public groups affected by the project, and any interested person had no input into the project (M. Clark, Defenders of Wildlife, personal communication, March 3, 2010).

Following the FONSI's release, DHS immediately commenced construction of the barriers. The Defenders of Wildlife challenged the project, using the BLM's appeal process and received a 10-day preliminary injunction from the U.S. District Court for the District of Columbia. But by then, significant topographic alteration had already been done, including trenching with bulldozers and the removal of vegetation to improve the existing dirt road (M. Clark, Defenders of Wildlife, personal communication, March 3, 2010).

It was during this time period that DHS issued its third waiver, eliminating the compliance requirements to construct the barriers within SPRNCA. The waiver allowed DHS to continue building the pedestrian fence and vehicle barrier, and it removed the legal grounds Defenders of Wildlife had used to obtain the injunction.

In response to the waiver, Senate Committee on Homeland Security and Governmental Affairs Chairman Joseph Lieberman wrote to DHS Secretary Chertoff about the committee's concerns regarding the issuance. Lieberman requested answers to critical questions including: What are the criteria used by DHS to determine when it is appropriate to use the waiver provision in the REAL ID Act to pursue border fence construction? What evidence does DHS have that the section of fence in the SPRNCA meets those criteria (Lieberman 2007)? Specific to wildlife management, Lieberman discussed concerns over the impacts of the project on wildlife movement and DHS coordination and cooperation with wildlife management agencies including the USFWS.

Other environmental groups, including the Sierra Club, became involved in the opposition movement to evaluate the validity of the environmental waiver. Concerns arose in regard to the project's potential impacts to the San Pedro River's water quality and general hydrology from increased erosion and runoff. Criticisms were rampant, not only toward the actions of DHS, but also to BLM, which manages SPRNCA. Critics cited that BLM failed to protect the lands under its jurisdiction by overlooking essential portions of the environmental review process, preparing an incomplete EA, and allowing DHS to move forward with its plans without an adequate evaluation.

Supporters of the barrier project responded by emphasizing that DHS issued these environmental waivers on the premise that it was "necessary to waive certain laws, regulations, and other legal requirements in order to ensure the expeditious construction of barriers and roads in the vicinity of the international land border of the United

States" (Department of Homeland Security 2008, p. 1). However, a thorough EA or EIS would have also considered the negative impacts of the "No Action Alternative" to the environment such as human and vehicle traffic associated with illegal immigration, and public health and safety concerns associated with drug trafficking. The DHS insisted that to achieve the necessary level of national security along the U.S.–Mexico border, its ability to issue environmental waivers to expedite supporting projects was critical, and that the country's lack of security along its southern border was a significant problem (U.S. Department of Homeland Security 2008). In their efforts to justify the issuance of these waivers, DHS and supporters cited the prevalence of the negative impacts to the environment as a result of illegal border traffic. The completion of the environmental review may have in fact determined that the project in SPRNCA would have been the appropriate solution.

Chertoff defended his decisions by stating that he had been given the authority to issue waivers when necessary to ensure that border security projects would proceed without the delays caused by administrative processes and litigation, while simultaneously maintaining that DHS was committed to responsible environmental stewardship. He stated that for the larger percentage of the border length involved in projects covered by the waiver, DHS had prepared draft EAs or EISs, and subsequently for a significant number of that majority, it had been determined that the negative impacts of the projects to the environment were insignificant (U.S. Department of Homeland Security 2008).

Countering DHS, Defenders of Wildlife stated that the document development "read like an EA with topics and format, but there was not adequate time for consultation with interested parties and they skipped over the public process," (M. Clark, personal communication, March 3, 2010). The DHS did hold public meetings, but concerns were given to a stenographer. "It was like a black box, we gave them our input, but there was no response" (M. Clark, personal communication, March 3, 2010).

Together, the Defenders of Wildlife and the Sierra Club challenged REAL ID, stating the waiver's provision was unconstitutional because it violated separation of powers (*Defenders of Wildlife v. Chertoff* 2007) and petitioned the U.S. Supreme Court. "The waiver was designed so that it could only be legally challenged on a constitutional nature," Clark said. Ultimately, the Supreme Court decided not to hear the case.

Environmental Planning

The environmental waiver allowed DHS to move forward with multiple projects including the installation of barriers at the SPRNCA, without working through the established environmental review process. In an effort to mitigate impacts and cumulative effects caused by the action, DHS created its own process to substitute the requirements set forth by NEPA. The DHS worked with USFWS to develop Best Management Practices (BMPs) to address natural resource management issues and reduce the impact of the barriers (Customs and Border Protection 2008). In doing so, the development of the BMPs and subsequently Environmental Stewardship Plans (ESPs) that included the results of environmental reviews and the associated mitigation measures, may have helped further minimize resistance from land management agencies, private landowners, and local communities who, because of the waiver, did not have the established venues for input and scoping specific to the proposed action.

The DHS developed the BMPs and ESPs in short timeframes and the documents were completed while the footprint of the project was being designed. The project's timeline based upon the mandated completion date did not allow for species-specific surveys using scientifically sound protocols to detect threatened and endangered species or for adequate analysis of environmental effects. "Most surveys undertaken were for habitat at a landscape level" (S. Sferra, USFWS, personal communication, March 23, 2010).

DHS's efforts to produce and implement BMP and ESP were criticized by environmental groups, and while the agency coordinated its activities with USFWS, it did not query the USFWS for the same level of review typically provided for a consultation. The timeframe under which the action was undertaken did not allow for adequate time for the USFWS to conduct a thorough review of all proposed fence segments and impacts on federally listed species (S. Sferra, USFWS, personal communication, July 13, 2010). According to Defenders of Wildlife, they only came up with a laundry list of effects and did not analyze them or quantify them (M. Clark, personal communication, March 3, 2010). According to Clark, the recommendations and mitigations were written in a vague manner or were incorrect all together. The Department of Homeland Security was in the process of updating the ESPs in 2010, correcting the proposed footprint of the fence and roads to the actual footprints. As of July 2010, the USFWS has not reviewed the revised ESPs.

A significant challenge for agencies preparing environmental documents, including ESPs, is the means of evaluating effects. The use of qualitative versus quantitative evaluation is not standard between EAs and EISs for example. All environmental documents are produced for review by government agencies, the public, and other stakeholders. As such, the different approaches to reviewing such documents present challenges for the agency responsible for considering and addressing them. Specifically, the means for collecting and analyzing information can vary significantly and achieving a standard method for doing so can become the crux of the problem, as may have been the case for DHS and the documents produced in support of the SPRNCA barrier project.

A total of $50 million was allocated for conservation throughout the border area as mitigation for the barriers, to be paid out over several fiscal years, according to a DHS Memorandum of Understanding with the USFWS. As of July 2010, DHS was still working on release of the first $6.6 million for mitigation projects out of the $27 million currently appropriated for release. The mitigation funds will only be spent on endangered species issues, as they are given top priority, and there were not enough funds allocated to cover the mitigation concerns of species not protected under the ESA.

Effects

The barriers in the SPRNCA were completed in early 2008, and the effects of the construction on the environment began to show shortly afterward. The pedestrian and vehicle barriers fragmented critical wildlife corridors for large mammals such as American black bears (*Ursus americanus*), deer (*Odocoileus* spp.), and mountain lions (*Puma concolor*), and endangered species including the jaguar (*Panthera onca*) and ocelot (*Leopardus pardalis*). In terms of the overall condition of the aquatic systems supported by the San Pedro River watershed, the barriers blocked desert washes that help maintain the San Pedro River and its floodplain, resulting in erosion.

U.S. Customs and Border Protection (CBP) made numerous fence modifications to address wildlife movement, but in some cases, the structural design of the border barriers and fences did not consider the effects to movement and genetic flow between wildlife populations on either side of the international border. "There is no over, under, around, or through for wildlife" (V. Supplee, Director of Bird Conservation, Audubon Arizona, personal communication, January 22, 2010), along the solid panel pedestrian wall on the east side of the San Pedro River. The fence on the west side of the river is a bollard-style fence with 10 cm between vertical posts. While this style of fencing (Figure 6.3) allows for the movement of small wildlife, including amphibians, reptiles, and small mammals, it is impermeable to larger species.

Another important effect of the barrier is potential degradation of the sensitive riparian habitat. Because of the barriers on either side of the river and within the river during times of low flow (the barriers are lifted with a crane and removed at times of high flow), the potential for human traffic to be funneled into the river itself is increased. Illegal immigrants, drug smugglers, and the agency officials trying to apprehend them on foot and in vehicles, trample vegetation, including the endangered Huachuca water umbel (*Lilaeopsis schaffneriana* var. *recurva*), which degrades habitat and increases disturbances to nesting avian species, such as the southwestern willow flycatcher, Bell's vireo (*Vireo bellii*), and yellow-billed cuckoo. Increased human traffic also increases

FIGURE 6.3 Collared peccary (*Pecari tajacu*) next to the border fence. (Copyright Matt Clark, Defenders of Wildlife. Used with permission.)

trash and the threat of fire, which would decimate the cottonwood–willow galleries. Increased traffic creates erosion, which increases sedimentation, leading to changes in stream morphology and the quality of aquatic habitat.

After construction, DHS altered some of the barriers, adding additional horizontal bars to the original Normandy fence design (Figure 6.4), which is made up of criss-crossed segments of railroad rails. The USFWS and other agencies have requested the additional bars be removed where they intersect with wildlife trails (S. Sferra, personal communication, March 23, 2010) but the USFWS has no legal leverage to override DHS's authority. With the waiver, there is no legal recourse to use by other governmental agencies such as the USFWS or by environmental groups to implement mitigation or to stop further project alterations that may affect the environment. Completing the NEPA review process would have most likely uncovered more effects, and quantified those identified, particularly cumulative effects.

As a result of Congress' mandate to increase border security, which included increasing the number of USBP agents, there are more people working and living

FIGURE 6.4 Normandy-style vehicle barriers in San Pedro riverbed, September 2010. (Used with permission from Lyda Suzanne Harris.)

within the rural San Pedro Valley, including additional law enforcement personnel, their families, and others providing support services. The increased human population has an effect on the water supply. "The human demand on water may turn out to be a bigger impact to the environment than the actual barrier" (V. Supplee, personal communication, January 22, 2010). Nearby U.S. Army Fort Huachuca has implemented strategies to minimize their effect on the water table after extensive litigation initiated by the Center for Biological Diversity. The Fort achieved a 53% reduction in ground water pumping over a 24-year period (U.S. Department of the Interior 2008). Supplee sees a dichotomy on the water draw-down issue because of the waiver: "The environmental review process would have brought to light the interdiction forces' water demand and strategies to create a water demand neutral situation" (V. Supplee, personal communication, January 22, 2010).

If a waiver had not been issued, the environmental review process would have evaluated these and other negative impacts and the benefits to wildlife and habitat by investigating the "No Action Alternative." The installation of different barriers along other sections of the border has resulted in positive environmental impacts. In California's Tijuana Estuary, a recognized nesting area for endangered avian species, Arizona's Organ Pipe Cactus National Monument, Cabeza Prieta National Wildlife Refuge, and Buenos Aires National Wildlife Refuge, pedestrian and vehicle barriers have significantly reduced illegal traffic (Ganster 2009), thereby reducing associated trash, unauthorized trails, human waste, and fires. There have been proven successes through consultation with stakeholder groups, such as the installation of vehicle barriers in New Mexico, where barriers were built with minimal damage to the environment, access road footprints were minimized, and staging/work areas were reclaimed (Ganster 2009).

Public involvement is critical in the environmental review process. Agencies are required to consider the input provided by private and public entities, and resource management agencies like the USFWS, which contribute invaluable information, ultimately enhancing the quality and completeness of the evaluation of cumulative effects. This may be one of the most significant issues related to the implementation of an environmental waiver, assuming that ESPs and BMPs are produced and coordinated as they are intended.

CONCLUSION

The DHS environmental waiver issued to build barriers within and adjacent to the SPRNCA is an example of a dichotomy in the social landscape. Under what circumstances does project completion circumvent environmental legislation? In this example, one side of the argument perceived an immediate need while the other side desired that the environmental process be carried through. In this case, the need trumped the established environmental review. Although a unique circumstance, now that a waiver of environmental due diligence has been used by the government, the DHS border barrier sets a precedence for the future. Only time will tell if the BMPs and ESPs were adequate to mitigate the effects of the barriers. The method by which DHS handled the situation may become a model of how to develop projects in the future, or it may become an example used to emphasize the importance of following established environmental laws. Until the power within and between the different governmental entities is clearly defined and appropriated among the acting agencies, the country will continue to struggle to attain equitable solutions.

The environmental review process and the concept of evaluating cumulative effects of federal actions must become standard and continue to consider all viable options. The vast differences in cumulative effects evaluations must be addressed to establish the necessary standards needed by preparers of EAs, EISs, ESPs, and other environmental documents. The use of qualitative versus quantitative evaluation methods must be tempered in concert with the specifics of the government action and the need for specific analysis. The ultimate achievement of satisfying agencies and the public in regard to the environmental review process will not be accomplished until the process of involving all stakeholders and considering their input is enacted. By nature, cumulative effects are broad in scope and in order to adequately identify and address them, the review and range of evaluation should be extended to the furthest extent feasible.

7 Piecemealing Paradise

Cumulative Effects on Scenic Quality in the Coronado National Forest

Debby Kriegel

CONTENTS

INTRODUCTION

Places where visitors can experience pristine natural viewsheds are becoming increasingly rare. "People need natural-appearing landscapes to serve as psychological and physiological 'safety valves' . . . Once plentiful natural-appearing landscapes are becoming more scarce" (U.S. Forest Service 1995, p. 14). The 12 "Sky Island" mountain ranges (i.e., mountain ranges, many over 3,000 m in elevation, surrounded by desert) in the Coronado National Forest are arguably the most important natural landscapes and providers of high-quality scenery in southeastern Arizona. Most effects to scenic resources are cumulative, boundaries for scenic resources are difficult to define, and numerous small project effects cumulatively add up to big effects. This chapter explores implications related to the successive loss of this treasured resource. A proposed transmission line is used as a case study.

THE CORONADO NATIONAL FOREST

The U.S. Forest Service (USFS) was established in 1905 and manages >78,000,000 ha of public land nationally. In Arizona, the Forest Service manages >4,500,000 ha of the state's most ecologically diverse lands. The Coronado National Forest (Coronado NF), located in southeastern Arizona and southwestern New Mexico, is made up of 15 Sky Islands. Elevations of the Sky Islands range from 1,000 to >3,000 m above sea level (U.S. Forest Service 2010). The Coronado NF is divided into 12 separate units of land, each containing at least one mountain range. Each unit is referred to as an Ecosystem Management Area (EMA) (Figure 7.1).

The Coronado NF offers a wide variety of scenic landscapes including deeply carved desert canyons, golden rolling grasslands, dense oak (*Quercus* spp.) woodlands, and mountaintop conifer forests. The Coronado NF Sky Island mountains also provide a visual backdrop to cities and roads in surrounding deserts (Figure 7.2).

Visitors who travel into the Sky Islands experience biological diversity similar to a trip from Mexico to Canada. Over 550 vertebrate species are found in Coronado NF, including unusual animals such as the coatimundi (*Nasua narica*), gila monster (*Heloderma suspectum*), and collared peccary (*Pecari tajacu*). Mountain lions (*Puma concolor*), bobcats (*Lynx rufus*), and black bears (*Ursus americanus*) make their home here, and in recent years, the elusive jaguar (*Panthera onca*) has been seen

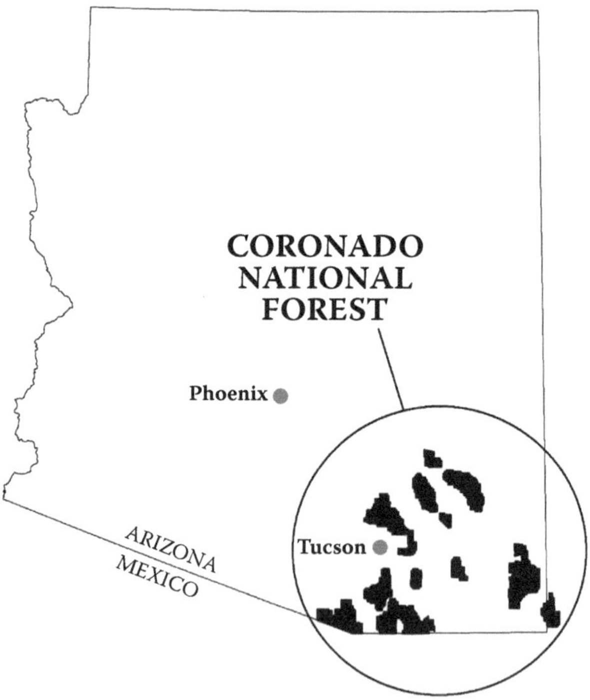

FIGURE 7.1 Sky Islands in the Coronado National Forest include 12 Ecosystem Management Areas in southeastern Arizona and southwestern New Mexico, each containing at least one mountain range.

FIGURE 7.2 The Santa Rita Mountains south of Tucson, Arizona. The high point is Mount Wrightson, with an elevation of 2,881 m. (Used with permission from D. Kriegel, Coronado National Forest.)

in the forest. Bird life is especially rich, with over 400 species found in southeastern Arizona. Madera Canyon and Cave Creek Canyon are world-renowned bird watching areas for rare species, including the elegant trogon (*Trogon elegans*) (Arizona Land and Water Trust 2010).

The Coronado NF hosts eight wilderness areas, more than 1,760 km of trails, and three scenic byways (i.e., roads and highways designated by state or national transportation departments in recognition of their outstanding scenic and recreational qualities). The forest offers year-round outdoor recreation opportunities, including hiking, camping, bird watching, driving for pleasure, picnicking, horseback riding, and mountain biking. Higher elevations are most popular during the summer, offering temperatures cooler than the desert. Low elevation recreation areas are popular during autumn, winter, and spring. Recreation settings vary from pristine wilderness to more than 100 developed recreation sites (e.g., campgrounds and picnic areas).

SCENERY MANAGEMENT

High-quality scenery is a benefit to people's lives, especially when it is related to natural-appearing forests (U.S. Forest Service 1995). Over 68% of visitors to the Coronado NF participate in viewing natural features (scenery), and this was the second most popular primary activity after hiking or walking. Additionally, over 25% of visitors to the Coronado NF used a scenic byway (U.S. Forest Service 2008).

Research indicates that people receive 87% of their information about the world through their eyesight alone (U.S. Forest Service 1995). Viewing natural scenery is psychologically and physiologically beneficial to people (U.S. Forest Service 1995).

Specifically, these benefits include people's improved well-being from viewing interesting and pleasant natural-appearing landscapes with high scenic diversity. People have positive responses (e.g., lower blood pressure, lower heart rate, reduced muscle tension) when viewing nature (Figure 7.3). When people feel better mentally and physically, they are more productive and society benefits. A 10% increase in nearby green space can decrease a person's health complaints in an amount equivalent to a 5-year reduction in that person's age (Arizona State Parks 2008). In addition, tourism is becoming the leading industry in many regions in the United States. In numerous communities near national forests, tourism and recreation are replacing the former leading roles of timber harvesting, mining, ranching, and farming. Scenic landscapes are key to the success of recreation and tourism (Outdoor Resources Review Group 2009).

Numerous federal laws (e.g., Wilderness Act of 1964, Wild and Scenic Rivers Act of 1968, National Trails System Act of 1968, National Environmental Policy Act of 1969, Environmental Quality Act of 1970, Forest and Rangeland Renewable Resources Planning Act of 1974, National Forest Management Act of 1976, Surface Mining Control and Reclamation Act of 1977, Public Rangelands Improvement Act of 1978) require that all federal land management agencies consider scenery and aesthetic resources in land management planning, resource planning, and project design, implementation, and monitoring. In addition, the USFS has routinely

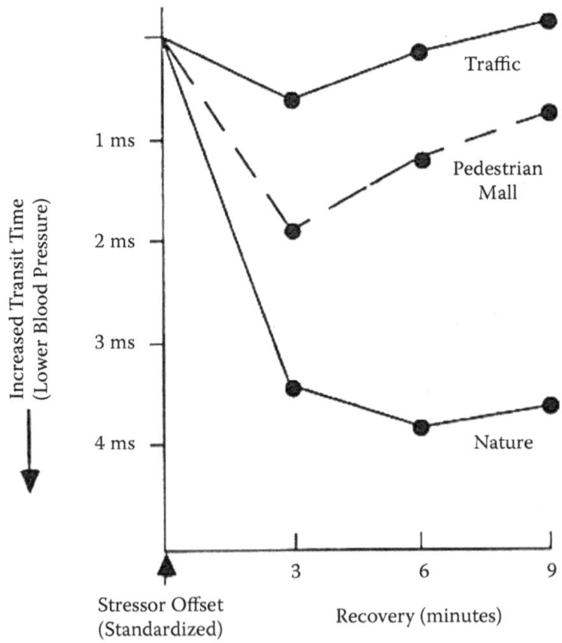

FIGURE 7.3 Positive response (lower blood pressure) of people recovering from stress, when shown photographs of nature, a pedestrian mall, and traffic. Research also shows similar results on heart rate and muscle tension. (Used with permission from U.S. Department of Agriculture Forest Service. 1995, p. 17.)

included scenery as part of the 1960 Multiple Use-Sustained Yield Act (U.S. Forest Service 1995, p. B1).

The Scenery Management System

The USFS Scenery Management System (SMS) is a tool for the inventory, analysis, and management of scenery on national forest lands. The SMS applies to every hectare of national forest and national grassland administered by the USFS, and to all USFS activities, including timber harvesting, road building, stream improvements, special use developments, utility line construction, recreation developments, and fuel breaks.

Although managing scenic resources is a combination of art and science, the SMS provides an organized system with standard components (Figure 7.4) that allows landscape architects to inventory and manage scenic resources consistently across national forest lands. Major components include describing the valued landscape's "sense of place" and uniqueness (i.e., Landscape Character and Scenic Attractiveness), evaluating the existing condition of "intactness" of the landscape and human-caused deviations (i.e., Scenic Integrity), defining travelways and use areas where the public has a high concern for scenery (i.e., Concern Levels), determining visibility of landscapes (i.e., Distance Zones), identifying the relative importance of aesthetics across a national forest (i.e., Scenic Classes), and establishing scenery goals across the forest (i.e., Scenic Integrity Objectives; scenic resource objectives in the current Coronado NF plan are called Visual Quality Objectives [VQO], because they were developed under the previous USFS Visual Management System (U.S. Forest Service 1974; the Coronado NF is currently revising the forest plan and will incorporate SMS). These elements are first used for forest-level mapping, and then applied at the project level to analyze project effects and determine mitigation measures. Management activities that contrast with or depart from the line, form, color, texture, or pattern of the valued landscape are usually negative impacts. While the SMS provides guidance for analysis of direct effects, it does not provide direction for cumulative effects analysis.

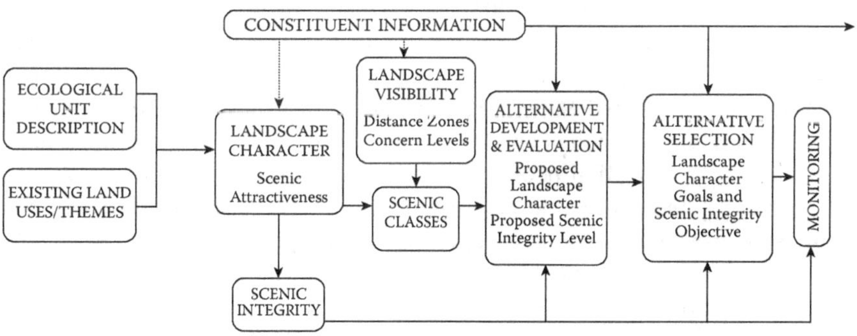

FIGURE 7.4 The U.S. Forest Service scenery management handbook provides guidance for managing scenic resources and uses standard components. This flow chart outlines the Scenery Management System process. (Used with permission of U.S. Department of Agriculture Forest Service. 1995, p. 6.)

CONSIDERATIONS IN EVALUATING CUMULATIVE EFFECTS

In my experience with the USFS working as an interdisciplinary team member on a wide variety of projects, I have noticed that the cumulative effects section in specialist reports and National Environmental Policy Act (NEPA) environmental analysis documents is frequently limited to one to two paragraphs, which is often insufficient. Equally often, the cumulative effects section summarizes or repeats direct or indirect effects instead of discussing cumulative effects. Cleary, many resource specialists and NEPA document authors do not fully understand cumulative effects.

The most profound effects on scenic quality are not from individual projects, but from multiple events and human activities over many years, including urban sprawl, the proliferation of cell phone towers, off-highway vehicle damage, and catastrophic wildfires. For projects that negatively affect scenic quality, a resource that is slowly being lost on the Coronado NF, each project may have a relatively minor effect, but cumulatively we are slowly losing once-vast wild places, resulting in a piecemealing of paradise.

As the Coronado NF revises the forest plan, USFS personnel asked for and received many comments from the public. One comment sums up what many others said: "What people are valuing now is the pristine quality of the Sky Islands. That is what these people (new residents) are looking for. But, that pristine quality is being affected by population growth and increased pressure on the resources" (U.S. Forest Service 2005, p. 15).

Just a few decades ago, southeastern Arizona had a modest population, human effects on landscapes were limited, and the Coronado NF was miles away from threats of urban sprawl. This is no longer the case. There are now a number of major past and present effects on scenic quality in the Coronado NF.

Among the important factors to consider during cumulative effects analysis for scenic quality on the Coronado NF are these seven: population growth, border issues, communication structures, astrophysical developments, off-highway vehicles (OHV), mining, and forest health.

1. *Population* growth. At statehood in 1912, Arizona's population was approximately 200,000 people. In 2005, the Arizona population increased to more than 6 million (Arizona State Parks 2008). Arizona can no longer be considered a sparsely populated state. Arizona has the fastest rate of population growth in the United States (Arizona State Parks 2008). The population of southeastern Arizona continues to grow. The Tucson metropolitan area is more than a million people (Pima Association of Governments 2009), other cities and towns and leapfrog developments near the Coronado NF are spreading across the landscape, and for decades southeastern Arizona's mild winters have attracted a seasonal migration of older Americans (i.e., snowbirds) from northern states. In many locations, housing, resorts, and commercial development have grown right up to the forest boundary (Figure 7.5). Arizona's population is expected to grow to more than 12 million people by 2050 (Arizona State Parks 2008).
2. *Mexican border activities.* Approximately 93 km of the Coronado NF border is adjacent to the international border with Mexico. Illegal immigration,

FIGURE 7.5 Urban growth in southeastern Arizona sprawls across the deserts, often up to the Coronado National Forest boundary. Here, Metropolitan Tucson, Arizona has grown up to the Santa Catalina Mountains in the background. (Used with permission from the Coronado National Forest.)

drug smuggling, and law enforcement activities affect scenic quality and alter forest landscapes. In every year since 1998, the Border Patrol's Tucson, Arizona, sector has had the largest number of apprehensions in the Southwest. In 2008, the Tucson sector accounted for 317,709 or 45% of all Southwest border apprehensions (U.S. Department of Homeland Security [DHS] 2008). The number of border crossers is likely much greater. In recent years, a large percentage of border crossers used Coronado NF lands because of increases in border security off-forest. Impacts on the Coronado NF from immigration and smuggling include numerous unauthorized roads and trails, and extensive trash and debris piles. Impacts from DHS (i.e., Border Patrol activities) infrastructure and operations include new roads, fences, walls, and surveillance towers. Effects from these activities now extend many kilometers into the Coronado NF, not just along the actual border.

3. *Communications and power lines and towers.* As the human population grows, additional communication towers and power lines are needed, and these facilities are increasingly proposed to be located on USFS lands. Construction of cell phone towers across southeastern Arizona's landscapes is only the most recent related activity that negatively affects numerous viewsheds. On the Coronado NF, heights for towers at a major electronic site in the Santa Rita Mountains were limited to 13.3 m to reduce scenic impacts, but in the late 1990s, a 66.6-m tower for a new television station was authorized,

and this tower is now visible on the ridgeline from many kilometers away. The proposed power line to be constructed by Tucson Electric Power is another example of the proliferation of utilitarian infrastructure that impacts scenic quality by introducing urban facilities into a natural landscape and requiring the construction of numerous new access roads.

4. *Astrophysical developments*. Southern Arizona has clear, dark skies, important conditions for viewing the heavens. Southern Arizona has come to be known as the astronomy capital of the world, and the Coronado NF Sky Island mountain ranges are ideal for astrophysical sites. These facilities are generally located atop mountain peaks where dust and light pollution are minimal, but unfortunately where the light-colored structures contrast with the natural landscape and are visible from many kilometers away. Access roads to the sites provide additional impacts to scenic quality (Figure 7.6).
5. *Off-highway vehicle use and nonsystem roads*. Off-highway vehicles (OHVs) are four times as popular as they were a decade ago, and in the West, OHV sales are double the national average, increasing over 1.5 times in 5 years (Arizona State Parks 2008). Off-highway vehicle use is an increasingly popular activity in many areas on the Coronado NF and OHV recreation is one of the fastest growing activities on public lands in the nation (Arizona State Parks 2008). While OHV use on many forest system roads is an appropriate recreation activity, some careless users do not stay on the designated motorized

FIGURE 7.6 The Mount Graham International Observatory in the Pinaleño Mountains, Coronado National Forest, Arizona. (Used with permission from A. Lynch, Coronado National Forest.)

routes. Both public OHV use and Mexican border activity contribute to the creation of unauthorized (unofficial, user-created "wildcat") roads. Vehicle use off-road damages fragile desert vegetation and soils, and desert soils tend to be much lighter in color than the vegetation, creating visual contrast. With little topsoil and arid conditions, scars do not heal quickly.

6. *Mining.* Rich mineral resources on the Coronado NF result in numerous mines and quarries, which often contrast sharply with line, form, color, and texture found naturally in the landscape. Reclamation of past mines is rarely sufficient (Figure 7.7). A limestone mine in the Santa Rita Mountains of the Coronado NF resulted in a large white scar that can be seen from across the Tucson Valley. In 2007 the Coronado NF received a proposal for a copper mine that would disturb some 1,485 ha of scenic national forest land (Rosemont Copper 2007).
7. *Forest and ecosystem health.* Although the SMS generally considers natural processes (e.g., wildfire, insect and disease outbreaks) as part of the natural landscape character, in recent years these disturbances have been outside their normal scale, intensity, and frequency on the Coronado NF. Fire suppression, drought, and climate change are likely causes (University of Arizona 2006). These events will sometimes affect scenic quality across entire viewsheds for decades. For example, two massive wildfires in 2002 and 2003 burned extensive areas of mature vegetation in the Santa Catalina Mountains, and since the 1990s, bark beetles have killed a significant percentage of conifer trees in the Pinaleno Mountains.

FIGURE 7.7 A quarry in the Dragoon Mountains, Coronado National Forest, Arizona. (Used with permission from D. Kriegel, Coronado National Forest.)

Cumulative effects analysis is critical to the management of USFS scenic resources. Every proposed project has its merits, and the USFS has a multiple-use mandate. Scenic resources are often the "losers" as individual projects frequently have effects that are not significant, and many projects that impact scenery are approved. Many other effects are outside of USFS control (e.g., border activities, development on private lands). Therefore, cumulative effects analysis is the only way to fully disclose the impacts of a project. "Evidence is increasing that the most devastating environmental effects may result not from the direct effects of a particular action, but from the combination of individually minor effects of multiple actions over time" (Council on Environmental Quality 1997, p. 1).

AN EXAMPLE: PROPOSED TRANSMISSION LINE

In 2000, Tucson Electric Power (TEP) applied to the Department of Energy (DOE) for a permit to construct a double-circuit 345-kV electric transmission line that would begin just south of Tucson, Arizona, cross the U.S.–Mexico border, and continue into the Sonoran region of northern Mexico. The DOE's Environmental Impact Statement (EIS) was completed in 2005.

Major scenic effects on the Coronado NF would occur from the introduction of new features into the landscape, including up to 196 43-m-tall towers, 12 transmission line wires, two neutral ground wires, and many new access roads for construction of the power line and maintenance of towers (U.S. Department of Energy 2005).

Three alternative routes crossing the Tumacacori EMA on the Coronado NF were considered: The Central, Western, and Crossover routes. The Central Route would follow the route of an existing underground gas pipeline corridor on the east side of the Tumacacori Mountains, would place 24.3 km of new power line on the Coronado NF, and would require minimal new road construction. The Western Route would place 47.5 km of new power line through a scenic and undeveloped valley on the west side of the Tumacacori Mountains (where many new roads would be required), then follow Ruby Road, a popular public sightseeing route, east to Nogales, Arizona. The Crossover Route would place 47.2 km of new power line through a scenic valley on the northwest side of the Tumacacori Mountains (and require new roads in this area), then cross through the Tumacacori Mountains (in an area so pristine it has been proposed as a new "wilderness") and continue south to the existing underground gas pipeline route on the east side of the mountains to Nogales, Arizona.

Analysis of the direct and indirect effects of the project was relatively easy. For the proposed TEP project, the construction would contrast with the valued landscape character, and impacts could not be fully mitigated. There would be negative effects on views along sensitive travelways and the project would not meet VQOs from the Coronado NF plan. Scenic Integrity on up to 7,307 ha would be permanently reduced. Simulations of the project helped disclose the direct effects and supported this analysis.

Analysis of cumulative effects was not as simple. The definition of a cumulative impact is "the impact on the environment which results from the incremental impact of the action when added to other past, present, and reasonably foreseeable future actions regardless of what agency (federal or non-federal) or person undertakes such other actions. Cumulative impacts can result from individually minor but collectively significant actions taking place over a period of time" (Council on Environmental Quality 1986, p. 28).

For the proposed power line, and other projects, at least three concepts need to be considered.

1. *Most effects to scenic resources are cumulative.* As projects (e.g., structures, mines) are added to a landscape, there is a gradual decline in scenic quality. Most people visit and value the Coronado NF because of its natural scenery and wild settings, and most human activities detract from this. Most projects are permanent or leave permanent scars, and landscape character is not restored unless facilities are removed or landscapes are naturalized. It is extremely rare that facilities are removed or lands fully reclaimed. The proposed TEP power line would create another permanent change and continue the decline in scenic quality.
2. *Scenic resource boundaries are difficult to define.* As people travel through a landscape on a road or trail, they experience a sequence of viewsheds. For the proposed TEP power line, there are two scales to consider: The Tumacacori EMA and the entire Coronado NF.

The Tumacacori EMA

Because the proposed TEP project would affect such a large part of the Tumacacori EMA, it was the first scale analyzed. Due to the basin and range geology of southeastern Arizona, Sky Island mountains rise above the desert and serve as focal points for travelers, often visible from many kilometers away and from numerous travelways inside USFS lands, so on the Coronado NF the landscape of an EMA is a logical scale to examine. Cumulative effects analysis for an EMA also includes effects on private lands adjacent to and within the forest boundary.

Past actions that impact scenic quality in the Tumacacori EMA include mining activities (especially at Ruby town site, small mines along California Gulch and Warsaw Gulch, and at the end of Rock Corral Road) unmanaged recreation (including wildcat roads and OHV damage), the existing underground gas pipeline that creates a linear clearing across the landscape, extensive impacts created by illegal aliens traveling north from Mexico (including many nonsystem trails and discarded debris), and resulting actions from U.S. Border Patrol enforcement activities (e.g., road improvements, cross-country travel, remote video surveillance towers). Present actions that impact natural landscapes in the area include the urban growth and development that continues to spread across the valleys and lower elevations (e.g., Nogales, Green Valley, other communities). A possible future action with similar impacts to visual resources was a proposal from the Power Company of New Mexico (PNM) to construct a 345-kV power line.

The Entire Coronado NF

Because the Coronado NF is the largest provider of natural public landscapes in southeastern Arizona, this is also an important scale to consider. Many of the public comments for the TEP power line related to scenic beauty of the region and the Coronado NF in general (U.S. Department of Energy 2005). It was helpful to examine all large blocks of natural landscapes across southeastern Arizona (i.e., the Sky Island landscapes), including National Park Service and Bureau of Land Management lands, other federal conservation lands, and even areas protected by state and local jurisdictions to analyze impacts influencing scenic quality.

National Park Service lands adjacent to Coronado NF include Saguaro National Park, Chiricahua National Monument, and Coronado National Monument. These lands provide natural, public landscapes, and their scenic resources are generally better protected than those on Coronado NF. However, Saguaro National Park West is relatively small and nearly surrounded by private development, and Chiricahua and Coronado National Monuments are relatively small, so at best these provide only limited protection of natural landscapes for the region. Bureau of Land Management (BLM) lands constitute extensive areas in southeastern Arizona, but are mainly low elevation and often less unique landscapes, and much BLM land is fragmented.

Other federal conservation areas include the San Pedro Riparian National Conservation area, which is a relatively narrow strip of protected land, Buenos Aires National Wildlife Refuge, which is being impacted by illegal immigrant activities along the border with Mexico, and military and Indian Reservation lands, which afford little protection for scenic resources and are not extensively managed for public recreation. Therefore, these lands provide limited long-term protection of natural landscapes available to for public enjoyment. There are also numerous State Trust lands in the area, which are generally not protected, and many local (city and county) lands that are fragmented. Thus, the analysis focuses back to the Coronado NF.

Past actions that impact scenic quality on the Coronado NF include astrophysical facilities on mountaintops, utility structures such as communication towers and power lines, mining activities, catastrophic wildfires, administrative facilities, wildcat roads, OHV-damaged areas, and developments on private lands within and adjacent to forest boundaries. Present actions that impact scenery include a continuation of past activities (some of which are growing worse). There are also numerous future actions listed on the Schedule of Proposed Actions that would negatively affect scenic quality (U.S. Forest Service 2010).

3. Seemingly small projects can have large effects on scenic quality, and small project effects can add up to big ones. Projects with relatively small footprints can impact vast viewsheds. For example, the Mount Graham International Observatory has a footprint of 3.5 ha, yet the large boxy white telescope structure is visible on the ridge of the Pinaleño Mountains from more than 60 kilometers away. Additionally, relatively insignificant individual projects across the landscape added over decades can ultimately result in enormous impacts.

For the proposed Tucson Electric Power construction, the footprint of permanent disturbances would be less than 15 ha, yet the project would negatively affect scenic quality across landscapes. Add that to past, present, and future projects, and the impacts increase.

Past, present, and reasonably foreseeable future projects for the two scales of analysis used for the proposed TEP project suggest how cumulative effects contribute to altered landscapes (Table 7.1). Past projects and actions continue to have negative impacts on scenic quality, the proposed power line would further impact scenic quality and not meet VQOs in the Forest Plan, and there are known projects that would further reduce scenic quality in the future. Therefore, cumulative effects, especially for the Western and Crossover alternative routes, are substantial.

Tucson Electric Power Powerline Status

The Coronado NF named a preferred alternative in its July 2005 Final EIS—the Central alternative—which runs parallel to the underground gas pipeline corridor on the east side of the Tumacacori Mountains except where it avoids an Inventoried Roadless Area. Decisions that could allow construction of the line are currently on hold. In December 2006, a new alignment was proposed by TEP, and this would

TABLE 7.1
Past, Present, and Future Project Influences at the Time of Analysis of a Proposed Tucson Electric Power Transmission Line on a Small (Tumacacori Ecosystem Management Area [TEMA]) and Large Scale (Coronado National Forest [Coronado NF]), Southeastern Arizona

TEMA	Coronado NF
Past Actions	
• Mining (Ruby, Rock Corral, California Gulch, Warsaw Gulch) • Wildcat roads and off-highway vehicles (OHV) • Underground gasline corridor • Illegal Mexican border activities (trails, debris) • U.S. Border Patrol enforcement activities (roads, towers)	• Astrophysical facilities • Utility lines/towers • Mining • Major wildfires • Administrative facilities • Wildcat roads/OHV damage • Development on private lands (adjacent lands and inholdings)
Present and Ongoing Actions	
• Continuation of past activities (some growing worse) • Urban development in the area (Nogales, Green Valley, etc.)	• Continuation of past activities (some growing worse)
Reasonably Foreseeable Future Actions[a]	
• PNM powerline	• Greaterville placer mining (Santa Rita EMA) • Greaterville road relocation (Santa Rita EMA) • Alpha Calcit mine (Dragoon EMA)

[a] Reasonably foreseeable future actions for this analysis were limited to projects on the Schedule of Proposed Actions at the time of analysis.

need additional NEPA analysis. Additionally, remedial equipment and capacity have largely resolved Nogales' power outage problems. For the Coronado NF to proceed with a Record of Decision, the Coronado NF would request DOE to consider updating the analysis to reflect changed conditions since the final EIS, including jaguar (*Panthera onca*) sightings and a new wilderness proposal (H.R. 3287).

REVISIONS OF THE CORONADO NATIONAL FOREST PLAN

In addition, to better cumulative effects analyses at the project level, the Coronado NF Plan revision is under way. The original Coronado NF Land Management Plan was approved in 1986 and expected to provide direction for 15–20 years. Major components of the revised plan will include desired conditions, objectives, and guidelines. Although plans can be amended and USFS lands are managed for multiple uses, the revised plan should help forest managers protect scenic quality and minimize some cumulative effects. One tool in the new plan will be land-use zones that establish appropriate land uses across the forest. For example, in the draft plan there is a land-use zone called "Wild Backcountry," which includes Inventoried Roadless Areas, areas adjacent to wilderness areas, and other relatively pristine areas across the forest. For this land-use zone, there is a proposed guideline: "New utility structures and power lines should not be allowed." This type of direction in a forest plan should help reduce direct and cumulative effects on scenic quality on USFS lands.

CONCLUSION

"Everybody needs beauty as well as bread, places to play in and pray in, where nature may heal and give strength to body and soul" (Muir 1912). We are slowly losing the scenic, natural landscapes in our national forests. Individual projects that impact scenic resources are often not significant: one more power line, one more mine, one more telescope. Each approved project affects viewsheds in only a portion of our public lands, but over the years and decades, the impacts add up. Additional impacts to scenic quality are outside the control of USFS decision makers. The purpose of NEPA analysis is to disclose project impacts, which helps decision makers make informed decisions. Good cumulative effects analysis and documentation are generally the best way to disclose impacts on scenic resources, and in some cases may provide an opportunity to enhance scenic quality. A larger problem occurs when scenic resources are not considered by decision makers to be important for health benefits, nor to contribute to society's well being, nor seen as necessary to meeting the needs of an increasing population.

8 Understanding the Cumulative Effects of Human Activities on Barren-Ground Caribou

Anne Gunn, Chris J. Johnson, John S. Nishi, Colin J. Daniel, Don E. Russell, Matt Carlson, and Jan Z. Adamczewski

CONTENTS

INTRODUCTION

In Canada, the techniques and methods necessary for measuring cumulative effects have developed slowly despite introduction of the concept in the 1970s when Justice

Berger referred to cumulative effects of the proposed Mackenzie Valley pipeline (Berger 1977). From a regulatory perspective, cumulative effects are the aggregate stresses from past, present, or future human activities on a valued ecosystem component. Although there are other definitions—and one might differentiate between cumulative effects and impacts—this is an intuitive concept (Johnson and St-Laurent 2010). Despite this simplicity, application of the concept to resource management and conservation continues to remain a "mystery to most EIA [environmental impact assessment] practitioners" (Duinker and Greig 2006, p. 157). Progress toward effective cumulative effects assessment (CEA) is being questioned, despite having been a requirement for environmental impact assessments in Canada for three decades (Kennett 1999; Dowlatabadi et al. 2004; Duinker and Greig 2006).

In the Northwest Territories (NT), the lack of progress on cumulative effects was apparent during public hearings for open-pit diamond mining. For example, the environmental assessment panel for the Ekati diamond mine concluded that ". . . further work is needed on the cumulative effects of exploration activities on wildlife in the region" (MacLachlan 1996, p. 68). Similar concerns were echoed during the assessment of Diavik, the second diamond mine in the NT (Canadian Environmental Assessment Agency 1999a). Since the mid-1990s, the diamond mines in the NT focused attention on project-specific EIA, and now government agencies must interpret these individual EIAs in a broader regional context. Additionally, people in northern communities express concerns about how even small-scale exploration projects such as diamond drill operations may affect caribou (*Rangifer tarandus*). These small land-use operations typically fall below the criteria for environmental assessment. Logically, they have to be considered as part of the cumulative footprint of human activities on caribou ranges.

The challenge of undertaking and applying CEA partly stems from lack of clear policies, regulations, and terms of reference (Chapter 3). This is the case in the NT, despite a multi-stakeholder, consensus-driven process that was initiated in 1999 to develop a framework for assessing cumulative effects (NWT CEAM Steering Committee 2007). The absence of systematic approaches to identify, evaluate, and respond to regional/territorial cumulative effects has been identified in recent regulatory reviews (Government of Northwest Territories 2009) and environmental assessment hearings especially where the management or conservation of caribou is an issue. Caribou are highly valued across northern Canada and Alaska, and the responses of caribou to mining and oil and gas development are highly visible and controversial. Disagreement surrounds the effects of even large and well-studied developments such as the Prudhoe Bay oil fields, and the indirect or cumulative effects of human developments, including climate change (Joly et al. 2006; Noel et al. 2006).

The "mystery" of cumulative effects is also a consequence of technical shortcomings. Environmental assessment in general, and cumulative effects assessment in particular, has been a slowly emerging field of applied ecological science and has lagged behind other areas of conservation biology and landscape ecology. There have been relatively few published efforts to design and test approaches to measure cumulative effects. With a few exceptions (Schneider et al. 2003; Johnson et al. 2005; Sorensen et al. 2008) past studies have dealt more with the "process" and policy reviews

(Chapter 3), or particular aspects such as the failure to include aboriginal traditional knowledge, different values, and world views (Usher 2000; Paci et al. 2002).

Studies designed to measure the influence of human activities on wildlife tend to deal with individual effects such as behavioral or physiological responses (Seip et al. 2007; Stankowich 2008; Thiel et al. 2008; Fahrig and Rytwinski 2009) or, less frequently, demographic responses such as changes in calf survival (Shively et al. 2005). Few authors have described responses to multiple disturbances or measures of population productivity (Nellemann et al. 2000, 2003; Johnson et al. 2005). Reimers et al. (2003) cautioned that interpreting shifts in animal distribution without understanding underlying ecological conditions is difficult. Behavioral, physiological, or distributional responses should be linked to population dynamics (Vistnes and Nellemann 2008) requiring measures such as energetic cost or change in reproduction and survival across a range of disturbance levels. These studies are methodologically difficult (Johnson and St-Laurent 2010), thus, the bias toward simpler response indices. We found only one published account that links the behavioral, energetic, and demographic responses of caribou to human disturbance (Murphy et al. 2000).

The technical challenge for defining and estimating cumulative effects is threefold. First, scaling up from project-specific to regional effects requires estimating the likelihood of additional industrial projects that are plausible but do not yet exist. Second, spanning the gap between assessing effects at the project-specific scale up to the regional scale requires identification of appropriate temporal and spatial scales and study boundaries (Vistnes and Nellemann 2008). Third, an assessment of cumulative effects requires pathways that integrate individual and population-level wildlife responses to single and multiple projects. Consequently, the science of cumulative effects assessment in northern Canada has often lagged with a reliance on a "check box" approach using qualitative summations of individual categorical ratings (simple addition of single effects) and little consideration of development scenarios and scale issues.

In this chapter, we describe a conceptual framework and supporting methods for assessing the cumulative effects of industrial development for the Bathurst caribou herd, a migratory herd of barren-ground caribou (*R.t. groenlandicus*) in the NT (Figure 8.1). We focused at the regional rather than the project-specific level because aboriginal communities and northern governments are struggling with process and information needs that occur across broad areas, and time frames that exceed the development of single projects. The lack of well-defined regional and strategic environmental assessment processes that support project specific assessments is one of the current CEA deficiencies in the NT (NWT CEAM Steering Committee 2007).

The chapter is organized into five sections. Following the introduction in the first section, we describe characteristics of industrial development in the Canadian central Arctic, briefly describe the regulatory and policy framework in the NT, and outline the significance of migratory barren-ground caribou to cumulative effects assessment. In the third section, we describe and introduce the concept of resilience because it conceptually grounds our understanding of cumulative effects for caribou. In the fourth section, we describe a collaborative research project designed to develop an integrated modeling framework that will help regulators, industry, and northern communities to better understand the potential cumulative effects of

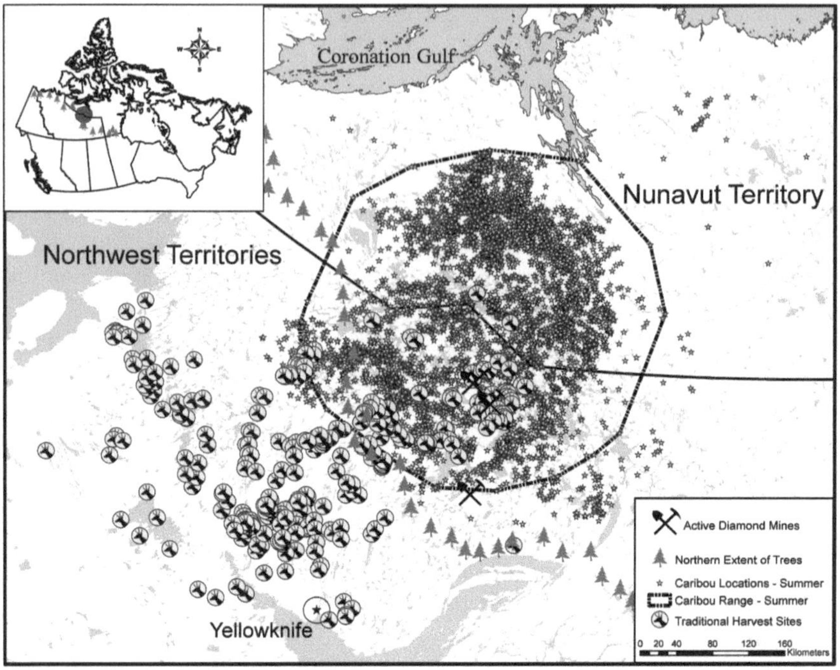

FIGURE 8.1 Extent of summer range, illustrated within inset map, and locations of individual caribou of the Bathurst herd collared from 1996 to 2008 in the Northwest Territories and Nunavut, Canada. Location and use of harvest sites by the Tlicho people was documented for the period 1935–1998. (From Legat, A. et al. 2001. Caribou migration and the state of their habitat. Final Report, Yellowknife, Northwest Territories, Canada.)

development for the Bathurst caribou herd. In the final section we provide a brief discussion and concluding remarks.

THE NORTHERN PERSPECTIVE

Characteristics of Mineral Resource Extraction in the Northwest Territories

The NT is a large area (1,346,106 km^2) with a small human population. Based on the 2008 national census (www.stats.gov.nt.ca), there were 43,283 people living in 29 communities and only Yellowknife has more than 4,000 people. The NT has a low density and clumped distribution of human activities including industrial exploration and developments. Compared to Canadian provinces, the NT is sparsely developed with few roads, and relatively little landscape-scale change due to agriculture or forestry. Mining exploration typically follows boom and bust cycles with highs in exploration activities in the 1970s (mostly uranium) and the 1990s (diamonds). The discovery of diamonds in 1991 at Lac De Gras resulted in the largest staking rush in Canadian history leading to the construction of two open pits and one underground mine. The collective operations of the three diamond mines produce 15% of the world's rough diamonds, and the annual production in 2007 was worth $1.4 billion (www.iti.gov.nt.ca).

Although developments such as mines are relatively few (i.e., about 35 mines since the 1930s) people have long memories; concerns about past practices and abandoned mines have influenced public perception of recent mining activities. The lack of all-season roads means that fly-in operations are typical, although the cold winters allow the use of seasonal ice roads to service remote camps and developments. Tourism is important for the NT economy; however, it is mostly confined to the larger communities serviced by roads or to lakes and rivers. On the central barrens, human activity other than mining exploration and development occurs mostly as guided hunting based out of seasonal camps. The majority of oil and gas development is found in the western Mackenzie Valley region while mining activities are more frequent in the central and eastern NT.

Regulatory and Policy Framework in the Northwest Territories

The NT regulatory regime has greatly changed since the early 1990s. Currently, governments (i.e., federal, territorial, aboriginal) participate in comanagement relationships, accepting shared responsibilities for resource development and management. The emphasis on comanagement of land and resources was formally recognized through enactment of the Mackenzie Valley Resource Management Act in 1998. The act was a result of commitments made by Canada during the negotiations of the Gwich'in and Sahtu comprehensive land claim agreements settled in 1992 and 1993, respectively (Donihee 1999).

The Mackenzie Valley Resource Management Act is counterpart to an environmental assessment act, and gives aboriginal people a greater say in resource development and management through an institutional framework that emphasizes comanagement, collaboration, and inclusion of indigenous knowledge (Armitage 2005; Ellis 2005; Christensen and Grant 2007). The Mackenzie Valley Resource Management Act resulted in the establishment of the Mackenzie Valley Environmental Impact Review Board (www.reviewboard.ca/), which is a comanagement board that has shared aboriginal and government roles, and is responsible for the environmental impact assessment process in the Mackenzie Valley, including areas used by the Gwich'in, Sahtu, Deh Cho, Akaitcho, and Tlicho aboriginal peoples. The Review Board considers cumulative effects during its assessments. The Mackenzie Valley Resource Management Act also established a requirement for monitoring cumulative impacts on the environment. Adopting a community-based approach, this work is directed by the NT Cumulative Impact Monitoring Program (NWT-CIMP) working group, which is a partnership among NT aboriginal governments, the Government of Canada, and the Government of NT.

Concerns about cumulative effects during the comprehensive review for the Diavik Diamond mine prompted the regulatory agencies to commit, in 1999, to a regional cumulative effects assessment and management framework (NWT CEAM Steering Committee 2007). The framework is meant to formally involve federal, territorial, and aboriginal governments, regulatory agencies, nongovernmental organizations, and industry in the design and implementation of a monitoring, management, and planning framework that addresses limits on regional cumulative effects (Chapter 3). The framework was slow in development as the requirement for consultations was time consuming. By February 2008, the CEAM Steering Committee recognized that cumulative effects were only one component of the framework, which more properly involved management of human activities through stewardship. Formally recognizing the broader role of the framework, it was renamed the NT Environmental Stewardship Framework.

The Significance of Migratory Barren-Ground Caribou

Caribou are of profound cultural, spiritual, and economic value to aboriginal peoples who have hunted the herds and depended on them for food and clothing (Kuhnlein and Receveur 1996; Legat et al. 2001). Indeed, the present-day relationships between northern peoples (aboriginal and nonaboriginal) and caribou is more accurately described as a complex, adaptive socioecological system (Berkes et al. 2003, 2009), where social capacity for responding to and shaping ecosystem dynamics is a powerful feedback mechanism (Folke et al. 2005). Consequently, one of the strongest public concerns expressed during the environmental review of diamond mines in the 1990s was for the migratory barren-ground caribou herds, especially the Bathurst herd that ranges across the majority of staked kimberlite deposits.

Within the context of formal environmental assessments, caribou are considered a valued ecosystem component due to their importance to northern people. The Bathurst herd is one of the seven herds of migratory barren-ground caribou in the NT and Nunavut. The winter range of Bathurst caribou is below treeline, and the herd can migrate over 1,000 km to the tundra for calving and summer habitats. The migrations are the caribou's evolutionary strategy to cope with variable environmental conditions (Bergerud et al. 2008), suggesting that barren-ground caribou may be especially vulnerable to human activities that interfere with or interrupt movement behavior. The migratory tundra caribou are also gregarious, especially during calving and postcalving, which can increase their vulnerability to human activities.

One of the difficulties for cumulative effects assessment is selecting and rationalizing spatial and temporal boundaries (Vistnes and Nellemann 2008). The boundaries are important because the definition and application of any thresholds for regional development will be largely dependent on scale. The seasonal and annual distribution of barren-ground caribou can help identify and justify boundaries for cumulative effects assessments across regional areas. The definition and application of any thresholds for regional development will be largely dependent on scale, and therefore the assessment has to be defined as or nested within the annual range of the study herd. A consequence of migratory behavior is the dilution and transfer of effects across regional areas that can complicate CEA between neighboring jurisdictions.

Less attention is paid to the logic for selecting temporal boundaries (Vistnes and Nellemann 2008). Typically, the timescale for less intensive activities such as exploration is years and for fully developed mineral deposits is often 20–30 years, although failure of reclamation can extend the time period. The abundance and distribution of caribou changes at the decadal scale (30–60 years) with relatively regular phases of increase, decrease, and low numbers (i.e., cyclic behavior; Gunn 2003;

Zalatan et al. 2006). Measures of individual and demographic vigor vary between different phases of the cycle that has implications for baseline, monitoring, and mitigation and describing responses to human activities. For migratory barren-ground caribou, then, the temporal scale for cumulative effects is decades and tied to the period and amplitude of the cycle that represents repeatable changes in abundance and distribution.

Ecological Characteristics of Migratory Barren-Ground Caribou

Understanding the ecological characteristics of caribou ranges is necessary because "cumulative effects have to be identified and assessed within the framework of a variable natural environment" (Cameron et al. 2005, p. 7; Wolfe et al. 2000). Natural environmental variability is an ecological driver of population dynamics and influences the resilience of caribou in the context of additional stresses that can be imposed directly through natural predation and hunting mortality, or indirectly through anthropogenic disturbance or displacement associated with industrial development and infrastructure (Figure 8.2).

There are several important ecological characteristics of the tundra and taiga seasonal ranges used by barren-ground caribou that can interact directly or indirectly with human activities. First, a highly seasonal pulse of annual plant productivity occurs during the relatively short and warm summer, interspersed with long cool winters when most ecological processes are dormant or slow (Bliss et al. 1973).

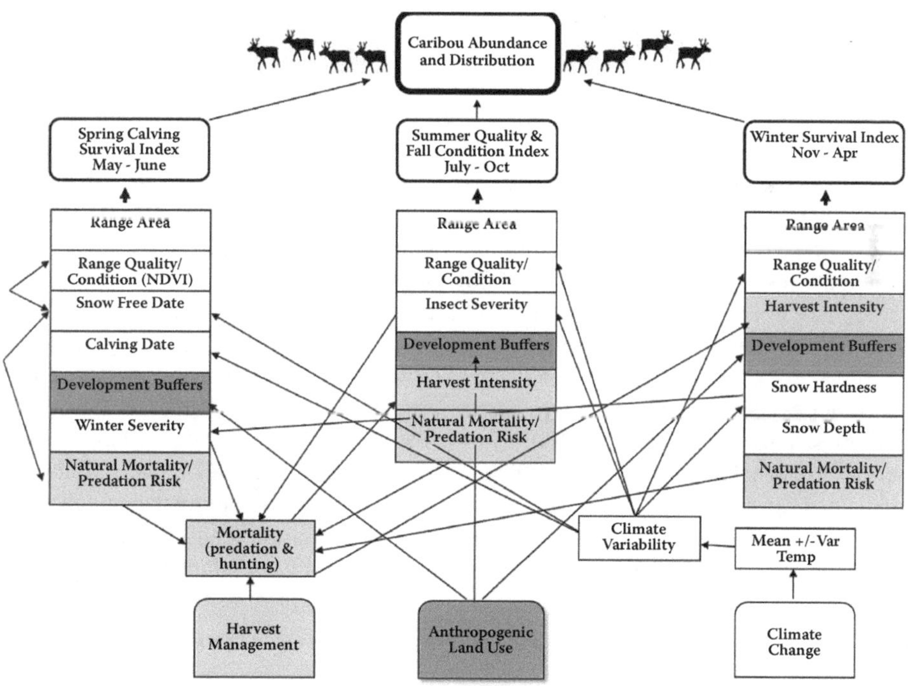

FIGURE 8.2 General factors and associated interactions influencing the abundance and distribution of barren-ground caribou populations found across North America.

Second, weather patterns that affect timing and length of the plant growing season are annually variable and unpredictable. The timing of new plant growth in spring, and forage production through the growing season are essential elements in the survival and growth of calves, and in the ability of females to meet the demands of lactation and regain body condition sufficiently to breed in autumn (Russell et al. 1993). Summer weather also influences the abundance of parasitic and biting insects, which in turn influences caribou body condition (Russell et al. 1993; Hagemoen and Reimers 2002).

Third, climate variability influences the frequency and size of fires on the taiga range that may burn large areas of winter foraging habitat, and effectively remove lichen biomass for decades (Thomas et al. 1996; Rupp et al. 2006). Finally, snow cover (i.e., depth, hardness, and duration) and freeze-thaw cycles reduce access to forage and may limit energetic and nutritional intake of caribou during winter (Adamczewski et al. 1988).

Barren-ground caribou are hunted by wolves (*Canis lupus*), grizzly bears (*Ursus arctos*), and less frequently by wolverine (*Gulo gulo*) and lynx (*Lynx canadensis*). Newborn calves are especially vulnerable to avian predators such as golden eagles (*Aquila chrysaetos*). Many of the evolutionary strategies of caribou, including migration, are shaped by predation (Bergerud et al. 2008). Caribou are gregarious and migrate in groups that range from a few individuals to tens of thousands. Group size can affect the average behavioral response level of individuals in the group (Roberts 1996; Manor and Saltz 2003). Gregarious behavior of caribou varies seasonally. During calving, all breeding females of a herd congregate on the calving ground, and during postcalving, caribou form large aggregations of thousands to tens of thousands of individuals. Thus, sampling designed to measure cumulative effects must be robust to variation in the relatively low probability that caribou will encounter a site of human activity, while each encounter can include a large proportion of a herd.

Caribou responses to human activities are likely influenced by their reaction to predators (Frid and Dill 2002). Indirectly, hunting may modify the responses of caribou to other human activities including habituation (Haskell and Ballard 2008) or avoidance responses (Coleman et al. 2001; Reimers et al. 2003). Prior to European settlement, aboriginal peoples accessed barren-ground caribou by anticipating and following the movements of the migratory herds on foot and establishing seasonal hunting camps; their traditional knowledge of water-crossing sites, over-winter areas and other aspects of caribou migratory movements were a key aspect to aboriginal hunting strategies (Legat et al. 2001; Stewart et al. 2004; Parlee et al. 2005). Today, people use trucks, off-highway vehicles (i.e., snowmachines, all-terrain vehicles), boats, and aircraft to access barren-ground caribou from late summer through winter. Access for caribou hunting from winter or all-season roads associated with industrial exploration and development is a key aspect for cumulative effects.

RESILIENCE IN THE CONTEXT OF CUMULATIVE EFFECTS

Our approach to cumulative effects is premised on the concept of resilience. Resilience captures the ability of caribou (individuals and populations) to cope with natural and anthropogenic environmental variation and stressors. For caribou, ecological resilience is measured by the amount of disturbance that is absorbed (coped with) before the individual (or herd) changes behavior (Holling 1973, 1986; Gunderson 2000). Natural environmental variation such as level of insect harassment can reduce or increase resilience of a caribou, which then changes the impact of the same individual's response to human activities. Gunn et al. (2001) explored the concept of resilience for caribou to integrate responses to human activities and the effect of insect harassment and snow conditions. The application of resilience as a concept allows us to integrate project-specific CEA and range-wide monitoring to offer testable predictions about the cumulative effects of human activities on the Bathurst caribou herd.

Considering human–caribou interactions, the concept of resilience also applies to the socioecological system's ability to build and increase the capacity for learning and adaptation by people (Berkes et al. 2003). Resilience in socioecological systems is closely tied to the concept of sustainability and the challenge of meeting current demands without degrading the potential to meet future requirements (Ludwig et al. 1997; Walker and Salt 2006). The concept of resilience shifts perspective from the anthropogenic desire to control change in systems assumed to be stable, to sustain and enhance the capacity of socioecological systems to adapt to change (Folke et al. 2002). This definition of resilience is similar to how Tlicho elders view respect and knowledge as a cornerstone of their relationship with caribou (Legat et al. 2001).

DEVELOPING AND DEMONSTRATING A SPATIALLY EXPLICIT DEMOGRAPHICS MODEL FOR MIGRATORY CARIBOU

Since 1980, the efforts of governments, industry, and independent researchers have contributed greatly to the understanding of the distributional and population dynamics of barren-ground caribou. With the development of oil reserves on Alaska's North Slope and the discovery of diamondiferous kimberlite deposits in the Canadian central Arctic, much of the recent emphasis on research and monitoring has been placed on understanding the impacts of human activities (Cronin et al. 2000; Johnson et al. 2005; Joly et al. 2006). For the Bathurst herd, three diamond mines and associated exploration activities have served as the impetus for a number of innovative cumulative effects studies (Gunn et al. 2001; Legat et al. 2001; Johnson et al. 2005). None of these works, however, captured the full range of hypothesized interactions between industrial development and the long-term dynamics of caribou. Likely the greatest limitation of these studies was their inability to interface effectively with decision-making frameworks focused at herd management, regulatory approval for new development, and strategic land-use planning.

Despite a considerable amount of research and a number of overlapping federal and territorial review and approval processes, aboriginal communities remain concerned about the impacts of development on caribou. The precipitous decline of Central Arctic and other caribou herds (Vors and Boyce 2009) has increased the pressure on government agencies and comanagement boards to better understand and, if possible, halt the decrease in the number of caribou. Recognizing the limitations of past research and approaches, we suggest a suite of interacting and complementary methods that provide a more complete perspective on the distribution and population dynamics of terrestrial mammals. Considering the current declines in caribou populations and the increasing level of development across the Arctic, this approach is timely and well illustrated using barren-ground caribou as an example species. This is especially the case for the Bathurst population, where nearly 15 years of distribution data, decades of past studies and knowledge of the biology, population status, and traditional uses of the herd, and recent concerns around a new and expanding industry suggest that further advances in cumulative effects analyses are warranted and likely fruitful.

We are developing an integrated and adaptive modeling framework that draws on the learning, data, and approaches developed for the Bathurst and other migratory caribou herds. This includes distribution data of Bathurst caribou collected at the mines through aerial census and across the annual range using satellite collars; activity data as part of a number of behavioral studies; and traditional ecological knowledge describing the long-term distribution of caribou. These data are being applied to a set of simulation and statistical models that have shown good utility for Bathurst caribou and other taxa, but to date have not been integrated to understand changes and interactions in (1) caribou distribution and (2) abundance in the context of (3) long-term and large-scale anthropogenic activities and environmental variation.

A team of experts in spatial, nutritional, and population models, caribou biology, and aboriginal knowledge of the historical distribution and behavior of caribou are working collaboratively to develop a set of models that will provide some perspective on future outcomes in the distribution and abundance of the Bathurst caribou herd in the context of global change and more localized development scenarios. This project is meant to be larger than a 2- to 4-year research endeavor, serving as a long-term planning framework that is adaptive and reflexive to changing knowledge, development pressures, government and community needs. As the starting point, we envision a collection of interconnected models that are premised on the distribution and avoidance responses of caribou to human disturbances (Figure 8.3). This includes spatial avoidance of human features, with inherent consequences for habitat use and the behavioral and ultimately nutritional costs of such decisions. These habitat relationships are then integrated within a mechanistic nutrition-population model to forecast the demographic consequences of avoidance behaviors and habitat change associated with the current human footprint. Finally, the products of these models will interface with a regional cumulative effects simulator (A Landscape Cumulative Effects Simulator [ALCES®], Schneider et al. 2003) that will allow the research team to engage local communities, government, and industry in a formal discussion of the possible implications of future land use and environmental change across the range of the Bathurst herd.

One of the initial objectives of the project was to illustrate the utility and benefits of the integrated modeling process. As such, the scope of the project is limited to the summer range of the Bathurst herd, the area with the greatest concentration of industrial activities. Through successive iterations, the research team will increase the detail and specificity of the models, accuracy and precision of input data, and the total area of application, including the annual range of Bathurst caribou and other herds of caribou in the Central Arctic. Thus, we are pursuing an iterative modeling process with no set end-point. Model predictions will be posed as tentative hypotheses that help engage the public, stimulate discussion and consideration of regional development thresholds

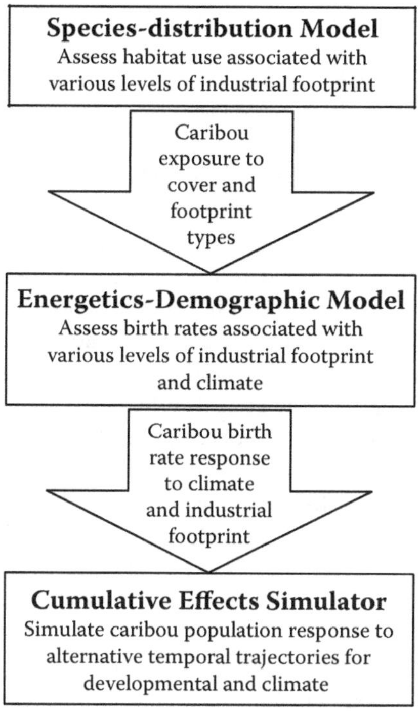

FIGURE 8.3 Integrated modeling framework for documenting and understanding the impacts of human developments for barren-ground caribou found across northern Canada. The analytical role of each model and information flows between models are presented.

for caribou, and direct further learning of limiting or regulating factors for the distributional and population dynamics of the species (Starfield 1997). Below, we discuss the conceptual framework and supporting research for developing and interpreting the three foundational elements for the cumulative effects demonstration project: distribution model, nutritional-demographic model, and cumulative effects simulator.

Understanding the Distributional and Avoidance Responses of Caribou

The first step in developing the integrated modeling framework was to use species-distribution models to describe the distribution and avoidance response of caribou relative to sites of industrial exploration and development. Defined by Guisan and Thuiller (2005, p. 994) as "...empirical models relating field observations to environmental predictor variables, based on statistically or theoretically derived response surfaces," species distribution models are now a well-accepted technique for quantifying and spatially representing the response of plant or animal species to variation in important resources.

Mace et al. (1996, 1998) were pioneers in applying species distribution models to the question of cumulative impacts. Working with grizzly bear locations, they demonstrated that bears had a lower probability of occurrence in areas with a high density of roads and that human activity resulted in a cumulative reduction in the availability of bear habitat. Similarly, Carroll et al. (2001) used species distribution models to quantify the impacts of human-caused landscape alteration on the distribution of carnivore species across the Rocky Mountain region of western North America.

Species distribution models are useful for identifying zones of influence. These zones represent the static area and perhaps time of year when wildlife demonstrate measurable responses as a result of an existing development such as an avoidance response, altered behavior in the vicinity of a facility, or changes in the types or quality of habitat used by animals. The zone of influence can determine the area of effect, serve as a metric for regional measures of cumulative effects, or help guide monitoring and mitigation strategies. For example, Nelleman et al. (2003) reported a zone of avoidance of 2.5–5 km for reindeer (*R. t. tarandus*) responding to powerlines, resorts, and roads.

Although an intuitive concept, the zone of influence and measures of significance are difficult to quantify (Quinonez-Pinon et al. 2007). This is especially apparent where multiple developments interact. Also, the zone of influence should be premised on the type of response that is measured, and there may be multiple zones depending on the source of effect. Direct mortality via road access, for example, is normally restricted to the area in the immediate vicinity of the road corridor or road density across a larger area. Habitat alteration or avoidance responses relative to noise or human presence may occur over a larger spatial extent. Recent research has focused on developing techniques that indicate statistically meaningful responses of animals to human activities or facilities that can then be translated to zones of influence used in regulatory frameworks (Bennett et al. 2009). When empirical data are absent, expert opinion is used to estimate probable zones (AXYS and Penner 1998). Often, the processes to collect such ecological data are flawed (Johnson and Gillingham 2004), making a strong case for the application of formal and repeatable species distribution models for such purposes.

Species distribution models and their associated outputs are easily adapted and applied to resource management or conservation models. These multimodel approaches often integrate maps, illustrating the location and amount of selected habitats, with predictive movement models, population viability analyses, or habitat supply models. Johnson et al. (2005), for example, used maps of the distribution of high-quality habitats for a number of Arctic species, including caribou from the Bathurst herd, to quantify the impacts of hypothesized development scenarios on the distribution and availability of habitats and abundance (Johnson and Boyce 2004). Similarly, Carroll et al. (2003) linked species distribution and spatially explicit population models to understand the relative value of a range of reintroduction strategies for wolves under current and predicted future landscape conditions.

Building on previous research (Johnson et al. 2005), we are using Resource Selection Functions (RSF), one type of species distribution model, to investigate the responses of Bathurst caribou to broad-scale vegetation patterns (i.e., habitats) and human disturbances. We hypothesize that after statistically controlling for variation in the distribution of plant communities, caribou will demonstrate a decreasing avoidance response as distance from human facilities increased. An RSF produces a series of coefficients that quantify the strength of avoidance or selection for specific habitat covariates. When considered additively (Equation 8.1), the series of coefficients indicate the relative probability of caribou using any location from across the study area (Johnson et al. 2006).

$$w(x) = \exp(\beta_1 x_1 + \beta_2 x_2 + \ldots + \beta_i x_i) \tag{8.1}$$

When normalized, the RSF score $w(x)$ equals the relative probability of the occurrence of caribou; this is a function of the weighting coefficients (β_i) and the magnitude of the covariate at that site (x_i). Using simple image arithmetic, the RSF equation can be applied to GIS data for each covariate resulting in spatially explicit habitat predictions; these maps can then be used to further develop population, movement, or habitat supply models.

We are using one form of a broad range of models capable of quantifying species–habitat relationships (Elith and Leathwick 2009). Alternative species distribution models may provide a quantitative perspective on caribou distribution. The exponential form of the RSF, however, is inherently flexible, accommodating a range of habitat covariates; is grounded in statistical theory and has a number of proven methods for validation; results in a predictive metric, relative probability of use, that is easily understood and transferable; and is insensitive to data imprecision and choice of formulation (Johnson et al. 2006; Johnson and Gillingham 2005, 2008).

An RSF is constructed using point data that illustrate the spatial location of an animal across some defined area and time period, a comparison set of random locations that represent the amount and distribution of resources or features, and a number of habitat or disturbance covariates that describe or model the observed pattern of animal locations relative to the set of random locations. For this project, we drew upon 13 years of location data collected by satellite collars deployed on 67 female caribou (Gunn et al. 2002). Recognizing longer-term dynamics in caribou distribution and the need to directly involve aboriginal communities in the project, we derived additional point locations from an extensive Traditional Use Study focused on the harvesting of caribou (Legat et al. 2001). Here, Tlicho elders documented locations where they had hunted caribou since 1932 and trails used by caribou. Each hunting location was considered as a separate datum, and the trails were converted into point locations with a 5-km interval to ensure independence. Because the traditional ecological knowledge information was located on the western extremity of the study area, these data were used to model vegetation covariates only (Figure 8.1).

We are using the RSF to investigate two vegetation and three human disturbance covariates. Because of the large study area, vegetative habitat variables were derived from two independent mapping projects based on Landsat Thematic Mapper images, but different legends and classification routines. Variation in green plant biomass and phenology influences caribou distribution (Griffith et al. 2002). Thus, we used the Normalized Difference Vegetation Index (NDVI) to measure the response of caribou to seasonal changes in plant productivity. We selected a number of types of human disturbance features that would aggregate cumulatively to influence the distribution of caribou. Drawing from the methods of Johnson et al. (2005), the distance of each caribou and random location was calculated from existing diamond mines and a gold mine. Recognizing that not all mines were in operation during the study period, distances were relative to active mines, on an annual time scale. We also calculated the distance to areas of the summer range where mineral exploration activities occurred. This included known locations of exploration camps and activities, buffered by 10,000 m, and broader areas for which an active mineral lease was on record. Previous work (Johnson et al. 2005) suggested that Arctic wildlife may avoid outfitter camps. We buffered camps by 500 m to represent the broader area of influence of such activities. Outfitters often hunt caribou from lake shorelines; therefore, we buffered lakeshores 5 km inland when situated within 20 km of a hunt camp.

Through the modeling, we are identifying combinations of resource and disturbance variables that serve as hypotheses that might explain patterns in the distribution of Bathurst caribou. We are generating candidate models for three time periods of distinctive behavior across the summer range: post-calving (June 14–July 5), early summer (July 6–July 18), and late summer (July 19–August 22). Then, we use an information-theoretic approach to guide model development and selection (Anderson et al. 2000).

Understanding Changes in Abundance of Caribou with an Energetics–Demographic Model

The second model explores how caribou integrate their behavioral responses to human activity and environmental variation (e.g., insect harassment, foraging conditions) from the individual to the herd scale (Nicolson et al. 2002; Kruse et al. 2004). The model (Russell et al. 2005) predicts the change in daily body mass and body composition of a female caribou, her milk production, and the daily body mass change of her calf as a function of milk intake. The variables driving these outcomes are daily activity budgets, forage quality, and forage quantity. The energetics model consists of two submodels (Figure 8.4 and Figure 8.5). The first is the energy submodel (Figure 8.4), which predicts daily changes in a female's metabolizable energy intake (MEI) by calculating her food intake and then simulating the functioning of the female's rumen and her digestive kinetics on an hourly basis. Using MEI as the index of change, specific objectives of the energy submodel are as follows: to show effects of environmental conditions and movement patterns as reflected by changes in activity budgets, forage quality, and forage quantity; to evaluate effects of human and natural disturbance such as mining activities and insect harassment; and to evaluate winter severity as reflected by snow depth.

The MEI predicted by the energy submodel is transferred to the second model. The growth submodel calculates the female's energy expenditure, her energy balance, and the subsequent daily change in her mass, milk production and hence the daily change in mass of her calf. The growth submodel evaluates effects of changes in seasonal activity budgets and MEI on the energetic and reproductive status of a female caribou. The growth submodel's specific objectives: to evaluate the impact of changing activity costs, maintenance costs, and MEI on the female's energy balance and subsequent change in body composition and growth; and to evaluate effects of the female's energy balance on the growth of her fetus during pregnancy and her calf during lactation.

We are using nine different scenarios to explore the sensitivity of the body condition model predictions to varying levels of human development and environmental

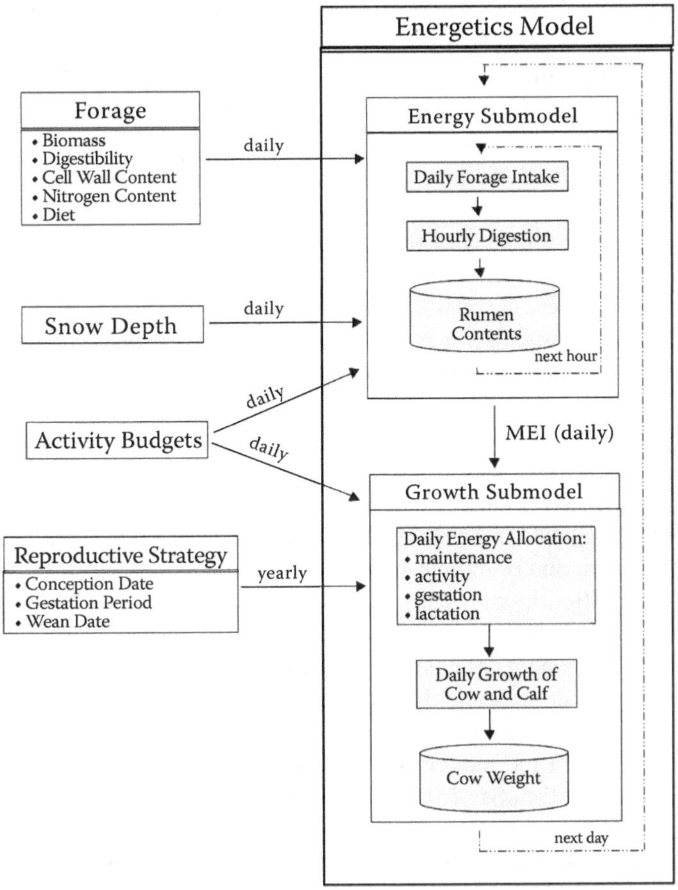

FIGURE 8.4 Generalized structure of the barren-ground caribou energetics model developed by Russell et al. (Used with permission from the Canadian Wildlife Service.)

change. Each scenario has an assumption regarding the level of development (none, current, or double current development) and climate (average, worst case, and best case). Within each scenario, we incorporate two zones of influence: habitats and behavior adjacent to the diamond mines and the area outside the zone of influence. With respect to the level of development, "current" scenarios represent current mining activity on the Bathurst summer range. About 6% of the summer range is within the 30-km zone of influence of mine sites for these scenarios; the no development scenarios assume that no development occurs on the summer range, while the two-times current development scenarios assume that the total area within the development zone is double that of current conditions (i.e., 12% of summer range).

For the climate scenarios, the average scenario represents current average climatic conditions. The worst-case climate scenario represents the worst possible combination of climatic conditions for caribou (i.e., high winter snow levels, high summer insect harassment, and a short green-up period for plant biomass). Similarly, the best-case climate scenario represents the best possible climatic conditions (i.e., low winter snow levels, low insect harassment, and long green-up).

To determine the time-series of forage biomass available to caribou every day, the model requires the annual maximum biomass ($kg{\cdot}ha^{-1}$), for plant groups within each of the 10 habitats taken from the RSF analysis (i.e., moss, lichens, mushrooms, horsetails, graminoids, deciduous shrubs, evergreen shrubs, forbs, standing dead, *Eriophorum* heads). Next, the phenology (green-up) of each plant group is characterized using three dates: start of plant emergence, date of maximum biomass, and end of plant senescence. These dates vary as a function of the climate year-type (i.e., average, worst-case, best-case), representing early and late green-up.

The model also requires estimates of seasonal activity budgets for the herd, specifying the proportion of time spent by the animal each day in foraging, lying, standing, walking, and running. The proportion of total foraging time is further broken down into the proportion time spent eating and time spent pawing. We specify activity budgets for each possible climate year-type to account for the differing effects of snow depth and insect harassment on caribou activity. Activity budgets are also varied within and outside the development zone, to reflect changes in caribou activity patterns as a function of human-related disturbance.

The body condition model requires an estimate of the proportion of time spent in each landscape stratum within the summer range, where landscape stratum refers to a combination of habitats and development zone. For the post-calving, early summer, and later summer seasons, this proportional use of each landscape stratum is estimated by simulating caribou movement based on the relative probabilities taken from the RSF. We run the model through each of the three seasons to predict the autumn mass (i.e., on October 15) of a lactating adult female for the nine scenarios. Having run the body condition model for this suite of scenarios, the last step in the analysis is to relate predicted changes in body condition to changes in one or more demographic parameters such as birth rate the following spring, as determined using data from the Central Arctic and Porcupine herds (Cameron and Ver Hoef 1994).

Understanding Implications of Landscape and Environmental Changes on Caribou: A Landscape Cumulative Effects Simulator (ALCES®)

We use ALCES (www.alces.ca), a landscape simulation model, to explore the cumulative impacts of land-use scenarios for barren-ground caribou. Scenarios are plausible, but structurally different descriptions of how the future might unfold (Duinker and Greig 2007; Mahmoud et al. 2009). Although computer-based scenario simulations do not provide quantitative predictions or forecasts of conditions in any particular year, they can be used to assess the influence of assumptions or management approaches, and to explore uncertainties and strategies for mitigating cumulative effects (Schneider et al. 2003; Carlson et al. 2007; North Yukon Planning Commission 2009).

The model ALCES is capable of simulating and tracking changes in land cover types caused by anthropogenic land uses and natural ecological processes at a regional scale (Hudson 2002). The cumulative effects simulator can represent natural disturbance regimes, human land-use trajectories, and climate drivers according to user-defined inputs. Resource development (e.g., mining, hydrocarbon extraction, forestry and agriculture), the ecological composition of the land base, and climate are translated into biological indicators such as the area of a particular plant community that may serve as habitat or the actual abundance of a wildlife species. For example, relationships between rates of increase of boreal caribou and functional habitat loss due to natural and anthropogenic landscape disturbance (Sorensen et al. 2008) have been incorporated in ALCES to simulate and evaluate implications of alternative management strategies for woodland caribou (*Rangifer tarandus caribou*) in northern Alberta (West Central Alberta Caribou Landscape Planning Team 2008, Athabasca Landscape Team 2009).

Effective and transparent application of ALCES for strategic-level cumulative effects modeling of barren-ground caribou systems requires that the model is able to simulate multiple plausible impacts that potentially affect caribou, and the impact hypotheses are grounded in relevant expert knowledge and empirical research. Our goal is to parameterize the model to integrate and simulate the effects of resource development and land use, climatic variability, and hunting and predation for barren-ground caribou (Figure 8.5). As a first step, a population dynamics

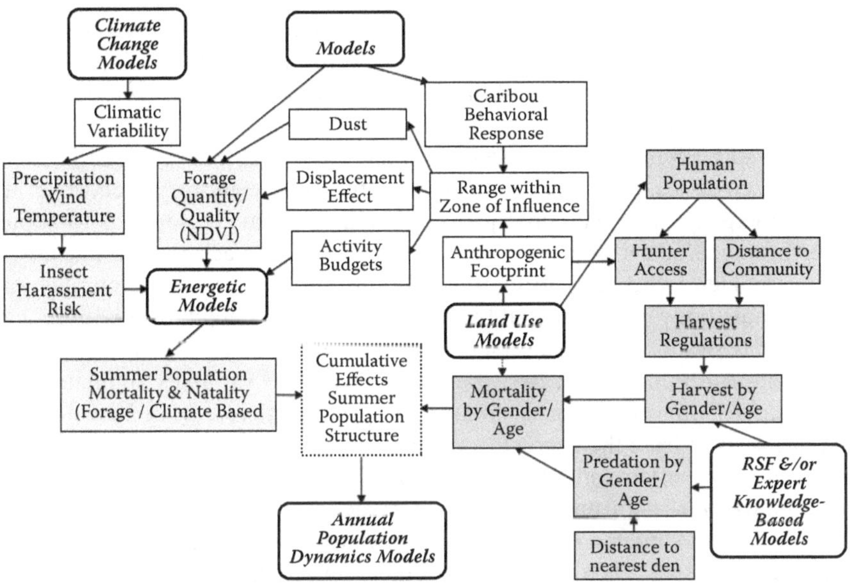

FIGURE 8.5 Structure of a cumulative effects simulation model for barren-ground caribou on the summer range. Submodel components—shown as dark outlined boxes—serve as data sources (e.g., RSF/expert models) and data processors (e.g., energetic and population dynamics models). Regular boxes reflect data themes with arrows showing linkages and flow of information among submodel components. Lightly shaded boxes outline factors that may be generated from climatic variability and are used as inputs to the energetics model, darkly shaded boxes represent factors that directly influence mortality via predation and hunting, and unshaded boxes represent factors that determine the anthropogenic footprint and associated responses of caribou.

submodel was added to ALCES to provide the link between direct and indirect effects of landscape change, and natural and anthropogenic stressors to population performance of caribou. The submodel also facilitates simulation of the direct effects of predation and hunting to sex and age-class specific mortality of a caribou population (Figure 8.5).

Our current focus on resource selection and energetic models is a way to define and incorporate empirically based functional relationships that describe responses of barren-ground caribou to changes in landscape composition that may occur due to resource development. Through an additional series of simulation experiments, we will use the resource selection and energetics models to define a suite of functional response curves that link pregnancy rate to varying levels of anthropogenic land use and natural stressors (e.g., summer insect harassment). For example, the RSF is first applied to simulate the distribution of females during the calving and summer seasons, including the amount of time spent in each cover type and the proportion of time spent within a 30-km zone of influence of an active mine. The energetics model is then used to estimate the implications of that avoidance scenario for the body mass of average female caribou and, ultimately, birth rate. This modeling approach is used to estimate birth rates associated with a number of land-use scenarios across a continuum from no active mines to several times the current area of active mines. Results from these scenarios will be used to define a relationship between proportion of the study area within the zone of influence of active mines and caribou birth rate. This relationship will be integrated into the population dynamics submodel within ALCES as a modifier to birth rate.

Because ALCES is able to simulate variation and trends in temperature and precipitation regimes based on user-defined climatic parameters for an ecozone, it is possible to explore potential consequences of climate change scenarios to population performance of caribou. We are developing an approach that would use the energetics model within a factorial simulation experiment, to relate climate variables to female body condition and birth rate. Simulations conducted with ALCES would track relevant climate variables using random variation around user-defined expected climate trends. Based on the array of climate variables simulated for a given year, ALCES would then select the appropriate birth rate from the database of birth rates derived from the energetics model.

The cumulative effects simulator, ALCES, will serve as the integration tool, projecting empirical output from the RSF and energetics model over long time periods (e.g., 50–100 years) under a range of change scenarios, including future climate and land-use development. Such scenarios will be premised on historical or anticipated rates of development, predicted levels of climate change relative to key caribou parameters (e.g., insects, green-up of spring forage), and questions that come directly from the users and managers of caribou. Predicting the possible outcomes of these cumulative impacts for the Bathurst caribou herd will allow for a better understanding of how resilient the herd is to anticipated levels of landscape change that occur over time periods (e.g., 50–100 years) that exceed the current temporal and spatial frameworks of site-specific CEA. Such information is crucial for long-term strategic planning in the context of cumulative effects and may help address many of the shortfalls of the current regulatory and monitoring processes (Chapter 3).

ADDRESSING THE CUMULATIVE EFFECTS QUAGMIRE—CARIBOU AND BEYOND

This research is an ongoing collaborative project between researchers, government, and a comanagement board, and was designed to meet the particular concerns of people in the NT and criticisms about CEA (Kennett 1999; Duinker and Greig 2006). Those concerns and criticisms (Duinker and Greig 2007) also relate to projects that fall below screening requirements, so we developed a practical approach to incorporate potential small-scale effects arising from exploration sites and camps on caribou range. Our integrated modeling approach lends itself to land-use planning that can be an effective approach for considering cross-sectoral developments strategically across regional areas (Chapter 3). Understanding the range of caribou responses to development, from the behavior of individual animals to the demographic changes of the herd, will allow communities, comanagement boards, and government to better manage or plan for cumulative effects. Additionally, this improved understanding will lead to a clearer appreciation of trade-offs between goals for sustainable hunting and persistence of healthy caribou populations in the context of broader goals for landscape management that include industrial development and resource extraction. An example of the trade-offs between sustainable harvesting and industrial development is apparent in the management of the Central Arctic herd that interacts with the Prudhoe Bay oilfield. Here, harvest levels are kept low (<2%) to offset possible detrimental effects of disturbance during calving (Lenart 2007).

As suggested by Vistnes and Nellemann (2008), pre- and post-development studies provide a powerful framework for environmental assessments and for understanding cumulative effects. The benefit of simulation models is that they can be used as learning tools to compare pre- and post-development scenarios to develop testable hypotheses that are derived from a current understanding of caribou ecology and responses to human activities. The goals of current and future modeling work in this project are to better understand the cumulative effects of a range of stressors on the long-term viability of the Bathurst caribou herd. These effects, however, can be considered in the context of economic development for northerners, including aboriginal communities.

If trade-offs are to be considered and integrated into the decision-making process, managers will need to move from command and control strategies (Holling and Meffe 1996) to those that recognize and accept that natural systems are complex and dynamic. Furthermore, people will need to be considered as part of these systems, which requires a shift in thinking and the development of decision support and learning tools. Such a shift has occurred in the NT as the management and regulatory process is premised on comanagement and community involvement. There is still, however, much room for improvement. Agencies and biologists must move from simply including traditional knowledge in science and management to a broader acceptance of other world views and the development of adaptive capacity in these complex socioecological systems (Paci et al. 2002).

Understanding the influences of natural environmental variation is fundamental to moving the science of cumulative effects from correlation to causation (Perdicoúlis et al. 2007). Causality is the link between human activities (actions) and their environmental impacts, although more often than not causality refers to hypotheses rather than facts (Perdicoúlis et al. 2007). Consequently, there is a need to select a design/methodology for CEA that can accommodate natural and industrial changes. Resilience allows one to integrate natural environment variability with the responses of animal populations to human activities. Applying the concept of resilience also emphasizes that there are limits to what caribou (or ecological systems) can cope with and that this limit will depend on naturally varying environmental conditions. This means that we can anticipate thresholds in development levels above which caribou ecology, distribution, or population dynamics will shift to a different state. We have to weigh those impacts in the context of the benefits of the development for local and national economies.

The impact of industrial development on wildlife is a frequent and worldwide concern, especially for migrant species whose traditional routes can be threatened by infrastructure or landscape change (Berger 2004). Although we designed our approach to cumulative effects specifically for migratory tundra caribou in the NT, the approach likely has application to migratory tundra caribou elsewhere. Caribou, such as those in the Bathurst herd, have ecological similarities to other open habitat, gregarious and migratory ungulates in Africa and Asia that face similar threats (e.g., Mongolian gazelles [*Procapra gutturosa*]; Ito et al. 2005). Thus, we suggest the general approach we have developed may be applicable to other species of wide-ranging herbivores.

When asked in 2008 about whether industry and caribou could coexist in the north, Fred Sangris (Chief, Yellowknives Dene) said, "When the buffalo went from the plains, the people of the plains, the Cree, the Dakota—their culture died, their spirit died. Here, we have a chance to save it" (Canadian Arctic Resources Committee 2007, p. 22). His comparison of bison and caribou underscored two important points: there is a strong cultural link between aboriginal peoples, wildlife, and the land, and abundance of wild animals today is no guarantee of their future survival. Bergerud et al. (1984, p. 19) also linked bison and caribou when they concluded: "But, adaptable as the caribou is, it still has the same problems as the buffalo—overharvest and the need for space." Seasonal migration is an adaptive strategy of caribou to their predators, parasites, and availability of forage. Caribou herds will lose their ability to cope with environmental changes and human activities if their ability to find space is compromised or restricted.

CONCLUSION

The future for caribou populations across the Canadian central Arctic is uncertain. We are currently witnessing historic lows of all populations with extreme curtailments on harvest for all northern residents, including aboriginal communities. The Bathurst herd is a case in point with numbers dropping from a peak of 472,000 ± 72,000 (SE) in 1986 to 128,000 ± 27,300 in 2006, and then a historic low of 31,900 ± 5,300 in 2009 (Government of Northwest Territories Environment and Natural Resources 2009). Continued decline will have consequences for ecosystem integrity and the livelihoods of northern residents with strong cultural and economic ties to caribou (Forchhammer et al. 2002).

Uncertainty for the future of barren-ground caribou arises from the cumulative effects of the types and levels of land-use activities that we permit across their seasonal ranges. Climate change with its positive and negative effects also confronts the use and management of caribou (Brotton and Wall 1997; Gunn et al. 2009). Given these cumulative impacts and uncertainties, governments, caribou users, and industry will have to collaborate to maintain the space that caribou need to cope with landscape changes. Our integrated modeling will help people to work together, and the model outputs provide perspective on the potential risks of development scenarios. Our goal is to move cumulative effects research from the realm of scientists to comanagement and government decision-makers as a step toward sustainable development.

9 The Cumulative Effects of Suburban and Exurban Influences on Wildlife

Paul R. Krausman, Sonja M. Smith, Jonathan Derbridge, and Jerod Merkle

CONTENTS

INTRODUCTION

Urbanization is the primary cause of species endangerment and a leading threat to biodiversity in the contiguous United States (Czech et al. 2000). It represents one of the most severe forms of unregulated cumulative effects impacting wildlife and their habitats. Rural, suburban, and exurban development in the United States consumes ~1,000,000 ha of land/year (Milder et al. 2008) and contributes to increased anthropogenic mortality of wildlife from collisions with vehicles and windows, dog and cat predations, malicious gun shot, monofilament line injuries, tar, oil, fly paper contact, and pesticide toxicities (Burton and Doblar 2004). Diseases also contribute to mortality factors of urban wildlife (Krausman 2002): house finch conjunctivitis, trichomoniasis, salmonellosis, sarcoptic mange, and rabies (Krausman 2002; Burton and Doblar 2004). Cities, towns, and villages in the United States make up approximately 5% of the land mass, but nearly 80% of the population lives in them and the surrounding suburbs (U.S. Census Bureau 2000). Thus, cumulative effects are continual and largely unmitigated. This anthropogenic development results in the modification of landscapes so severe that native habitat for wildlife is altered and eliminated (McKinney 2002). Most alteration occurs around cities, and the influences of development decrease as human activity decreases from the core through suburban and into rural landscapes. This expansion of residential areas into rural landscapes is urban sprawl (Lindstrom and Bartling 2003), which creates new habitats from the native landscapes: cemeteries, parks, gardens, golf courses, lakes, and tons of asphalt and concrete. Although some species adapt well to urban sprawl (e.g., some amphibians, reptiles, small mammals, birds), many more are negatively impacted (Randa and Yunger 2006) due to abrupt habitat boundaries, road construction (Hawbaker et al. 2006), introduction of exotic flora and fauna, degradation of landscapes by humans causing long-term habitat loss, and increased extinction rates (McKinney 2002) leading to international biotic homogenization (McKinney 2006). As "cities expand across the planet, biological homogenization increases because the same 'urban-adaptable' species become increasingly widespread and locally abundant in cities across the planet" (McKinney 2006, p. 247). These "global homogenizers" combined with native species that adapt well to suburban environments (i.e., early successional plants and edge animal species, ground-foraging, omnivorous and frugivorous birds that use gardens, forest fragments, and other suburban landscapes) provide some developed areas a rich assemblage of flora and fauna even though many native species declined. This has likely received little attention because many humans that live in suburban and exurban habitats (of all income levels) become increasingly disconnected from local indigenous species and their roles in the natural ecosystem (McKinney 2006).

In the United States and Canada, on average, the greatest levels of biotic homogenization were predicted for plants (22%) and fishes (14%), followed by reptiles and amphibians (12%), mammals (9%), and birds (8%). Homogenization is predicted to be greatest for fish in the southwestern and northeastern United States, highest in eastern North America for plants, greatest for birds and mammals along the west coast of North America, and peaks in the Southeast for reptiles and amphibians (Olden et al. 2006).

Similar effects occur as rural landscapes are altered with livestock and agriculture. In both situations, the concentration of anthropogenic influences is not generally beneficial to wildlife and is creating serious challenges for wildlife biologists and managers. Wildlife species that are able to tolerate humans within their habitat are often undesirable to many citizens leading to human–wildlife conflicts; these conflicts can range from minor annoyances to property damage and the loss of life. Until planners consider wildlife at the development stage and plan for cumulative effects, habitats will continue to be altered, populations will be reduced and eliminated, and the overall quality of life will be reduced. The area of the western United States dominated by anthropogenic features is ~13%. Areas by rivers are generally influenced more by the human footprint than are lakes (Leu et al. 2008). The disproportional influence of humans on the western landscape creates a challenge to biologists and managers.

Unfortunately, urban sprawl has only recently been addressed as a serious issue. In this chapter, we concentrate on one of the most pristine regions of the country (i.e., the inland Northwest) to demonstrate how cumulative effects of suburban and exurban development impact wildlife. No one believes this area will be immune to the negative issues that are facing wildlife throughout North America. Exurban development will likely persist in the Rocky Mountains as people search for scenic and secluded lifestyles (Romme 1997). How does this growth affect wildlife, their movements, habitats, and connectivity? When negative impacts are identified, what practices are being applied to avoid, minimize, or mitigate the impacts? How effective are the mitigations? These data will provide an understanding of the current state of how urban sprawl is influencing wildlife. Plant communities have been examined with gradient analysis for decades (Whittaker 1967), and the technique has been used with avifauna, mammalian, and aquatic studies. Most recently, it has been used to examine the complexity of urban ecology (McDonnell et al. 1993).

Development of Wildlife Conservation along the Rural–Urban Gradient

In the United Kingdom, the shift to urban lifestyles began earlier than in the United States, and ecologists began studying urban ecology in the early 1900s (Shenstone

1912; Shaw et al. 2009). Fitter (1945) published the first book on urban ecology as development in and around London progressed (Shaw et al. 2009). Urban ecology became popular in the United States in the 1960s and 1970s, associated with the environmental movement (Shaw et al. 2009). However, Leopold (1933) actively searched for ways to minimize anthropogenic influences on wildlife (Miller and Hobbs 2002). Leopold's efforts marked a shift in wildlife management to include nongame species and a conservation-based approach to management (Hadidian and Smith 2001). The shift was accepted by the urbanizing public because they were interested in attracting wildlife to their homes and understanding the distribution of animals in developed areas (DeStefano and DeGraaf 2003; Shaw et al. 2009).

With the shift from "old conservation" (i.e., exploitation) toward "new conservation" (i.e., clean water, air, open space, outdoor recreation, quality of human environments; Dasmann 1966; Shaw et al. 2009), a suite of laws was enacted (e.g., Endangered Species Act), and nongame management programs were established in state fish and wildlife agencies (Shaw et al. 2009). The first conference on the urban environment was sponsored by the U.S. Fish and Wildlife Service in 1968, and the National Institute for Urban Wildlife was founded in 1973 (Adams 2005). These activities and others led to an explosion of interest and research in attracting wildlife to backyards (Shaw et al. 2009), how wildlife coexisted with humans (Destefano and DeGraff 2003), bird–habitat relationships (Emlen 1974; Campbell and Dagg 1976; Lancaster and Rees 1979; Beissinger and Osborne 1982; DeGraaf 1991; Blair 1996; Germaine et al. 1998; Melles et al. 2003), values of wildlife watching, surveys of public attitudes, and planning and management (Lyons and Leedy 1984; Shaw et al. 1985; Shaw et al. 2009). However, urban ecology remained at the fringes of mainstream ecology because many considered anthropogenic alterations to areas as biological degradation and suggested that studies in undisturbed natural areas were of more value (Miller and Hobbs 2002). During the 1980s, urban wildlife research concentrated on ways to mitigate human–wildlife conflicts (Loker et al. 1999; Shaw et al. 2009) and, throughout the 1980s and 1990s, interest in urban ecology surged with growing interest from management agencies, universities, working groups, and professional societies (Shaw et al. 2009). In the 1990s, the significance of urban and exurban landscapes to the conservation of biodiversity was widely recognized; conservation that focused primarily on wildlands was not adequate to maintain a full range of biodiversity (Shaw et al. 2009). Furthermore, urbanization was fragmenting habitats, and urban ecology moved "... quickly from the periphery of ecological science to the mainstream as concerns about unprecedented suburban growth and the attendant loss of open space motivated research agendas aimed at informing regional planning and conservation..." (DeStefano et al. 2005; Shaw et al. 2009, p. 120). Universities hired urban wildlife specialists (Decker et al. 2001), and Baltimore, Maryland, and Phoenix, Arizona, were added to the National Sciences Foundation's Long-term Ecological Research programs (Kingsland 2005). By 2000, there were nongame wildlife programs in every state wildlife agency (Shaw et al. 2009).

Restoring and maintaining habitats in urban areas to conserve indigenous species by reducing the impacts of urbanization and reducing human–wildlife conflicts in urban areas are the dominant fields of study in urban ecology (McKinney 2002; DeStefano and DeGraff 2003; Shaw et al. 2009). Both arenas have involved extensive involvement with passionate and often controversial conflicts; most urban centers are on public lands dominated by humans. The public has to be part of any method used to enhance wildlife or wildlife habitat, minimize habitat alteration, or reduce human–wildlife conflicts. "In urbanized areas, developing a more ecologically informed public can be the most effective way of promoting conservation of native species and reducing human-wildlife conflicts (McKinney 2002)" (Shaw et al. 2009, p. 121).

Our objective is to review relevant primary and gray literature that addresses how suburban and exurban growth influences wildlife and fish. Because this is a relatively new area of study, there are few specific examples from the inland Northwest, so we will present examples from other areas that may be applicable to understand how cumulative effects influence wildlife and their habitats so that planning can progress with minimal negative influences to wildlife.

DEFINITIONS

Because urbanization is a relatively new field in wildlife management (Krausman 2002), we present definitions of terms used in this chapter.

1. *Biophysical-behavioral forces.* Biological and ecological properties associated with wildlife including population dynamics, reproduction, habitat, and behavior (Kellert and Clark 1991).
2. *Exurban development.* Approximately 6–25 homes/km² and includes urban fringe development on the edge of cities and rural residential developments in rural areas that have natural amenities (Hansen et al. 2005). "... Exurban development occurs beyond incorporated city limits, and the surrounding matrix remains the original ecosystem type" (Odell and Knight 2001).
3. *Institutional-regulatory forces.* These are factors that allow wildlife management to operate successfully: financial base, law enforcement, states' rights for residential wildlife, expansion of wildlife values, public trust, federal regulation of migratory species, and inclusion of habitat when defining wildlife (Kellert and Clark 1991).
4. *Rural areas.* An area with <193 people/km² (U.S. Census Bureau 2002).
5. *Social–structural forces.* The role of power versus property relationships that reflect the rights and privileges to use and control wildlife resources (Kellert and Clark 1991). Wildlife in the United States is common property but access can become complicated on private lands, especially where federal and private lands merge.
6. *Suburbs.* The patchwork of residential, commercial, municipal, and industrial land uses and related transportation and utility corridors often adjacent to urban centers (Knuth et al. 2001).
7. *Synurbization.* Adjustment of wild animal populations to specific conditions of the urban environment (Luniak 2004).
8. *Unnatural food.* Any food for wildlife made available by humans (Peine 2001).
9. *Urban areas.* An area with a high-density core of ≥386 people/km² (U.S. Census Bureau 2002).

10. *Urbanization.* Changes in landscape caused by urban development (Luniak 2004).
11. *Urban stream syndrome.* The consistently observed ecological degradation of streams draining urban lands. Symptoms "include a flashier hydrograph, elevated concentrations of nutrients and contaminants, altered channel morphology, and reduced biotic richness, with increased dominance of tolerant species" (Walsh et al. 2005, p. 706).
12. *Urban wildlife.* Species that find habitat in urban and suburban landscapes (e.g., deer [*Odocoileus* spp.], raccoons [*Procyon lotor*], coyotes [*Canis latrans*]; Krausman 2002).
13. *Valuation forces.* The general manner in which society considers wildlife. These forces can be categorized as economic, ecological, and social–physiological (Kellert and Clark 1991). Valuation forces are essentially the manner in which society views wildlife (Peine 2001). For example, tourist operations may use wildlife to entice customers (economic), biologists encourage biodiversity for the necessary roles in ecological processes (ecological), while Native Americans revere some wildlife spiritually (social–physiological).

REPTILES AND AMPHIBIANS

Most of the available data on anthropogenic influences on reptiles and amphibians is categorized under biophysical–behavioral forces. An early paper (Minton 1968) summarized the importance of altered aquatic habitats. Between 1949 and 1958, two species of salamanders, six species of anurans, six species of turtles, and seven species of snakes were recorded on the boundary of Indianapolis, Indiana. Six years (1963–1964) later, only two species of anurans, one species of turtle, and four species of snakes were recorded. In the earlier survey, ≥11 species bred within the area but, in the 1960s, there was no evidence of amphibians breeding (Minton 1968).

> Insofar as amphibians and reptiles are concerned, clearing of the land with removal of ground cover and underbrush affects all terrestrial species but is particularly severe for salamanders and some snakes. Modification of aquatic habitats by drainage, dredging, pollution, or removal of vegetation has serious effects on all amphibians except those whose egg and larval stages are spent on land or in small or transient collections of water. Aquatic turtles are more or less severely affected, whereas snakes suffer indirectly through deprivation of aquatic refuge and prey. Road construction and subsequent traffic can be devastating when it denies a population free access to its hibernating or breeding areas; otherwise, traffic kill has little effect, particularly on smaller species. Direct killing by man affects chiefly the larger snakes and occasionally turtles. It can be a major factor in extermination of species that aggregate for breeding or hibernation but, more frequently, it delivers the *coup de grace* to a population severely endangered by alteration of its habitat. Changes in food supply and in predator–prey relationships incident to urbanization are difficult to evaluate. They apparently cause some small species of amphibians and reptiles to increase in numbers in urban and suburban sites, at least temporarily. *Natrix kirtlandi* and *Thamnophis butleri* are good examples in the Midwest. On the other hand, urbanization may deprive other species of suitable prey or leave them vulnerable to excessive predation. Pesticides undoubtedly have the

> potential for doing severe harm to amphibian and reptile populations. Evaluation of their effect is difficult without toxicologic studies on exposed populations. (Milton 1968, p. 115)

The results are consistent with other early reports on the influence of development on reptiles and amphibians in the Midwest (Seibert and Hogen 1947; Conant 1951) and in San Francisco County, California (Banta and Morafka 1966). These results also reflect the negative consequences of continued urbanization on reptiles and amphibians throughout the United States. Lizards, snakes, turtles, salamanders, and anurans have all decreased in the face of anthropogenic activities.

Lizards were surveyed along a gradient ranging from a mean density of 3.26 homes/ha to native, undisturbed landscapes in Tucson, Arizona (Germaine and Wakeling 2001). Low–moderate development densities supported lizard abundance, species richness, and evenness but, as house density and paving increased beyond moderate levels, lizard assemblages decreased rapidly, and the tree lizard (*Urosaurus ornatus*) was dominant (Germaine and Wakeling 2001).

In New Hampshire, snake abundance, occupancy, and richness were examined on habitat patches (range = 0.2–120.0 ha) in a landscape undergoing substantial land use (Kjoss and Litvaitis 2001). The species–area hypothesis (MacArthur and Wilson 1967) suggests that populations on small patches would be volatile and characterized by few species relative to larger patches. This hypothesis was supported in the study of Kjoss and Litvaitis (2001). Because populations that are dependent on early successional and shrub-dominated habitats have declined to levels where they are subjected to regional extinction, Kjoss and Litvaitis (2001) advocate the maintenance of patches >10 ha.

Suburban and urban developments are also detrimental to turtles, especially those developments with dense road networks and abundant populations of generalist predators (e.g., raccoon). Reductions in turtle populations are likely caused by reduced recruitment caused by habitat alterations (e.g., dense road networks) that will reduce or eliminate local populations (Marchand and Litvaitis 2004).

Similar results are documented for the southern two-lined salamander (*Eurycea cirrigera*) and northern dusky salamander (*Desmognathus fuscus*) in North Carolina (Price et al. 2006), spotted salamanders (*Ambystama moculatum*) in Rhode Island (Skidds et al. 2007), and California newts (*Taricha torosa*) in California (Riley et al. 2005). In North Carolina, the southern two-lined salamander and northern dusky salamanders have decreased 21%–44% from 1972 to 2000 as urban lands have influenced stream ecosystems (Price et al. 2006). Salamanders in Rhode Island and central Pennsylvania are negatively influenced (i.e., reduced reproduction) by urbanization (Skidds et al. 2007).

Simply maintaining habitat components for salamanders is not enough. Connectivity to other populations is important. Allelic richness and heterozygosity were lower in isolated populations of eastern red-backed salamanders (*Plethodon cinereus*) due to urbanization in southern Québec compared to populations located in continuous habitat (Noël et al. 2007). In an urbanized area north of Los Angeles, California, streams in more developed watersheds have fewer native species such as California newts compared to streams not influenced by urbanization (Riley et al. 2005).

More research has been conducted on urban influences on anurans than other reptiles and amphibians. Unaltered landscapes without exotic fish that include aquatic resources and associated uplands are important for the maintenance of many amphibians (Pearl et al. 2005). However, urbanization alters these landscapes that may make them unsuitable for some species. For example, eastern spadefoot toads (*Scaphiopus holbrookii*) require wetlands and upland habitat to complete their life cycle; wetlands for reproduction and larval development, and uplands for feeding and burrowing (Jansen et al. 2001). When urbanization alters the characteristics of burrowing sites, adequate habitat is eliminated and precludes the persistence of the eastern spadefoot toad in Florida (Jansen et al. 2001).

Other anurans are at risk due to the creation of road networks in and adjacent to their habitat. Vagile species such as the northern leopard frog (*Rana pipens*) were negatively affected by traffic density within a 1.5 km radius of their habitat (Carr and Fahrig 2001) in Ontario, Canada. Roads serve to fragment habitats for anuran communities. When urbanization occurs, it is important to consider interconnected wetlands and upland habitats to avoid isolation and potential population reduction or local extinction (Pillsbury and Miller 2008). In central Iowa, all seven species of anurans studied exhibited a negative association with urbanization (Pillsbury and Miller 2008) due to roads, alteration of upland habitats, predation, and increased hydroperiod. Anuran species' richness was also significantly lower in breeding ponds in urban landscapes compared to forested and agricultural landscapes in and around Ottawa, Ontario and Gatincau, Québec, Canada (Gagné and Fahrig 2007). These studies are consistent with the findings of Lehtinen et al. (1999), Ficetola and DeBernardi (2004), Riley et al. (2005), and Rubbo and Kiesecker (2005). In an urban area north of Los Angeles, California, when development exceeded 8% of the watershed, exotics increased and native species such as California tree frogs (*Hyla cadaverina*) decreased (Riley et al. 2005).

Although each of these studies related to reptiles and amphibians and urbanization are categorized under biophysical–behavioral forces, the social–structural and valuation forces have led to detailed management plans in at least one area in the United States, Pima County, Arizona. The county wants to minimize impacts of earthmoving, alteration of seasonal waters where amphibians breed, and other negative influences associated with urbanization. To that end, the county surveyed amphibians in the Tucson, Arizona, metropolitan area including distribution and life history characteristic data. When urbanization was to occur, translocation was used to mitigate its negative effects. The suite of 13 amphibian species was translocated in a salvage–rescue translocation project for impacted areas. Over 600 amphibians of 4 species (i.e., Couch's spadefoot [*Scaphiopus couchii*], Mexican spadefoot [*Spea multiplicata*], Great Plains toad [*Bufo cognatus*], and Sonoran Desert toad [*B. alvarius*]) were salvaged and translocated.

The management plan also reviews mosquito ecology and biological control relevant to amphibian conservation to determine how urban wetland communities might be structured to avoid public health hazards. Two approaches are suggested: incorporate mosquito-eating native fish into summer rain-pool ecosystems, and manage for populations of beneficial mosquito-eating tadpoles and aquatic invertebrates.

The program in Pima County, Arizona, also suggests techniques to enhance ecological restoration in infrastructure and parks, and stresses the importance of monitoring (Rosen and Funicelli 2008). This is a comprehensive management plan that if implemented will minimize the influence of urbanization on this sensitive wildlife assemblage.

Biologists and managers concerned with amphibian and reptile conservation need to be aware that relatively limited development of watersheds (≥8%) can alter the habitat and viability for some species; development in the watershed alone, not necessarily in the riparian area itself or even directly upstream, can negatively influence amphibians. Monitoring for amphibians and exotics should be a common practice in landscape development (Riley et al. 2005).

AVIFAUNA

Most of the research on the influence of urban and exurban development related to avifauna is categorized under biophysical–behavioral forces. However, studies range from small areas of interest (Smith and Sharp 2005) to anthropogenic influences across countries (Pidgeon et al. 2007) and continents (Clergeau et al. 1998; Clergeau 2008). Most of the studies examine various aspects of community composition and density along the rural to urban gradient, and others relate to specific impacts (i.e., noise) related to development. Fewer studies provide specific management recommendations (Marzluff and Ewing 2001).

Many of the studies were designed to measure various characteristics associated with avifauna communities and altered landscapes from urbanization and most covered large areas. In all cases, native avifauna benefited from native vegetation in undeveloped land or surrounded by undeveloped lands. In Portland, Oregon (Hennings and Edge 2003), Tucson, Arizona (Germaine et al. 1998), Pitkin County, Colorado (Odell and Knight 2001), Jackson Hole, Wyoming (Smith and Wachob 2006), and Tulsa, Oklahoma (Boren et al. 1999), urbanization influenced avifauna communities.

In Oregon, breeding bird and plant communities were surveyed along 54 streams in the Portland metropolitan region. Total and exotic bird abundance was higher in narrow forests, and native bird abundance was greater in narrow forests surrounded by undeveloped lands. Native species richness and diversity were greater in less-developed areas (Hennings and Edge 2003). Neotropical migrant abundance, richness, and diversity were greater in areas with fewer roads and open-canopied landscapes (Hennings and Edge 2003). The negative influence of roads on avifauna was also documented in Spain by Palomino and Carrascal (2007).

In a gradient study in Tucson, Arizona (Germaine et al. 1998), residential development was beneficial to nonnative birds. Nonnative birds are at a disadvantage when competing for resources in natural habitats with native birds (Green 1984), and their resource requirements are best met in urban environments (Emlen 1974). Native breeding birds increased with increases in the array of land cover types, but Germaine et al. (1998) did not determine how birds were distributed among cover types or how cover types met the resource needs of birds. However, they did determine that anthropogenic influences altered physiognomic, floristic, and spatial habitat alterations that influence bird assemblages (Germaine et al. 1998). If some species (e.g., loggerhead shrikes [*Lanius ludovicianus*]) are left unmolested and sufficient open areas are maintained within the urban environment, the species may not be negatively influenced by urbanization (Boal et al. 2002).

In Pitkin County, Colorado, the avian densities reported were not statistically different between high (1.04 ± 0.67 houses/ha) and low (0.095 ± 0.083 houses/ha) density development but were statistically different from undeveloped sites. The American robin (*Turdus migratorius*), black-billed magpie (*Pica pica*), brown-headed cowbird (*Molothrus ater*), European starling (*Sturnus vulgaris*), house wren (*Troglodytes aedo*), and mountain bluebird (*Sialia currucoides*) had higher densities in developments of high housing density, and eight species had significantly reduced densities in high-density housing densities: black-caped chickadee (*Poecile atricapillus*), blue-gray gnat catcher (*Polioptila caerulea*), black-headed grosbeak (*Pheucticus melanocephalus*), dusky flycatcher (*Empidonax oberholseri*), green-tailed towhee (*Pipilo chlorurus*), orange-crowned warbler (*Vermivora celata*), plumbeous vireo (*Vireo plumbeus*), and Virginia's warbler (*Vermivora virginiae*).

In another study conducted in the West, Smith and Wachob (2006) sampled bird community parameters and habitat variables at microhabitat, macrohabitat, and landscape scales along a residential development gradient within the Snake River riparian corridor in Jackson Hole, Wyoming. Overall, species richness and diversity declined as residential development increased. The most negatively impacted species were neotropical migrants that declined in proportional representation on forested plots as residential development densities increased (Smith and Wachob 2006).

Changes in vegetation cover also altered avian communities in Tulsa, Oklahoma (Boren et al. 1999). The loss of neotropical migrants and increased number of generalist species in high-density rural populations was related to decreased native vegetation, road development, and increased landscape fragmentation. Other studies that documented avifauna changes related to urbanization were consistent with the above works.

In Colorado, grasslands along an urbanizing region at the western edge of the North American Great Plains, rough-legged hawk (*Buteo lagopus*) populations declined by nearly 75% between 1971 and 2003, and red-tailed hawk (*B. jamaicensis*) populations nearly tripled as the human population steadily increased over the 33 years. Rough-legged hawks avoided human development in preference for treeless grassland. Red-tailed hawks adjusted to utility poles and areas closer to buildings and roads, which allowed an increase in population size. The rough-legged hawk and other grassland species such as mountain plover (*Charadrius montanus*), long-billed curlew (*Numenius americanus*), burrowing owl (*Athene cunicularia*), common night hawk (*Chordeiles minor*), loggerhead strike, and lark bunting (*Calamospiza melanocorys*) declined as a result of urbanization and represent the challenge facing managers as landscapes are manipulated (Schmidt and Bock 2005).

Open space grasslands can support sizeable populations of diurnal raptors (e.g., red-tailed hawks, Swainson's hawks [*Buteo swainsoni*], American kestrel [*Falco sparverius*], bald eagles [*Haliaeetus leucocephalus*], golden eagles [*Aquila chrysaetos*]) if prey populations are available. In Denver, Colorado, black-tailed prairie dog (*Cynomys ludovicianus*) towns received heavy use by wintering raptors (Weber 2004). However, some species are highly sensitive to landscape urbanization (e.g., bald eagle, ferruginous hawk [*Buteo regalis*], rough-legged hawk, prairie falcon

[*Falco mexicanus*]) when <8% is urbanized (Berry et al. 1998). Raptor numbers and the number of species were lowest in a study in Washington near Spokane, but both consistently increased as distance from the city center increased (Ferguson 2004). Bird species also declined in riparian woodlands along a gradient of urbanization in Santa Clara Valley, California. Decreases were attributed to increased development (i.e., buildings, bridges), decreased width of riparian habitat, and decreases in native vegetation (Rottenborn 1999). Landscapes directly influenced by urbanization caused a decline in bird density, but lands adjacent to urbanized areas also were influenced (Rottenborn 1999). These responses were documented also in Santa Clara County, California, where bird ecology was examined across a gradient of undisturbed land to highly developed landscapes including biological preserves, recreational areas, golf courses, residential neighborhoods, office parks, and business districts. Birds in undisturbed sites were predominately native species but were replaced by exotic and invasive species in the business district (Blair 1996). These patterns were related to shifts in habitat structure that occurred along the gradient. The area had been dominated by oak (*Quercus* spp.) woodlands. The patterns described by Blair (1996) as woodlands were transformed to urban sites and suggests that any land-use development is detrimental to avifauna. "Even minor perturbations, such as the grazing that formerly occurred in the open-space recreation area, apparently lead to a loss of species. Moreover, species that disappear at the lightly disturbed sites do not reappear at some more highly disturbed site" (Blair 1996, p. 517).

Contrasting bird abundance across continents also demonstrated the influence of human development. Avifauna was compared in the cities of Québec, Canada, and Rennes, France, that had landscapes that changed from rural to urban (Clergeau et al. 1998). Diversity of avifauna declined across the gradient as development increased in both cities. Because the surrounding landscape did not explain the variation of species in the city, Clergeau et al. (1998) suggests that urban environments be viewed as new ecological landscapes rather than degraded environments. If that approach is adopted, birds could be regrouped into two major categories: omnivorous species adopted to the urban environment and its food resources (i.e., garbage), and the species that find resources in the urban environment, normally exploit in their usual habitat (Clergeau et al. 1998).

Most of the researchers that examined bird communities across the rural to urban gradient used a variety of plots randomly placed within the gradient, and systematically documented bird composition and density within plots to obtain their data. These designs are appropriate for relatively small study areas, but other techniques are used for larger landscapes.

To examine forest bird species richness associated with housing across the United States, Pidgen et al. (2007) used the North American Breeding Bird Survey (Sauer et al. 2003) as their data source for bird distribution and abundance. This survey consists of an annual monitoring system that censuses birds on permanent monitoring plots administered by the U.S. Fish and Wildlife Service, and provides data on the relative abundance of birds across the 48 conterminous U.S. states and southern Canada. The survey has been conducted since 1966. Data on humans and development were determined by examining the bounding rectangle that encompassed the survey route. Bounding rectangles were centered on survey routes and developed by extending half the length of the route to define 1,200 km² landscapes. The human development within these boundaries was determined from the 1990 and 2000 U.S. Decennial Census (i.e., human density, housing density, household density, seasonal housing density) and the National Land Cover Data (i.e., landscape composition). From these data, Pidgeon et al. (2007) investigated species richness of all forest birds versus the predictor variables of housing density in 2000 and the abundance of forest, seminatural, and intensive use land cover. These data were used to develop models that showed housing density and residential land cover as significant predictors of forest bird species richness. Results were the same for smaller-scale studies; urbanization decreased species richness.

Other approaches have been used to examine avifauna related to urbanization (i.e., capture–recapture models; Cam et al. 2000), but the results are generally the same; human settlement at some levels may limit avifauna by reducing resources, increasing nest predation, competition for resources, and brood parasitism (Marzluff and Ewing 2001).

However, in exurban situations, some species increase in abundance from increased environment heterogeneity and biotic resources (Fraterrigo and Wiens 2005). More urban studies provided evidence of negative relationships between human development and species richness (Batten 1972; Emlen 1974; Huhtalo and Järvinen 1977; Beissingeo and Osborne 1982; Bezzel 1984; Rapport et al. 1985; Jakimäki and Suhonen 1993; Zalewski 1994; Clergeau et al. 1998). Others provide evidence that intermediate levels of urban development cause peaks in avifauna species richness (Jakimäki and Suhonen 1993; Blair 1996). In the urban-gradient studies, the decline in avifauna was often associated with broad-scale vegetation patterns, and ground-level habitat patterns that were not associated with the development gradient studied in an exurban landscape (Fraterrigo and Wiens 2005). The bird community results of the study in exurban development of north-central Colorado (Fraterrigo and Wiens 2005) were consistent with the concept that "... human settlement can act as an intermediate disturbance on the landscape, ... and that habitat heterogeneity can enhance avian diversity" (Fraterrigo and Wiens 2005, p. 271). Species that increase with increases in building density are habitat generalists (i.e., house sparrows [*Passer domesticus*], common grackles [*Quiscalus quiscula*]) that may benefit from resource supplements provided by development. Habitat specialists did not increase with the exurban environments (Fraterrigo and Wiens 2005).

Urbanization influences life history characteristics of wildlife in numerous ways besides abundance and distribution. Several authors have examined the influence of urban noise (Wood and Yezerimcec 2006), nest predation (Blair 2004), and reproduction (Millsap and Bear 2000; Morrison and Bolger 2002; Thorington and Bowman 2003; Phillips et al. 2005) in relation to urbanization and birds.

Sound pressure levels from anthropogenic activities have the potential to mask bird songs, which could alter the ability of males to attract mates. Song sparrows (*Melospiza melodia*) singing at locations with higher sound pressure levels exhibited higher-frequency low notes and had less amplitude in the low-frequency range of their songs (1–4 kHz) where most anthropogenic sound pressure levels occurred. How this will influence the population is under study but is an environmental variable that warrants further study (Wood and Yezerinac 2006).

Urbanization also influences nest predation rates. Predation causes most nest failures in birds (Thorington and Bowman 2003). In south-central Florida, Thorington and Bowman (2003) examined the influence of nest predation using artificial nests in a suburban matrix. Nest predation was highest at high housing density (<1 house/ha) and lowest at low housing density (<0.5 house/ha); nest predation may increase with human housing density (Thorington and Bowman 2003). Others have documented predation decreasing on artificial nests with increasing urbanization (Blair 2004). However, these data did not reflect the nesting success of birds that did not increase with urbanization. Blair (2004) classified birds along the urban gradient as urban exploiters or urban avoiders from the individual, species, community, landscape, and global perspective. Urban exploiters successfully reproduce, invade locally, have multiple broods, use heterogeneous patches, and ubiquitous species invade, respectively, leading to homogenization of communities. On the other hand, urban avoiders do not successfully reproduce, become locally extinct, have single broods, require homogeneous patches, and require maintenance of native species leading to conservation.

Blair's (2004) generalization needs to be considered cautiously as species and habitats are important considerations when discussing or predicting fragmentation effects. For example, in Southern California, Morrison and Bolger (2002) examined whether rufous-crowned sparrows (*Aimophila ruficeps*) were subjected to higher nest predation at the edge of urban sites compared to sites within the habitat interior from 1997 to 1999. Total reproductive output did not differ between edge and interior in any year. However, wood thrushes (*Hylocichla mustelina*) in southern Ontario (Phillips et al. 2005) and burrowing owls in Florida (Millsap and Bear 2000) had reduced reproductive rates due to urbanization. Wood thrushes breeding in woodlots with embedded houses experienced higher rates of parasitism by brown-headed cowbirds than wood thrushes in woodlots with homes less than 100 m of the forest edge, or undeveloped woodlots. However, nest predation did not increase in developed woodlots in Ottawa (Phillips et al. 2005). The effects of housing developments on wood thrushes may be region specific or depend on cowbird density.

Burrowing owl populations benefited from high prey densities around homes in Florida, but this advantage was offset by increased human-caused nest failures and declines in the number of young fledged. Nest site density of owls increased until 45%–60% of lots were developed before decreasing, and the number of young fledged decreased as development increased above 60% (Millsap and Bear 2000).

Threshold distances from urbanization that avifauna can tolerate without alterations in their life history characteristics would be useful in urban planning, but this has not been studied by many. However, results from a study in central Spain (Palomino and Carrascal 2007) suggested that "As a general rule, the significant threshold distances in the models averaged 400 m for cities, and 300 m for the roads, although these figures varied among different bird populations" (Palomino and Carrascal 2007). These buffer distances in a densely developed landscape matrix suggest severe fragmentation of suitable habitat for native avifauna.

It is not surprising that habitat alteration influences bird communities even in small patches such as gardens. Gardens were examined in ten suburbs of Hobart, Tasmania, Australia (Daniels and Kirkpatrick 2006) and, although use varied, native birds used exotic plants but exotic birds largely used exotic plants. Because gardens can be designed and managed to favor particular species, gardeners have a potential role in the conservation of native birds. However, larger landscape alterations such as golf courses may not be as beneficial for native birds. Number of species, number of neotropical migrant species, and degree of conservation concern of the species present were higher in less-altered golf courses in coastal South Carolina and were significantly influenced by percent of forested area (Jones et al. 2005). Enhancing avian habitat is possible through increasing the amount of forest and reducing the amount of managed turf grass (Jones et al. 2005). Others have also demonstrated that the size of the area considered, amount of natural vegetation, and percent of urbanization and natural land are all important factors in explaining winter bird use in Ontario, Canada (Smith 2007). Unfortunately, estimates of threshold sizes for habitat islands have not been determined (Beissinger and Osborne 1982). It has been established, however, that the habitat structure and population-suppressing factors in developed areas create habitat for only a few bird species (Beissinger and Osborne 1982). Fortunately, when some altered landscapes (i.e., grasslands) are restored, the habitats they provide contain bird assemblages similar to those in native prairie habitats, suggesting that restored grasslands may provide similar habitat for most grassland birds (Fletcher and Koford 2002).

Management for Avifauna in Altered Landscapes

The problems with urbanization and avifauna have been clearly established, but fewer recommendations have been made as to how urban, suburban, and exurban centers can exist and minimize their influence on birds. Many authors who address avifauna in urban landscapes provide suggestions to enhance native species, but they are unproven suggestions that may or may not improve avian assemblages in the face of landscape alteration. Suggestions for management include educating residents about their impacts (Millsap and Bear 2000; Fraterrigo and Wiens 2005), developing buffer zones around sensitive areas (Millsap and Bear 2000; Rottenborn 1999), maintaining native plant communities (Germaine et al. 1998; Boren et al. 1999), maintaining and enhancing riparian corridors (Germaine et al. 1998; Hennings and Edge 2003), maintaining native patches of vegetation >1 ha, and managing for sensitive species (Germaine et al. 1998). Because natural landscapes are important, Blair (1996) and Odell and Knight (2001) recommend that cluster developments will have less of an impact than development that traverses the landscape. Clustered development has been suggested as an important conservation tool to maintain and enhance wildlife habitats. Theobald et al. (1997) conducted a study in Summit County, Colorado, which demonstrated that clustered developments reduce the negative impacts on wildlife habitat. However, this is a new concept that has received limited attention. Others contrasted the influences of clustered and dispersed housing (1 house/2–16 ha) with undeveloped areas in Boulder, Colorado (Lenth et al. 2006). They contrasted densities of songbirds, nest density, and survival of ground-nesting birds, presence of mammals, and percent cover and proportion of native and exotic flora as indicators to assess conservation value. Both types of housing had significantly higher densities of exotic and human-commensal species (e.g., dogs, common grackles, European starlings, American robins, red-winged blackbirds, doves, killdeer [*Charadrius vociferus*]) and significantly lower densities of native and human-sensitive species (e.g., western meadowlarks [*Sturnella neglecta*], vesper sparrows [*Pooelcetes gramineus*], field mice [*Peromyscus* spp.]) than undeveloped areas. Additional research is needed to determine the ecological value of clustered housing on broader scales (Lenth et al. 2006).

Because the majority of the nation's forests are in private ownership, national conservation plans need to incorporate housing in their management strategies. "The recent development of several conservation plans at the national and international scale demonstrates the general buy-in of the conservation community to this idea" (Pidgeon et al. 2007, p. 2008).

To maximize golf courses as habitat for avifauna, wildlife should be considered before development begins to increase forest patches, reduce managed turf grass, maintain as much native vegetation as possible, plant native vegetation in disturbed sites, reduce mowing, and maintain natural waters (Jones et al. 2005).

Clearly, management needs to be considered, and Marzluff and Ewing (2001) offer 15 specific recommendations that would improve the suitability of reserves for birds.

1. Increase the foliage height diversity within fragments.
2. Maintain native vegetation and deadwood in the fragment.
3. Manage the landscape surrounding the fragment (matrix), not just the fragment.
4. Design buffers that reduce penetration of undesirable agents from the matrix.
5. Recognize that human activity is not compatible with interior conditions.
6. Make the matrix more like the native habitat fragments.
7. Actively manage mammal populations in fragments.
8. Discourage open lawns on public and private property.
9. Provide statutory recognition of the value of complexes of small watersheds.
10. Integrate urban parks into the native habitat reserve system.
11. Anticipate urbanization and see creative ways to increase native habitat and manage it collectively.
12. Reduce the growing effects of urbanization on once-remote natural areas.
13. Realize that fragments may be best suited to conserve only a few species.
14. Develop monitoring programs that monitor fitness.
15. Develop a new educational paradigm.

Eight research needs are suggested by Marzluff and Ewing (2001) to guide the way toward enhancing habitats for birds in urban areas.

1. Are corridors used by dispersing birds and do they facilitate the functioning of metapopulations?
2. Does increasing native vegetation of the matrix help?
3. How does the pattern of housing affect avian population viability in surrounding fragments?
4. How do we design effective buffers that shield birds in fragments from the disturbance of the matrix?
5. How does urbanization affect insect communities?

6. Is it possible to use some nonnative plant species to reduce invasions by species known to be disruptive to ecosystem function?
7. Can fragments of native habitat in urbanized landscapes make tenable contributions to avian conservation?
8. What are effective means of encouraging citizens to conserve birds and their habitats and reduce their impacts?

MAMMALS

Small Mammals

Urbanization and exurbanization vary in their influence on small mammals. Cottontail rabbits (*Sylvilagus audubonii, S. floridanus*) were counted in plots grazed by livestock embedded in housing developments, or both, or neither in southwestern Arizona (Bock et al. 2006a). The number of cottontails documented was positively correlated with the number of homes near plots. Cottontails benefited from exurban development due to increased cover provided by structures and landscaping especially in open grasslands with limited natural cover (Bock et al. 2006a). Rodents, however, were negatively influenced by livestock grazing or its effects on vegetative ground cover. Exurban development did not influence rodent abundance or diversity. If housing densities are low and embedded in a matrix of natural vegetation with limited grazing, exurban development can maintain rich assemblages of grassland and savannah rodents (Bock et al. 2006b). In contrast, native rodents as a group were captured more often in interior plots compared to edge plots in Boulder, Colorado. Plots that had the highest capture rates of native species were in landscapes <10% suburbanized. The authors concluded that proximity to suburban landscapes had a negative influence on the abundance of native rodents in open-space grasslands (Bock et al. 2002).

The study was not designed to understand why rodents were relatively scarce near suburban edges on open-space grasslands. However, they did eliminate habitat, competition, and patch size as likely causes. Probable causes for the differences reported are predation by domestic cats (Bock et al. 2002).

Species abundance and richness has been the most common metric used to document how urbanization influences small mammals. Behavior has also been used to measure separate and combined impacts of anthropogenic actions at the individual and population levels. With the use of giving up densities, Bowers and Breland (1996) determined that gray squirrels (*Sciurus carolinensis*) living close to humans were more limited by food or less sensitive to risk of predation than squirrels in more natural areas.

In Pennsylvania, small mammal (e.g., eastern chipmunks [*Tamias striatus*], white-footed mice [*Peromyscus leucopus*], deer mice [*P. maniculatus*], and woodland jumping mice [*Napaeozapus insignis*]) species richness was lowest in parks containing manicured landscapes surrounded by human-modified landscapes. Only 1–2 species of small mammals were in mowed landscapes (Mahan and O'Connell 2005). These authors recommended leaving strips of 10 to 15 m that are not mowed along streams and planting native trees along stream corridors to encourage small mammals in suburban and urban parks. Similar results have been reported worldwide. For example, in Melbourne, Australia, small ground-dwelling mammals were the group of mammals most negatively affected by urbanization; only 2 out of 15 species has a >10% probability of persisting (van der Ree and McCarthy 2005). Because Melbourne will continue to expand, van der Ree and McCarthy (2005) recommend that state and local governments design and adopt a comprehensive strategy to manage habitat networks with cross-jurisdictional boundaries and include Melbourne to enhance viability of mammalian populations.

Chiroptera

Many of the studies designed to examine the influence of human development on bats have approached the question by contrasting bat use of habitat within rural to urban gradients. In residential areas in California, *Myotis yumanensis* preferred large trees ($\overline{\chi}$ diameter = 115 cm) for roosting, close to water with forest cover in the surrounding 100-m radius. As urbanization encroaches into these areas, managers need to preserve large trees and forested parkland, especially along stream corridors to help maintain bat populations in urbanizing landscapes (Evelyn et al. 2004). Indiana bats (*Myotis sodalis*), a federally listed endangered bat, also prefer woodlands over developed habitats for roosting and foraging (Sparks et al. 2005).

Compared to agricultural landscapes, urban habitats were not used as much for foraging by evening bats (*Nycticeius humeralis*) or big brown bats (*Eptesicus fuscus*) in Indiana. However, big brown bats roosted in human-made structures (i.e., urbanization), but evening bats roosted in tree cavities in woodlots. Evening bats are likely more sensitive to suburban development near roosts than big brown bats (Duchamp et al. 2004).

Urban settings do not always represent negative habitats for bats. Five species of bats and the *Myotis* group were monitored in the Chicago metropolitan area. All bats identified exhibited positive relationships within the urban matrix (Gehrt and Chelsvig 2004). They were also detected more frequently in urban areas than in more rural habitat fragments. "Urban areas may represent islands of habitat for some bats within larger landscapes dominated by intensive agriculture. Thus, the nature of the relationship between urbanization and bats is probably dependent on context at the macrogeographical scale and local habitat quality within the urban matrix" (Gehrt and Chelsvig 2004, p. 625).

Mesopredators

Mesopredators are common urban inhabitants, but little is known about their demographic response to urbanization (Prange and Gehrt 2004). Because raccoons exploit anthropogenic resources efficiently, they have become common residents of urban areas (Prange and Gehrt 2004). Opossums (*Didelphis virginiana*) and striped skunks (*Mephitis mephitis*) were also common in developed areas. In northeastern Illinois, density of raccoons was greater in urban plots than rural plots throughout the year and, at urbanized sites, raccoons may have had larger litter sizes. Adult female survival was also higher in urban sites in the absence of disease (Prange et al. 2003).

Raccoons in urban areas had fewer mortality sources (disease was greatest), and those residing in suburban and rural sites had the most (roadkills were the greatest). There also may have been greater site fidelity at urbanized sites (Prange et al. 2003). Because raccoons are capable of quickly repopulating an area after the resident population has been reduced, control methods will have to be continuous.

As raccoons and other mesocarnivores become abundant, society often views them as pests. They create nuisance-related problems and may transmit diseases and parasites to humans and domestic animals. Increased survival, higher annual recruitment, and increased site fidelity contribute to increased densities of raccoons in urbanized areas. To efficiently manage the population, direct control measures (e.g., trapping and removal) and reduction or elimination of anthropogenic food sources will be required (Prange et al. 2003). However, direct removal of overabundant species may not be economically feasible and could be socially unacceptable (Goodrich and Buskirk 1995).

Domestic cats are an often overlooked influence on fauna in urban, suburban, and exurban landscapes, but they have a significant impact on native rodents and birds (Hawkins et al. 2004). In one study in southeastern Michigan, 25% of landowners owned outdoor cats, which killed at least 1.4 birds/cat/week. Over 23 species were killed including species of conservation concern (Lepczyk et al. 2004). There were at least 3,100 cats across the landscape that killed at least 47,000 birds during the breeding season (~1 bird killed/km/day). Even taken conservatively, these data (Lepczyk et al. 2004) suggest that cat predation plays an important role in fluctuations of bird populations.

Devices have been developed that can reduce the efficiency of domestic cats as predators (i.e., collar-mounted pounce protectors, the Cat-Bib™). However, they need to be worn consistently by cats and most cat owners were reluctant to maintain this cat apparel (Calver et al. 2007).

Overabundant wildlife and cats in urbanized landscapes are clearly an increasing problem. More effective programs have to be developed for the management of urbanized species (DeStefano and DeGraaf 2003).

Larger Predators

The Problem

Addressing the influence of suburban and exurban development on other large predators (e.g., wolves [*Canis lupus*], mountain lions [*Puma concolor*], wolverines [*Gulo gulo*], grizzly bears [*Ursus arctos horribilis*]) has not generated as much concern as other predators because they are primarily wilderness species (Leopold 1933) that do not benefit from close associations with humans. However, as the human population increases, cumulative effects increase, which continually alters landscapes; thus, habitats for larger predators are influenced. Successful translocations of larger predators into historic habitats will be challenging especially when they encroach upon exurban developments. Successful conservation is compounded because larger predators require large landscapes that often traverse public and private lands and international borders (i.e., gray wolf, jaguar [*Panthera onca*]). Maintaining habitats and connectivity between habitats is challenging (Harrison and Chapin 1998).

For example, a highway in Spain is a significant barrier to wolf movement, which may be isolating two subpopulations (Rodriguez-Freire and Cricente-Maseda 2008). An interstate freeway through Tucson, Arizona, also serves as a barrier for mountain lion movement (K. Nicholson, graduate research assistant, and P.R. Krausman, unpublished data). Mountain lions in California that dispersed used corridors associated with natural travel routes, cover, and underpasses to avoid high-speed road crossings and limited artificial lighting with less than 1 house/16 ha (Beier 1995).

When examining resiliency of larger predators, Weaver et al. (1996) examined plasticity in diet and food availability, demographic compensation to mitigate increased exploitation, and dispersal to maintain connectivity with fragmented populations. Wolves "possess resiliency to modest levels of human disturbance of habitat and populations" (Weaver et al. 1996, p. 964). Grizzly bears and wolverines possess less resilience because of requirements for forage, low productivity, and philopatry of females to maternal home ranges (grizzly bears). Humans have altered these life history characteristics causing widespread declines and, as exurban development spreads, conflicts will increase.

Controversy escalates as predators kill livestock or threaten human safety (Geist 2008). As a result, numerous authors call for management by maintaining refugia on public and private lands (Carroll et al. 2003) that encompass the full spectrum of required habitats that are connected to other refugia through landscape linkages (Weaver et al. 1996).

Valuation and institutional-regulatory forces have provided the mechanism for restoration of larger predators, and Rasker and Hackman (1996) proposed the hypotheses that the protection of refugia that sustains wild carnivores does not have a detrimental effect on local or regional economics. Environmental protection and economic development are complementary goals (Rasker and Hackman 1996). Agencies, organizations, and the public need to work together for successful conservation (Harrison and Chapin 1998), especially of larger carnivores.

Influences of exurban development have not been examined as thoroughly with larger predators, but lessons learned from other conflicts can certainly be used as models as these species and human habitats merge on the landscapes.

Coyotes

The Problem

Coyotes have established populations in habitats dominated by humans across North America (Atkinson and Shackleton 1991) likely because of habitats and unnatural food provided. Establishment and increases in population sizes are especially prevalent in suburban and exurban areas that contribute to habitat fragmentation. Unfortunately, life history characteristics of coyotes in human-dominated landscapes are not well documented but, when coyotes and humans use shared habitats, attacks on humans increase, creating concern for human safety and property. Similar problems exist for American black bears (*Ursus americanus*) and other carnivores often because of human encroachment into their habitat. Anthropogenic development often creates habitat for coyotes and serves as an undesired attractant.

Much of the available literature on coyotes addresses and contrasts their life history characteristics with coyotes in rural areas and wildlands, and addresses problems coyotes cause in human-dominated landscapes. Few studies have thoroughly investigated mechanisms to minimize coyote–human interactions.

Biophysical-Behavioral Forces

Information about coyotes in urban settings has increased from limited descriptions and observations (Gill 1965; Andelt and Mahan 1980) over 3 decades ago to more detailed studies of habitat, activity, survival, diet, and reproduction in human-dominated habitats. Most of these studies are strictly biological and provide limited data as to how problems between coyotes and humans can be minimized or mitigated.

Home-Range Sizes

Most studies of home-range size in animals report varying results due to techniques used to calculate sizes, habitat, sample size, and other factors (Table 9.1). The 95% minimum convex polygon home-range size of resident coyotes in western North America varies from 1.1 km^2 in urban areas, Los Angeles, California (Shargo 1988), to 118 km^2 in Washington. Mean home-range sizes for coyotes in habitats that include suburban and exurban landscapes ranged from 7.7 ± 4.2 (SE) km^2 in Canada (Atkinson and Shackleton 1991) to 26.8 ± 5.1 km^2 in Tucson, Arizona (Grubbs and Krausman 2009a).

Habitats in Urban Areas

Although coyotes continue to increase in urban areas and expand their range, they still heavily rely on cover. Of the several habitat studies of coyotes in urban areas conducted, most have involved monitoring radio-collared animals, but others have used public observations and scent stations to evaluate habitats. Broader studies have examined carnivores across a spectrum of habitats that essentially constitute an urban-to-rural gradient (Crooks 2002; Randa and Yunger 2006).

In an examination of 29 urban habitat fragments in coastal southern California ranging from those surrounded by human-modified landscapes to mesa-top habitat and others dominated with ornamental plants, Crooks (2002) examined carnivore distribution based on track plots. Coyotes were present in 26 of 29 sites. The estimated area for detecting a 50% probability of occurrence was 1 ha; larger-bodied carnivores require areas that will eventually disappear if not connected to other patches (Crooks 2002). The only patches examined that did not have coyotes were 12 and 2 ha in size. Due to their behavioral plasticity, allowing coyotes to exist in a variety of disturbed sites limits their utility as an indicator of connectivity (Crooks 2002).

In a similar study, Randa and Yunger (2006) examined carnivores across 47 sites in a rural–urban gradient in the Chicago metropolitan area counting carnivore tracks at scent stations. Sites varied from those with higher amounts of human influences (e.g., people, industry, commercial, high-road densities; urban) to those further removed from the city (e.g., agriculture; rural). Coyotes were common across the entire gradient: in 6 of 18 urban sites and 18 of 29 rural sites and in all habitats (i.e., mowed lawn, nonnative grassland, old field, prairie, row crops, shrubland, woodland, woodland edge) except lawn (Randa and Yunger 2006). The ability of coyotes

TABLE 9.1
Mean Home-Range Size of Resident Coyotes That Use Suburban and Exurban Landscapes in Canada and the United States

Home-Range Size (km^2)	Method of Calculation[a]	Sex	Location	Landscape	Source
7.7 ± 4.2 SE)	95% MCP	M	Lower Fraser Valley, BC, Canada	Rural–urban gradient	Atkinson and Shackleton (1991)
17.0 ± 20.7	95% MCP	F	Lower Fraser Valley, BC, Canada	Rural–urban gradient	Atkinson and Shackleton (1991)
10.8 ± 11.2	95% MCP	M/F	Lower Fraser Valley, BC, Canada	Rural–urban gradient	Atkinson and Shackleton (1991)
12.6 ± 3.5	95% MCP	M/F	Tucson, Arizona	Rural–urban gradient	Grinder and Krausman (2001)
26.8 ± 5.1	95% FK	M/F	Tucson, Arizona	Urban	Grubbs and Krausman (2009*a*)
15.4 ± 6.3	MCP	M/F	Tucson, Arizona	Suburban	Bounds and Shaw (1997)
2.97–23.48	ADK	M/F	West-central Indiana	Rural–urban gradient	Atwood et al. (2004)

[a] MCP = minimum convex polygon, FK = fixed kernel, ADK = adaptive kernel method.

to make large-scale movements has likely usurped the effects of habitat fragmentation (Randa and Yunger 2006).

Results of these broad-scale studies (Crooks 2002; Randa and Yunger 2006) are generally supported by specific examinations of coyote habitats in urban areas. In a series of papers, Quinn (1991, 1995, 1997) reported that coyotes in Seattle, Washington, preferred undisturbed habitats and used forest and shrub habitat for hiding cover, and used densely mixed vegetation (with forest and shrub vegetation) more than other habitats. These areas were adjacent to rural areas, provided escape, prey, and cover. Quinn (1991, 1995, 1997b) used radio-collared coyotes and public sightings to describe coyote habitat associated with urbanization due to the abundant patches of vegetation.

In west-central Indiana, Atwood et al. (2004) examined habitat use of coyotes along a rural–exurban–suburban gradient consisting of forest, grassland, fencerows, grassy drainage ditches, agriculture matrix, and human development including commercial and residential. Twelve percent of the 250 km^2 study area included human developments with 70% in agriculture production. Within their home ranges, coyotes used fence, ditch, and grassland. Forested habitat was used more than the agriculture or urban matrix. All coyotes preferred corridors when present (Atwood et al. 2004).

Habitat of coyotes in Tucson, Arizona, was examined in two different studies (Grinder and Krausman 2001; Grubbs and Krausman 2009b). In the first study, coyotes were captured and radio-collared throughout the city including areas adjacent to wildlands (Grinder and Krausman 2001). In this situation, coyotes concentrated in >30% of at least one of three habitats: natural areas (i.e., state and federal parks, private open space, and cropland; less than 1 house/ha), parks (i.e., schools, military grounds, cemeteries, zoos, golf courses, small parks, stables), and residential areas (i.e., more than 1 house/ha). These patch types contained food, cover, and breeding sites.

The study of Grubbs and Krausman (2009a) used different habitat classifications because all collared animals were entirely within the city and did not use adjacent wildlands. These coyotes used washes, medium- (2–7 residences/ha), or low-density (less than 1 residence/ha) residential areas more than expected by chance alone. All coyotes avoided high-density residential areas (greater than 7 residences/ha), commercial areas, and roads. Washes and medium-density residential areas offer an abundance of shade and cover—important items in times of high human activity. The presence of washes throughout Tucson, and especially in dense residential areas, is likely a reason coyotes persist throughout the city (Grubbs and Krausman 2009a).

Diet

Coyotes in urban areas exhibit the same catholic diets as animals in wildlands. In the Lower Fraser Valley, British Columbia, coyotes primarily consumed small rodents (70.2%), lagomorphs (8.12%), and other mammals (4.7%) (i.e., raccoon, opossum, muskrat [*Ondatra zibethicus*], and black-tailed deer [*Odocoileus hemionus columbianus*]). They also consumed birds (2.0%) and domestic stock (4.3%) (i.e., sheep, cattle, pigs, chickens). Plant material (10.3%) included plums, apples, grasses, and holly. Other items documented in the diet were insects, paper, cloth, plastic, and rubber (Atkinson and Shackleton 1991).

Across a rural to urban gradient in Western Washington, Quinn (1997a) reported changed foods, particularly mammals, as human-caused land-use patterns changed. Fruits and mammals represented the largest classes of food items in all habitats. Voles (*Microtus* spp.) were the most abundant mammalian food item (41.7%) in mixed agricultural–resident habitat, which shifted to house cat (13.1%) and squirrel (7.8%) in residential areas. Cats were a common source of food for coyotes in Tucson, Arizona, also (Grubbs and Krausman 2009b).

Coyote diets were also examined across a rural–urban gradient in the Santa Monica Mountains, California (Fedriani et al. 2001), and their data supported results of Quinn (1997a); consumption of anthropogenic foods by coyotes varied according to human density. The area with more people provided 14%–25% of diet to coyotes as anthropogenic foods (i.e., trash, livestock, and domestic fruit). The area with the fewest people provided 0%–3% of items, and the intermediate level of human activity provided 4%–6% of anthropogenic foods. Similar findings were reported by McClure et al. (1995) in Tucson, Arizona; anthropogenic foods replaced natural foods. Others also documented increases in anthropogenic foods in coyotes using urban areas (MacCracken 1982; Shargo 1998).

The diets of coyotes in urban areas have been studied rarely, possibly because tools to discriminate coyote feces from those of domestic dogs have not been available. Criteria that have been used to discriminate between dog and coyote scats include size (i.e., scats more than 5 mm in diameter as coyote; Atkinson and Shackleton 1991) and collecting scats in areas of relatively high coyote density (Quinn 1992, 1995, 1997a). However, these techniques are not accurate (Krausman et al. 2006) and, to ensure accuracy, DNA analysis should be used. The role that available food plays in foraging strategies of coyotes needs to be examined thoroughly to understand its ecology and relationships to dogs and humans in urban areas. DNA analysis used for scat identification provides a mechanism to distinguish canid scats (Krausman et al. 2006).

Activity

Activity patterns are plastic and not catholic. Some researchers documented coyote activity to be crepuscular with extensive nocturnal movements (Gipson and Sealander 1972; Andelt and Gipson 1979; Laundre and Keller 1981; Woodruff and Keller 1982). Others have documented coyote activity to peak at different times of day and night (Major and Sherburne 1987; Morton 1988; Brundige 1993; Patterson et al. 1999). Activity patterns are dependent on a variety of factors; Atwood et al. (2004) suggested that diel activity was typical of unexploited coyote populations. Others studying coyotes in urban areas documented a shift in activity to periods when prey is more active (Atkinson and Shackleton 1991; i.e., nocturnal). This was also observed in Jackson Hole, Wyoming, when coyote activity was contrasted between suburban and agricultural landscapes and undeveloped areas. Coyotes in the suburban and agricultural areas reduced activity during diurnal periods and increased activity during crepuscular and nocturnal periods (McClennen et al. 2001). Nocturnal activity increased in Saguaro National Park, Tucson, Arizona, likely due to easy access to anthropogenic foods (Bounds and Shaw 1997). The general shifts in activity in urban areas compared to rural landscapes have been suggested as a response to human activity. Humans are generally more active during the early hours of evening than at other times of the night and remain inactive longer than coyotes in more rural areas to avoid contact with humans (Grinder and Krausman 2001; McClennen et al. 2001; Atwood et al. 2004; Grubbs and Krausman 2009a). For example, in Tucson, Arizona, when human activity was high and coyote activity low in mid-day, coyotes were in areas of low human activity with cover. As evening approached and human activity decreased, coyotes moved to golf courses, washes, and residential areas. As night progressed, coyote activity increased, and they moved through high-density residential and commercial areas. As dawn approached, human activity increased, coyote activity slowed, and they returned to familiar day resting spots (Grubbs and Krausman 2009a). Others, however, have suggested that coyotes in Banff, Alberta, Canada, used habitats available to them regardless of human activity (Gibeau 1998). Gibeau (1998) concluded that coyotes were not attracted to urban environments.

Survival

Very few studies have examined survival of coyotes in exurban–suburban landscapes. In Tucson, Arizona, the annual survival rate of coyotes was 0.72. They were exposed to viral, bacterial, and parasitic infections common to many coyote populations, but humans were the major source of mortality (95%): vehicles ($n = 16$), trapping ($n = 2$), unknown ($n = 1$); Grinder and Krausman (2001). Human-caused mortality was also common in a later study of coyotes in Tucson, Arizona. Of eight monitored animals, three were killed by vehicles, one drowned, and two died of unknown causes (Grubbs and Krausman 2009a). In Southern California, the survival rate for coyotes was 0.742 and did not vary with urban association (Riley et al. 2003). Vehicles, other carnivores, and anticoagulant rodenticides were the main causes of death for coyotes. Toxincaused mortality was related to urban association. Rat poison was also responsible for the death of coyotes in the north edge of Boston, Massachusetts (Way et al. 2006).

Human–Coyote Conflicts

As urbanization encroaches into coyote habitat and vice versa, an undesirable human interface with coyotes has resulted (Howell 1982). Numerous attacks on adults, children, and pets have been reported from homes, yards, parks, and golf courses (Baker and Timm 1998). The most severe was in August 1981 when a 3-year-old girl in Glendale, California, was killed in her front yard by a coyote. These attacks and tragic deaths, plus hundreds of observations of coyotes in urban areas, caused valuation forces to demand action. The City of Glendale, California, provides a successful plan of action to minimize coyote–human interaction (Howell 1982).

Case Study: Glendale, California

The incidence of coyotes in urban areas is enhanced by humans. Humans modify habitats to provide prey for carnivores and subsidize diets with anthropogenic foods (Baker and Timm 1998) and provide cover and den sites (Grubbs and Krausman 2009a). When coyotes and humans clashed in Glendale, California, the County Board of Supervisors authorized the Agricultural Commissioner to contract with the City of Glendale to selectively trap and remove coyotes where conflicts occurred. In addition to direct removal, the commissioner initiated a public education program to minimize human–coyote conflicts. Furthermore, an ordinance was passed that prohibited feeding predatory animals and rodents in unincorporated areas of Los Angeles County.

The program had early success. In 80 trap days within a 0.8-km radius of the residence where the child was killed, 55 coyotes were trapped or shot (Howell 1982). The City of Glendale still has one of the best deterrent programs in the country. It is currently administered by the police department based on coexistence as the only long-term solution for coyote–human conflicts. When problems are identified, citizens are educated about fencing, habitat modification, human and wildlife behavior, coyote biology, city wildlife antifeeding ordinances, and the use of pepper spray. Trapping and killing coyotes are used when all recommended solutions fail. Success is attributed to reinstating the fear of humans and urban areas into coyotes (Baker and Timm 1988). All trapped coyotes (No. 3 Victor Soft Catch) are euthanized; relocation is not biologically sound, humane, or sometimes legal (Baker and Timm 1998). When translocation is attempted, it is usually to pacify opposing views of control: those who demand that something be done and those who are against control. Regardless, the issue with translocation is that the problem is also being translocated. Furthermore, translocated animals may create intraspecific conflict and disease transmission. The program in Glendale, California, has been successful; attacks and pet losses have been reduced.

The City of Glendale also informs citizens about coyote behavior characteristics and human safety risks (Baker and Timm 1998).

1. An increase of lost pets at night.
2. An increase in observations of coyotes on streets and yards.

3. Crepuscular observations of coyotes on streets, parks, and yards.
4. Observations of coyotes chasing or killing pets.
5. Attacking leashed pets, chasing joggers, bikers, and walkers.
6. Observing coyotes in and around play areas and parks in midday.

Management

The emerging picture of coyotes in urban areas suggests that they fulfill their daily requirements by shifting activity periods to times when humans are least active and by using areas where coyotes can avoid humans. Unfortunately, many urban dwellers enjoy seeing coyotes as part of their daily life (Kellert 1985) and often feed them directly or indirectly. Feeding causes attacks. Even the small girl killed by coyotes in California had been encouraged by her father to feed coyotes (Grinder and Krausman 1998). From a survey of 188 U.S. national parks, Bounds and Shaw (1994) reported that, where aggressive coyotes were reported, feeding by humans was significantly more commonplace than in parks without aggressive coyotes. Altering human behavior (Peine 2001) and public education is necessary to prevent wildlife–human conflicts (Timm et al. 2004).

Coyotes likely benefit from habitat alteration of urban areas (Grinder and Krausman 2001). Thus, in fragmented landscapes, cover likely facilitates occupancy of areas with relatively high human activity (Atwood et al. 2004). For example, forested habitat provides cover, forage, and travel corridors (Atwood et al. 2004). Society needs to understand how habitat modification to suburban development can progress without increasing mortality or dispersal rates (if that is a desired objective; McClennen et al. 2001). When the decision is made to coexist with coyotes, society will need more information about coyotes in urban areas, so conservation can be promoted by preserving open space, reducing use of rodenticides, providing usable crossing points under freeways and roads, and driving more slowly (Riley et al. 2003).

However, when problems arise, lethal control, education, hazing, and other methods have reduced conflicts (Timm et al. 2004). The Arizona Game and Fish Department recommends 12 ways to discourage coyotes in urban areas.

1. Do not feed coyotes.
2. Eliminate sources of water for coyotes.
3. Remove bird seed.
4. Edible garbage should not be available.
5. Do not keep pet food where coyotes have access.
6. Trim and clean shrubbery to the ground that provides cover for coyotes.
7. Fence the yard.
8. Install a battery-operated electric fence.
9. Do not leave small children outside unattended.
10. Keep pets restrained.
11. Assertively discourage coyotes by making loud noises and throwing rocks when they appear.
12. Ask neighbors to follow the same steps.

A further step could be broader-scale educational programs. In Vancouver, British Columbia, Canada, the Stanley Park Ecology Program (http://www.stanleyparkecology.ca/) has been a useful tool to reduce human–coyote conflicts.

American Black Bears

The Problem

As society encroaches further into habitats of American black bears, there are numerous opportunities for human–bear conflicts because bears are common in the United States and are increasing in many areas, can tolerate humans, and are attracted to human food (Spencer et al. 2007). As a result, American black bears have presented a serious challenge to biologists, especially where anthropogenic activities occur within bear habitat. An overview about bear populations, levels of complaints about bears, the type of human–bear interactions, management strategies, and documentation was provided by Spencer et al. (2007) via a mail survey completed by American black bear biologists. The survey was sent to wildlife agencies throughout North America. Although few details were provided about the survey design, Spencer et al. (2007) received responses from all 39 states surveyed (states without bears were not surveyed, and 8 Canadian provinces and Mexico responded to the survey). Most responding agencies (44 of 48; 91.6%) provided estimates of black bear population size (n = 747,000). All agencies reported bear complaints. Garbage and food attractants were the most common types of human–bear conflict followed by general sightings. Agriculture damage (i.e., apiary, orchard, crops) and human encounters were similar in rank. Attacks by bears on humans and livestock were the least common type of conflict reported. In Montana, human–black bear conflicts decreased from garbage and food attraction to human attacks: garbage and food attraction, human–black bear encounters, agriculture damage, general sightings, campsite encounters, livestock attacks, and human attacks. In Montana, 67% of human–black bear conflicts occurred in spring and summer. Most agencies (77%) did not have a damage fund to pay for the losses due to black bears (Spencer et al. 2007).

Most agencies (89%) had defined protocols for field personnel when responding to human–bear conflicts: site visit, capture and relocation, and euthanasia. Kill permits and use of hunters were ranked the lowest means in protocols. When trapped, culvert traps were the most common traps used followed by leg-hold snares and use of dogs.

Most agencies (75%) relocate problem bears; public pressure drives this decision many times (44%). Few agencies (15%) reported that relocation was the best management approach. On-site release was used by 42% of respondents; of these 65% of respondents maintained a database of released bears.

Most agencies marked some captured bears, but only 50% marked captured bears all the time to monitor results of aversive conditioning and relocation. Aversive conditioning (i.e., rubber bullets and loud noises) was a common tool used on bears involved in conflict (64%). Bear-resistant containers were used by 50% of respondents, and 8 of 24 respondents used agency funds to purchase them.

States, policies, or laws allowing fines for creating depredation situations were reported by 47% of respondents. Garbage management and fines were preferred deterrents to black bear problems over aversive conditioning and relocation. Education programs (i.e., brochures, press releases, radio, TV, workshops) were reported for 81% of agencies.

The survey was instrumental in providing an overview of American black bear issues in North America. Based on these data, Spencer et al. (2007) recommended that agency responses to human–black bear conflicts can be strengthened in at least three ways.

1. Develop protocols for marking and monitoring all bears captured related to conflicts and maintain a database.
2. Develop proactive garbage management.
3. Develop effective education programs.

Deterrents to Reduce Human–Black Bear Conflicts

Removal of Unnatural Food

Garbage and food attractants were the most common type of human–black bear conflicts reported in North America (Spencer et al. 2007). Most wildlife biologists agree that, by eliminating unnatural foods, human–black bear conflicts will be minimized (McCullough 1982; Breck et al. 2006). Bear-proof garbage cans, sanitation, and only putting garbage out on collection day are obvious solutions to minimize this problem, but the problems persist. As a result, other deterrents have been developed to keep bears from concentrated sources of food.

Translocation

Capturing and translocating problem bears is a common wildlife management practice in North America (Spencer et al. 2007). However, the process is generally expensive, time consuming, not always successful, and does not address the situation that caused the nuisance behavior (McArthur 1981). In a casual review of 179 American black bears (more than 2 years old) translocated in 11 states and provinces, Rogers (1986) reported that 20% of bears returned to the capture site (within 8–20 km) when translocated more than 220 km. When translocated less than 64 km, 81% returned; when translocated 64–120 km, 48% returned; and when translocated more than 120–220 km, 33% returned to the capture site. Of documented sexes, 54% ($n = 36$) of males and 70% ($n = 32$) of females returned. Translocation did not greatly increase natural mortality among translocations summarized by Rogers (1986).

An advantage of translocating nuisance bears was the removal of the offending individual. These same individuals resumed their negative activities less than 65% of the time but usually the following year. This allowed some animals to breed and preserved many for fall hunting. Bears shot during the season were used for food or trophies, but few bears killed as nuisances were used for either.

In 1967–1977, 112 American black bears were captured and translocated in Glacier National Park (McArthur 1981). Transplants were considered unsuccessful if the bear returned to the capture area. Other results were recurring problems by the translocated bear elsewhere or being killed outside the park. Thirty-nine bears were transplanted successfully on the first attempt and 19 on subsequent transplants. Females were more likely than males to return to the capture area, and males were more likely to create problems elsewhere (McArthur 1981). Factors that contributed to successful transplants included timing, distance moved, number of ridges between the capture and transplant site, elevation, and barriers to movement. Transplants were moderately effective in addressing nuisance black bears in Glacier National Park (58 of 112 bears were successfully translocated). However, how transplanted bears influenced population dynamics and community ecology at the capture or release site are largely unknown.

Education

Educating the public about nuisance American black bears is another common wildlife management practice in North America (Spencer et al. 2007). Any means possible to heighten the public awareness of the human–bear conflict issue also provided factual information for public debate and was instrumental (McCarthy and Seavoy 1994; Ternent and Garshelis 1999) in enhancing human understanding of bears.

Shocking

Shocking devices have been successful in preventing American black bears from obtaining unnatural concentrated food sources. One device, the Nuisance Bear Controller (NBC), operates on two 6-V batteries wired to a disk that emits 10,000–13,000 V. Activation occurs when a bear (or another animal) contacts the disk (Breck et al. 2006). The NBC was tested in Minnesota at ten independent sites in 2004. The NBC was placed on protected and unprotected birdfeeders; no protected feeders were disturbed or destroyed, but 40% of the unprotected feeders were robbed or destroyed. The NBC can be useful to deter bears from concentrated forage (Breck et al. 2006).

Taste Aversion

Taste aversion techniques have also been helpful in minimizing nuisance black bear behavior. Thiobendazole (72–165 mg/kg bear) was administered to five nuisance bears in central Minnesota, so they would avoid prepackaged military food. In the next 122 days, they ignored 6 (15%), approached but did not taste 12 (29%), tasted but did not consume 14 (34%), partially consumed 9 (22%), and did not totally consume any of the 41 offered prepackaged military foods (Ternent and Garshelis 1999). The authors concluded that taste aversion could minimize nuisance bear behavior toward target foods if alternate unnatural foods were minimized. In Saskatchewan, thiobendazole used around bee yards reduced black bear depredations (Polson 1983). However, in Alaska, bears simply learned how to avoid thiobendazole placed in trash cans and still managed to obtain garbage (McCarthy and Seavoy 1994).

Dogs, Pepper Spray, Crackers, and Bullets

Deterring bears from urban areas has been attempted with a variety of other measures including dogs, pepper spray, crackers, and rubber slugs and buckshot (McCarthy and Seavay 1994; Beckmann et al. 2004). None of the techniques was effective in the long term (Beckmann et al. 2004) and, without addressing the bear attractants congruently, will have limited success at minimizing future conflicts (Leigh and Chamberlain 2008). Beckmann et al. (2004) suggested that they not be used when attempting to alter bear behavior for more than 1 month.

Alternative View

Because many nuisance black bears are associated with unnatural food, attempts have been directed toward eliminating the food or deterring bears from the food. However, McCullough (1982) maintains that it is more than the food that bears become conditioned to. The stimuli for unnatural foods include human scent, human presence, human structures, and equipment among others. So, even if bears do not detect food, they are still attracted by related stimuli and the conditioned behavior will remain. Furthermore, in many situations (e.g., parks) there are no opportunities for reinforcement of fear (i.e., hunting) in bears toward humans. To avoid human–bear conflicts, McCullough (1982) suggests a model of bear management be developed that instills fear of humans into bears. "Humans and bears have coevolved as adversaries; to expect peaceful coexistence is both unnatural and unwise" (McCullough 1982, p. 32).

Biophysical-Behavioral Forces

Much of the research related to American black bears within suburban and exurban areas concentrated on deterrents to nuisance behavior with little attention paid to life history characteristics. An early study examined nuisance bears relying on unnatural food by describing life history characteristics of 126 bears (Rogers et al. 1976). The study was conducted in northern Michigan. Most captured bears (67%) using garbage were males as reported by others (Erickson et al. 1964; Rogers 1970). Males have larger home ranges than females and have more opportunities to encounter unnatural food. Number of cubs from females consuming unnatural foods was higher than for females using natural foods (3.1 versus 1.9). Bears using unnatural food were also heavier than those using natural food.

As a resource, black bears were not used for food or trophies when destroyed as nuisance bears. As with other studies, Rogers et al. (1976) recommends that garbage be minimized by prompt removal, bear-proof garbage cans, and garbage dumps located more than 1.6 km from campgrounds or residential areas (Rogers 1970).

More recently, Beckmann and Berger (2003a) contrasted bear ecology at the urban–wildland interface and in wildlands at the interface of the Sierra Nevada Range and Great Basin Desert, western North America. Bears in the urban areas had densities 3 times as high as historic values, sex ratios were 4.25 times more skewed toward males, body mass was 30% larger, home range size was reduced 90% and 70% for males and females, respectively, and bears entered dens later than those in wildlands and remained in them for fewer days (Beckmann and Berger 2003a). Bears in the urban–wildland interface also shifted activities to nocturnal periods (Beckmann and Berger 2003b). Black bears also shifted activity patterns from crepuscular and diurnal to nocturnal when using unnatural food in Sequoia National Park, California (Ayres et al. 1986). The shift to nocturnal activity in campgrounds (with unnatural food) was related to the reduced human activity at night.

Females in the urban–wildland interface had more potential reproductive years and gave birth to three times the number of cubs as wildland females. American black bears in the Lake Tahoe basin experienced a distribution shift rather than a demographic increase in response to unnatural food.

Although generally fecundity and age at first reproduction are higher and earlier in urban areas, elevated mortality rates also exist. Subsequently, bear populations associated with urban areas may function with source-sink dynamics, where the urban areas act as sinks to bears from urban and wildland areas (Beckmann and Lackey 2008).

Regulating human-related mortalities of American black bears is another important aspect of adult survival based on modeling populations on the southeastern Coastal Plain (Freedman et al. 2003). Regulations will require limitations to legal and illegal harvests, habitat connectivity, construction of highway underpasses, and speed reduction on highways during peak bear activity. Each of these will gain stature with increasing habitat fragmentation and subsequent anthropogenic mortality.

To counteract these changing dynamics and reduce human–black bear interactions, Beckmann and Berger (2003a,b) recommended extensive public education about how anthropogenic activities affect bears and laws, ordinances, and regulations against feeding wildlife unnatural foods. Landowners and businesses should obtain and use bear-proof garbage containers.

"Populations located on large tracks of public land, and in particular those with access to bear sanctuaries, may be buffered from increasing human impacts. Nevertheless, other populations will be less fortunate as conditions fostering positive population growth become increasingly rare. The continued subdivision of prime bear habitat and accompanying reductions in sub-adult and adult survival will increase the likelihood of local extinctions and make recovery of at-risk populations more difficult to achieve" (Freedman et al. 2003, p. 61). Although this statement relates to black bears in the southeastern Coastal Plain, it will also be applicable to other areas if appropriate management is not applied.

Community Case Studies

When problem black bears are in urban settings, human safety can be a serious concern. Peine (2001) used Kellert and Clark's (1991) framework to describe the evolution of public policy related to black bears for four communities in the United States.

Juneau, Alaska

Juneau is an ideal area to examine American black bears on the wildland–urban interface. It extends for 54 km with about half the area's concentrated development adjacent to prime American black bear habitat. Most areas of human habitation are within 0.4 km of prime black bear habitat. Development of policies and programs concerning nuisance bears took 4 years and went through several stages.

1. Research established density of 3–7 bears/1.6 km^2 associated with unnatural food. Researchers (McCarthy and Seavay 1994) concluded that garbage removal was the best solution to the problem, and they attempted to alter bear behavior (i.e., a key biophysical-behavioral force) via physical

and chemical aversive conditioning (i.e., 12-gauge shotgun slugs, 12-gauge explosive cracker shells, hand-thrown seal control bombs, thiobendazole). These methods failed to alter behavior.

2. In 1987, there were numerous complaints and 14 bears had to be killed. Media coverage led to public demands for nonlethal solutions. Non-Alaskans also influenced policy by voicing their protectionist views and threatened to demonstrate at terminals for cruise ships in Juneau, which constituted a major contributor to the community's tourist-based economy (i.e., valuation forces of environmental stewardship and economic community development). These valuation forces led to the formulation of community policy.
3. In 1987, the Juneau City Assembly drafted an ordinance defining guidelines for storage and collection of garbage to discourage wildlife from getting into the city's garbage. Once the ordinance was passed, the local police served as the institutional force to facilitate the policy.
4. This was followed by education that emphasized the link between unnatural foods and bears including graphic scenes of bears being shot. The education (via various media) included warnings of fines, videos, radio jingles, coloring books, bumper stickers, buttons, pins, and fliers to get the word out.

Voluntary compliance was not widespread, and garbage cans were not bear-proof. In 1991, complaints of nuisance bears increased, and 15 bears were killed and two people were injured. This led to increased emphasis on limiting unnatural foods, and through public education led to public acceptance of a revised ordinance requiring the use of bear-proof garbage containers causing some convenience and financial-related sacrifices. The evaluation of policy driven by economic stability, public safety, and wildlife stewardship was successfully implemented through education (McCarthy and Seavoy 1994; Peine 2001).

Mammoth Lakes, California

This community is surrounded by the Sierra National Forest, east of Yosemite National Park, and is another example of the development of a community-based wildlife policy that took years to evolve. The initiative began in 1996 when up to three generations of American black bears entered town for unnatural food and also followed people, some who were pushing baby carriages (Peine 2001). As a result, the community passed an ordinance to ban feeding and hunting within the city limits and provided procedures to minimize nuisance wildlife complaints (e.g., fines from \$100 to \$500 and 6 months in jail). Garbage had to be indoors until the day of pickup, and bear-proof dumpsters were provided for commercial businesses and a common community dump. These regulations were driven by valuation forces (e.g., public safety, protection of property, and protection of wildlife).

The institutional forces used involved the police, who could apply aversive conditioning to chase the bear or kill the animal, and a single private wildlife management consultant who developed an aversive conditioning program. The latter was designed to influence the behavioral forces through aversive conditioning (i.e., capture, immobilizing, removal of blood and a tooth) for first-time offenders. This "trauma" often discourages the bears from returning. If they do return for unnatural food, the animals are attacked with liquid-filled and rubber bullets, buckshot, pepper spray, audio devices, pyrotechnic devices, flash-bang devices, and Karelian dogs trained to chase bears. In Mammoth Lakes, policy was influenced by this consultant working with local police and the California Fish and Game Department. This is one of the few communities in the area that has addressed the issue of nuisance bears around the park (Peine 2001). No information was provided on the duration of the project or the overall success.

West Yellowstone, Montana

Located at the western entrance to Yellowstone National Park, this community is surrounded by American black bear and grizzly habitat. Valuation forces driving public policy for this community were public safety and protection of property. As a result, a comprehensive garbage disposal ordinance was passed that required "storing garbage, refuse and other food of any type what-so-ever edible by bears" in a secure building or bear-proof garbage can. Feeding, approaching, or harassing bears is also prohibited.

The Chief of Police constitutes the institutional forces and approves garbage cans used, schedules garbage collection, and penalizes offenders ($500 and/or 3 days in jail). This ordinance has been enforced for over a decade, is the most comprehensive community ordinance designed to minimize human–bear conflicts, and can serve as a successful model (Peine 2001).

Gatlinburg, Tennessee

Gatlinburg is the primary gateway to Great Smokey Mountains National Park, which receives more visitors than any other park in the United States. As a result, there are numerous tourist activities and venues available, and anthropogenic development is rapid. The park provides habitat for bears, but the animals have accessed unnatural food in the community for more than 25 years. Biophysical-behavioral factors were responsible for policy to reduce this activity. In the early 1990s, the bear population rapidly increased due to an abundant source of natural food, and there was an active plan to reduce poaching. In 1997, the natural foods were reduced due to an unusual frost and drought causing a higher dependence on unnatural foods in Gatlinburg. The result was a higher number of human–bear conflicts (Peine 2001).

Due to social–structural forces, there was resistance toward developing wildlife policy to address the problem. Gatlinburg was the prime area to hunt bears in the county, and harvesting bears using unnatural food was a tradition among locals. Tennessee allows baiting to entice bears up to 10 days prior to and during the hunting season. Members of the City Council were protective of hunters' rights and, if they were not hunters, they knew hunters, or had relatives who hunted (Peine 2001). In 1997, a record 81 bears were legally shot in Seiver County, most within the city limits of Gatlinburg.

In 1997 and 1998, this record harvest during the peak of the autumn season and the increased number of bears wandering in search of unnatural foods were instrumental in bringing media attention to Gatlinburg. In addition, a tourist witnessed a bear being shot at a dumpster in town, and the largest bear on record from the park was found dead; it had been shot searching for garbage. These valuation forces clashed and the national media asked why the City of Gatlinburg did not have a policy to deal with these conflicts. Clearly, the city benefited from the presence of bears as a tourist attraction and the city also began to have concerns about liability due to potential personal injury. This concern intensified when a person was killed by bears in the park near Gatlinburg. Institutional forces began. Some homeowners obtained bear-proof garbage containers. Once the unnatural food was secure in these areas, bears moved to restaurants and hotels and foraged in their trash leading to a task force in 1989 established by the city to investigate the problem. They reported five key findings.

1. Garbage control was not adequate.
2. Some citizens wanted the bears, others did not.
3. Sevier County would not let bears be translocated outside the county.
4. U.S. National Park policy did not allow bears to be translocated into the park.
5. Trained personnel to deal with the problem were limited.

A second group was established in 1997 and met until 1999 to further study the problem. They recommended garbage control and prohibitions on feeding bears with $500.00 fines for violations. Still some business owners failed to comply because pan-handling bears attract customers. The hunting issue was not resolved because that was controlled by the Tennessee Wildlife Resources Agency (Peine 2001). This conflict is another that demonstrates the complexity of establishing wildlife policy in a timely fashion, and considering the numerous cumulative effects.

In each of the examples cited (Peine 2001), human–bear conflicts involved unnatural food-conditioned bears, communities were reluctant to develop policy, policy initiation was triggered by specific events (e.g., economy, human safety, property protection, protection of bears), and after various attempts, bear-proof garbage containers were accepted (with the exception of Mammoth Lakes). Policy formulation occurred over 10–25 years. The time lags were due to human behavior, which is resistant to behavioral changes unless people recognize the consequences of their collective actions on resources in specific situations that are valued (Peine 2001).

Developing ordinances are important to minimize human–American black bear conflicts. As in the case studies presented by Peine (2001), the California Department of Game and Fish recognized that most human–bear problems were generated from improper garbage storage in the San Gabriel Mountains. Policy was established to educate and recommended that unnatural food had to be removed, garbage should be kept inside until the pick-up day, pet food should be kept inside, barbecue grills should be cleaned, and ripened and dropped fruit picked up. However, none of these measures was enforced and even if some citizens complied, the problems persisted (Lyons 2005). The cost of converting to bear-proof cans was the main reason the policy was not enforced. Human behavior has to be adjusted if human–bear conflicts are to be minimized (Lyons 2005). Adjusting trash cans and collection methods will likely be significantly cheaper than losing multimillion-dollar, out-of-court settlements due to bear attacks (Peine 2001).

UNGULATES

The Problem

Deer in urban areas are widespread throughout the United States, and deer–human conflicts are occurring at much higher rates with deer expansion and increasing human populations (Conover 1995; DeNicola 2000). These conflicts include human health concerns associated with Lyme disease, deer–vehicle collisions (DVC), and economic damage to landscaping and crops from herbivory (Conover 1995).

There are several factors that enhance deer populations in urban areas. In rural landscapes, deer depend on transitional zones between forests, grasslands, and agricultural areas where combinations of food and cover are available. Similar edge habitats occur in suburban areas that provide high-quality food in the form of ornamental plants, gardens, or fertilized lawns (Swihart et al. 1995), while adjacent woodlands or natural patches provide daytime cover (DeNicola et al. 2000). Additionally, many suburban and exurban areas have firearm restrictions or protective laws that limit hunting throughout most of the year and lack natural predators, leading to behavioral adjustments (Conover 1995; DeNicola et al. 2000). Finally, residential homes provide shelter from severe weather (Grund et al. 2002), and suburban residents often feed deer (Swihart et al. 1995; DeNicola et al. 2000).

White-tailed deer (*Odocoileus virginianus*) are adaptable species in diet breadth and habitat use, and have broadened their ranges to include expanding urban and suburban environments (DeNicola et al. 2000). In areas where deer herds in urban areas are not actively managed, these populations are becoming locally overabundant, (i.e., deer numbers approach or exceed human tolerance levels; DeNicola et al. 2000). The restrictions for deer management in urban areas include real and perceived safety concerns, conflicting perceptions and social attitudes toward wildlife, restrictions on hunting and firearm-discharge, and public relations and liability concerns (DeNicola et al. 2000). Clearly, deer in urban areas are becoming an increasingly important issue for state wildlife agencies, and continuing management and monitoring is needed to sustain healthy deer populations and mitigate human–wildlife conflicts.

A thorough understanding of the ecological factors concerning deer populations in urban areas should be incorporated into management decisions to determine the scope of the solution and likelihood of success. The current boom in exurban development also raises issues in deer management as many of the conflicts and resulting mitigation for deer in urban, suburban, or rural areas differ in exurban areas. Also, most research concerning deer ecology in urban areas focuses on white-tailed deer in the eastern and mid-western United States. Similarly, most research regarding control and management of deer in urban areas is also geared toward white-tailed deer. Because of the varying biophysical-behavioral forces between white-tailed deer, mule deer (*O. hemionus*), and elk (*Cervus canadensis*), each will receive separate treatment concerning the biophysical-behavioral characteristics of the species and the cumulative effects of urbanization on them.

White-Tailed Deer

Densities

Deer densities vary across studies and locations but tend to be higher in urban landscapes, ranging from 19.8 to 105.6 deer/km^2 (Table 9.2). Throughout a 5-year culling and contraception program in Rochester, New York, that reduced the population from 99% carrying capacity to 48%, deer densities decreased from 19.8 deer/km^2 to 9.6 deer/km^2 (Porter et al. 2004). Beringer et al. (2002) measured deer densities in a rural area to be 4.0/km^2. White-tailed deer densities ranged from approximately 2.5–5.9 deer/km^2 in rural Illinois (Nixon et al. 1991). However, on two study areas of the Lower Yellowstone River in Montana, deer densities were higher, averaging 34.3 deer/km^2 and 39.3 deer/km^2 (Dusek et al. 1989).

Home Ranges

Home ranges of white-tailed deer in urban areas vary across studies, states, and housing densities but are generally smaller than home ranges in rural areas (Table 9.3). Kilpatrick and Spohr (2000) recorded a 43.2 ha annual home-range size for female white-tailed deer in suburban Connecticut. Piccolo et al. (2001) reported female home-range sizes averaging 25.8 ha and 60.8 ha in two forest reserves in urban Illinois. Beringer et al. (2002) examined posttranslocation home-range sizes for deer in urban areas in Missouri and reported home ranges for males and females increased 170 ha posttranslocation.

Several studies also report a seasonal variation in home-range size, with home ranges being smaller in the summer at the onset of fawning season and largest during winter and early spring (Swihart et al. 1995; Grund et al. 2002; Etter et al. 2002; Storm et al. 2007). This seasonal behavior has also been reported for white-tailed deer in rural areas (Nixon et al. 1991; Sparrowe and Springer 1970). For example, home ranges in suburban Minnesota increased from 50.4 ha to 85.3 ha summer to winter (Grund et al. 2002). Kilpatrick and Spohr (2000b) and Porter et al. (2004), however, reported no difference between summer and winter home-range size in Connecticut or New York. This might be due to the limited availability of undeveloped areas

TABLE 9.2
White-Tailed Deer Densities across Rural–Suburban Gradients, 1989–2004

Density (deer/km^2)	Location	Landscape	Source
34.3–39.3	Lower Yellowstone River, MT	Rural	Dusek et al. (1989)
2.5–5.9	Illinois	Rural	Nixon et al. (1991)
72.7	Bridgeport, CT	Suburban	Swihart et al. (1995)
28	Groton, CT	Suburban	Kilpatrick and Walter (1999)
31	Town and Country, MO	Suburban	Beringer et al. (2002)
4	Huzzah Conservation Area, MO	Rural (translocation area)	Etter et al. (2002)
0–105.6	Chicago, IL	Suburban	Porter et al. (2004)
9.6–9.8	Rochester, NY	Suburban post-culling/ contraception	

TABLE 9.3
Annual and Season Home Ranges for White-Tailed Deer across Rural–Urban Gradients, 1970–2007

Home Range (SE) ha							
Annual	**Winter**	**Summer**	**Method**	**Sex**	**State**	**Landscape**	**Source**
	699.3 (440)	259 (129.5)	100% MCP	M/F	SD	Rural	Sparrowe and Springer (1970)
	132 (18.3)	221 (19.0)	Hand-drawn	F	NY	Rural	Tierson et al. (1985)
	150 (18.3)	233 (23.4)		M			
1,630			100% MCP	F	MT	Rural	Vogel (1989)
340						Exurban	
	177 (14)	55 (7)	MMA[a]	F	IL	Rural	Nixon et al. (1991)
	440 (19)	323 (49)		M			
45			100% MCP	F		Enclosure[b]	Beir and McCullough (1990)
142				M			
158			100% MCP	F	IL	Exurban[c]	Swihart et al. (1995)
67						Suburban[d]	
560			100% MCP	M/F	MT	Rural	Dusek (1987)
	42	58		F			
	32	255		M			
400 (50)	130 (3)	230 (3)	100% MCP	F	MT	Rural	Dusek et al. (1988)
	75 (17)	45 (19)			MN	Urban	Grund (1998)
	28.5 (2.8)	32.4 (2.7)	95% MCP	F	SC	Suburban	Henderson et al. (2000)
43.2 (2.7)	35.7 (3.2)	32.9 (3.2)	95% AK	F	CT	Urban-suburban	Kilpatrick and Spohr (2000a)
60.8 (13.1)			95% MCP	F	IL	Urban	Piccolo et al. (2001)
25.8 (4.9)							
		86e	Kernal	F	MO		
51.4–72.6		250–430[e]				Suburban	Beringer et al. (2002)
	40.5–61.5	22.0–30.0	95% MCP	F	IL	Post-trans.	
	85.3 (5.8)	50.4 (6.8)	95% AK	F	MN	Suburban	Etter et al (2002)
	22.4 (2.7)	21.4 (3.4)	90% MCP	F	NY	Suburban	Grund et al. (2002)
			95% FK	F	IL	Suburban	Porter et al. (2004)

[a] = Lower-density housing, 1–2 ha lots.
[b] = Deer surrounded by high density commercial/residential housing.
[c] = Minimum mean area.
[d] = Home ranges for fawns and adults.
[e] = George Reserve.

within deer home ranges and habitat fragmentation that restricts home ranges from expanding (Kilpatrick and Spohr 2000a,b).

Home ranges in suburban areas tend to be smaller than in exurban areas (Cornicelli 1992; Cornicelli et al. 1996; Storm et al. 2007). Summer and winter exurban ranges of females in Illinois averaged 53.0 ha and 90.6 ha, respectively (Storm et al. 2007).

Survival

Survival of white-tailed deer in urban areas is higher than for deer in rural areas (Table 9.4; Etter et al. 2002) although Storm et al. (2007) reported higher survival rates for white-tailed deer in exurban environments (87%) than in suburban (62%–82%) and rural (57%–76%) areas. Porter et al. (2004) recorded survival rates of 89%. Deer–vehicle collisions (DVC) are the primary source of white-tailed deer mortality in many urban and suburban areas (Etter et al. 2002; Nielson et al. 2003; Porter et al. 2004), whereas recreational hunting tends to be the primary source of mortality in rural areas (Nixon et al. 1991; Brinkman et al. 2004). The primary source of mortality for Florida Key deer (*Odocoileus virginianus clavium*) fawns was drowning in ditches and death due to collision with vehicles (Peterson et al. 2004). In contrast, there was 67% mortality of white-tailed deer fawns for the initial 8 weeks of life in an exurban area in Alabama. Coyotes were responsible for 41% of mortality, and starvation due to abandonment was responsible for 25%. Vehicle collisions were not an important cause of fawn mortality in Alabama (Saalfeld and Ditchkoff 2007). Fewer DVC occurred in exurban areas due to fewer roads (1.5 km/km^2; Storm et al. 2007). Additionally, hunting in exurban areas is not permitted on all properties (19%; Storm et al. 2007).

Habitat Use

Habitat use of white-tailed deer in urban areas varies across studies and locations as white-tailed deer is widespread, and topography and vegetation vary throughout their range. Studies of white-tailed deer in urban areas indicate that deer readily habituate to human development and, although they sometimes appear to avoid residential areas when possible (i.e., summer fawn-rearing season), will exploit these areas if little other choice is available, and there is sufficient cover, especially during winter (Swihart et al. 1995; Kilpatrick and Spohr 2000a; Grund et al. 2002). Females in exurban Illinois selected grasslands outside of zones of human influence during summer fawning (Storm et al. 2007). In Connecticut, residential areas were important foraging sites for deer during winter (Swihart et al. 1995).

In Connecticut, housing density decreased from 48.3 houses/deer home range to 30.2 houses/deer home range from winter to the fawn-rearing season, and distance from core area to residential development increased from 41.4 m to 99.4 m from winter to summer (Kilpatrick and Spohr 2000a). Others also found a decrease in the number of homes/ha within deer home range and core areas from winter to summer in an exurban area. Dwelling density in home ranges of deer and core areas averaged 0.13 ± 0.03 dwellings/ha and 0.14 ± 0.05 dwellings/ha, respectively, during fawning and 0.18 ± 0.02 dwellings/ha and 0.16 ± 0.03 dwellings/ha in winter home and core ranges, respectively (Storm et al. 2007).

TABLE 9.4
Annual Survival of White-Tailed Deer across Rural–Suburban Gradients, 1990–2007

Annual Survival (proportion killed by vehicles)							
All	**Fawns**	**Yearlings**	**Adults**	**Sex**	**State**	**Landscape**	**Source**
			0.46	M	MN	Rural	Fuller (1990)
			0.69	F			
			0.73	M/F	IL	Suburban post-translocation	Jones and Witham (1990)
	0.44		0.34	F	IL	Rural	Nixon et al. (1991)
	0.95	0.62	0.71	M			
	0.88	0.38	0.39	F	CT	Suburban	Swihart et al. (1995)
	0.38	0.86	0.82	M			
	0.77		0.83	F	SD	Rural	Deperno et al. (2000)
			0.57	M/F	MO	Suburban	Beringer et al. (2002)
				M/F	MO	Post-translocation	Etter et al. (2002)
0.69 (0.68)				F	IL	Suburban	Brinkman et al. (2004)
0.30 (0.09)				M	MN	Rural	Porter et al. (2004)
0.82 (0.66)	0.85 (1.0)	0.82 (0.80)	0.83 (0.55)	F	NY	Suburban	Storm et al. (2007)
0.83 (1.0)				F	IL	Exurban	
0.76 (0.20)	0.84[a] (0.17)	0.97[b] (0.0)	0.75[c] (0.23)	F			
0.57 (0.44)	0.6	0.89	0.62	F			
			0.87 (0.14)				

[a] = Neonate fawns.
[b] = 7–12 mo.
[c] = >12 mo.

Movements and Dispersal

Overall, seasonal movements of female white-tailed deer in suburban areas are limited and, with one exception, all females in a Rochester, New York, study moved less than 1.1 km (Table 9.5A,B; Porter et al. 2004). White-tailed deer in Bloomington, Minnesota, did not migrate or shift their home range centers during mild winters but moved less than 1.2 km during an extreme winter. Less than 10% of females dispersed and less than 15% migrated in suburban Chicago, Illinois (Etter et al. 2002). Meanwhile, 50% of male fawns and less than 10% of yearling and adult males dispersed, while no males migrated (Etter et al. 2002). Dispersal and migration distances for females were 7.6 km and 4.0 km, respectively, and dispersal distances for males averaged 5.4 km (Etter et al. 2002). Porter et al. (2004) observed 56% of females migrating (although nearly all deer showed more than 50% overlap between summer and winter home ranges). Annual dispersal rates were 14.3% for yearlings and 8.3% for adults, with an average dispersal distance of 4.0 km (Porter et al. 2004).

Few studies of deer in rural areas report low dispersal rates, except for adult females (less than 5%; Hawkins and Klimstra 1970; Tierson et al. 1985; Aycrigg and Porter 1997). Meanwhile, dispersal rates of 13%–50% for fawn and yearling females and 37%–80% for fawn and yearling males have been reported for rural areas (Hawkins and Klimstra 1970; Tierson et al. 1985; Dusek et al. 1989; Nixon

TABLE 9.5A
Dispersal Rates (%) of Male and Female Fawn, Yearling, and Adult White-Tailed Deer across the Rural–Suburban Gradients, 1970–2005

Dispersal Rates (SE)							
Fawn	Yearling	Adult	Distance (km)	Sex	State	Landscape	Source
	80	7	1.3–7.7	M	IL	Rural	Hawkins and Klimstra (1970)
	13	0		F			
	40.5	13.5		M	NY	Rural	Tierson et al. (1985)
		3.8[a]		F			
	46		18.5	M	WY	Rural	Dusek et al. (1998)
	17		19.5	F			
51			40.9 (5.0)	M	IL	Rural	Nixon et al. (1991)
50			49.9 (4.8)	F			
	20		17.6–168.0	F	MN	Rural	Nelson and Mech (1992)
		2.7	4	F	NY	Rural	Aycrigg and Porter (1997)
50	9.1	5.2	2.9–8.8	M	IL	Suburban	Etter et al. (2002)
7.3	4.7	6.5	1.9–33.9	F			
	14.3	8.3	4.0 (1.3)	F	NY	Suburban	Porter et al. (2004)
	36.7		1.9–13.8	M	TX	Rural	McCoy et al. (2005)

[a] = Did not specify age class.

TABLE 9.5B
Migration Rates (%) of Male and Female Fawn, Yearling, and Adult White-Tailed Deer across the Rural–Suburban Gradients, 1970–2005

Migration Rates (SE)							
Fawn	Yearling	Adult	Distance (km)	Sex	State	Landscape	Source
	20.6	46.2		M/F	WY	Rural	Dusek et al. (1989)
		19.6[a]	13.0 (3.0)	F	IL	Rural	Nixon et al. (1991)
4.9	14.3	4.8	4	F	IL	Suburban	Etter et al. (2002)
			0.64 (0.25)[b]	F	MN	Suburban	Grund et al. (2002)
		56.3[c]	<1.1[d]	F	NY	Suburban	Porter et al. (2004)

[a] = Includes yearling females.
[b] = Seasonal shift between geometric centers consecutive seasonal HR.
[c] = Did not specify age class.
[d] = Only one female had completely nonoverlapping seasonal HRs.

et al. 1991; Nelson and Mech 1992; McCoy et al. 2005). Migration rates for white-tailed deer in rural areas are also low; deer on the lower Yellowstone River were generally nonmigratory; 27 of 34 (79%) of adult and 13 of 19 (68%) yearlings had overlapping summer and winter ranges (Dusek et al. 1989).

Fertility

Fertility rates in urban landscapes differ by age class but are generally lower in urban than in rural landscapes (Table 9.6). In suburban Connecticut, fetus:female ratios averaged 0.0, 0.60, and 1.20 for fawns, yearlings, and adult females, respectively (Swihart et al. 1995). Fertility rates increase after translocation and removal programs. Beringer et al. (2002) measured pre- and post-translocation fertility rates of deer in Missouri. Fawn:female ratios for fawn and yearling females averaged 0.0 and 0.86 in the suburban area and 0.14 and 1.22 post-translocation (Beringer et al. 2002). Nielson et al. (1997) measured fertility rates before and after an adaptive management program; fetus:female ratios increased from 1.33 to 1.85 post-removal. Porter et al. (2004) examined the effects of culling and contraception on fertility rates over a 5-year period. Fetus:female ratios for fawn, yearling, and adult females increased from 0.0, 1.17, and 1.41 from a high-density (99% carrying capacity) population to 0.27, 1.75, and 2.11 for a population at 48% carrying capacity (Porter et al. 2004).

Mule Deer

White-tailed deer are among the most well-known and widespread large mammal species in North America and have flourished in wilderness and metropolitan areas during the last century (Conover 1995; DeNicola et al. 2000). However, little evidence suggests that mule deer survive and thrive as well as white-tailed deer in urban environments (Reed 1981; Vogel 1989). Furthermore, predictions of the consequences of fragmentation from exurban development to mule deer were in their infancy (Kucera

TABLE 9.6
White-Tailed Deer Fertility across Rural–Suburban Gradients, 1991–2004

Fertility (fetuses:female)						
All	**Fawn**	**Yearling**	**Adult**	**State**	**Landscape**	**Source**
1.33[a]				IL	Rural	Nixon et al. (1991)
	0.0[a]	0.60[a]	1.20[a]	CT	Suburban	Swihart et al. (1995)
1.33					Suburban	Nielson et al. (1997)
1.85					Post-AM[b]	
	0.0[a]		0.86[a]	MO	Suburban	Beringer et al. (2002)
0.96[a]	0.14[a]		1.22[a]		Post-translocation	
	0	1.17	1.41	NY	Suburban—high density	Porter et al. (2004)
	0.27	1.75	2.11		Suburban—low density	

[a] = Indicates fawns:female instead of fetuses:female.
[b] = Adaptive management program (culling and contraception).

and McCarthy 1988). In a study of the effects of housing density on white-tailed and mule deer in Gallatin County, Montana, a species composition shift occurred in which white-tailed deer increased in abundance or expanded into areas historically occupied by mule deer (Vogel 1989). This could be the product of several ecological factors between the species. For instance, white-tailed deer have higher natality rates, lower fawn mortality, and an overall younger population structure than mule deer (Vogel 1983). Disturbance is usually tolerated in species that have greater numbers of young, shorter life expectancy, wider dispersal patterns, and more nocturnal habits (Geist 1971), a description better suited for white-tailed deer than mule deer (Vogel 1989).

Density

Although studies that investigate the ecology of mule deer in urban areas are limited, several reports indicate that densities of mule and black-tailed deer are lower in urban areas than rural areas, unlike those of white-tailed deer (Smith et al. 1989). McClure et al. (2005) calculated densities of approximately 6.3 deer/km^2 in urban sites in Utah and 7.1 deer/km^2 in adjacent rural areas, coinciding with Vogel's (1983, 1989) observation of fewer deer observed in more developed areas in Gallatin County, Montana. Similarly, Columbian black-tailed deer densities averaged less than 0.3/ km^2 in urban Vancouver, Washington, versus 2.7 deer/km^2 in surrounding rural areas (Bender et al. 2004b). In Montana, mule deer densities can be fairly high; low-density rural winter ranges in Montana contain 8.3–10 deer/km^2, whereas more restricted mountainous winter range areas in Montana experience mule deer densities ranging from 31 to 50 deer/km^2 in mild winters to 83 to 250 deer/km^2 in concentrated areas during more severe winters (Mackie and Pac 1980).

Survival

Bender et al. (2004b) recorded survival rates of 0.70 (n = 11) for female and 0.86 (n = 6) for male Columbian black-tailed deer in urban Vancouver, Washington. Fawn survival rates averaged 0.84 ($n = 26$; Bender et al. 2004b). The survival rates of females were low compared to Columbian black-tailed deer in rural habitats in the Pacific Northwest (i.e., 0.73 and 0.80 for females; McNay and Voller 1995; McCorquodale 1999). Male Columbian black-tailed deer survival was higher compared to rural studies; McCorquodale (1999) estimated survival rates of 0.50, while Bender et al. (2004c) reported rural male survival 0.50–0.52. Mortality of urban black-tailed deer was primarily attributed to DVCs or deer–train collisions (83.3%, Bender et al. 2004b). Two of the three deer killed by DVCs were observed being chased by domestic dogs just prior to the collision (Bender et al. 2004b). Primary causes of mortality for rural black-tailed deer are predation and hunter harvest (McNay and Voller 1995; McCorquodale 1999). In rural areas in Washington, hunting contributed to 56% of the male black-tailed deer mortalities (Bender et al. 2004c).

Habitat

Some development affecting mule deer habitat and populations has occurred in areas used as summer and fall range and, while more is likely to occur, the greatest threat comes from development on and adjacent to major winter ranges (Mackie and Pac 1980). Because mule deer distribute themselves and exhibit fidelity to specific sites, loss of these regions can have profound implications on mule deer occurrence in different areas and other seasons (Mackie and Pac 1980; McClure et al. 2005). For example, in the Bridger Range north of Bozeman, Montana, deer winter in densities of 25.5–34.0 deer/km^2 and summer at densities of 3.9–4.6 deer/km^2 (Mackie and Pac 1980). Thus, the loss of 1 km^2 of winter range may result in a minimum loss of 26–34 deer in the population; the equivalent of all adult animals within 25–31 km^2 of summer range (Mackie and Pack 1980). Additionally, some areas of winter range are more important than others—during more severe winters, ~77 deer can concentrate on less than 1 km^2 of range; loss of this primary area could result in a significant loss in the deer population (Mackie and Pac 1980).

The influences of urban developments on mule deer habitat are also complex because many mule deer are migratory and the alteration of habitat conditions from urban developments in mule deer habitat affect population sizes and the ratio of migratory to nonmigratory animals (McClure et al. 2005). McClure et al. (2005) examined the ratio of migratory versus nonmigratory mule deer and assessed differences between fawn recruitment in deer herds using adjacent urban and rural winter ranges in the Cache Valley, Utah. They found that 15 of 17 radio-collared deer in urban areas were migratory, opposed to 8 of 14 deer in rural areas. Additionally, spring migration of deer in urban areas commenced an average 2–3 weeks sooner than deer in rural areas in both years examined (McClure et al. 2005). Deer in urban areas traveled an average 31.5 km, and deer in rural areas traveled an average 14.5 km between winter and a shared summer range (McClure et al. 2005).

Winter habitat use between mule deer in urban and rural areas also differs (McClure et al. 2005). For instance, deer in urban areas selected habitats with concealment vegetation, whereas deer in rural areas exhibited more neutral behavior toward this characteristic (McClure et al. 2005). McClure et al. (2005) postulated that deer in urban areas were less likely to take risks associated with large-scale movements for better forage, which may explain their lower nutrition and recruitment and the earlier, more pronounced migration from the herd in the urban area. The home ranges of deer in urban areas were also smaller than deer in rural areas. Activities of deer in urban areas were more aggregated than their rural counterparts (McClure et al. 2005).

In a comparison of fawn:female ratios between mule deer in rural and urban areas, fawn ratios are conspicuously lower in urban areas (McClure et al. 2005). In 1995 and 1996, the fawn:female ratios for migratory urban mule deer were 0.88 and 0.71, respectively; for nonmigratory urban deer, the ratio was zero in both years, indicating that nonmigratory behavior is perishing within the urban deer herds (McClure et al. 2005). Fawn recruitment for mule deer in rural areas was the same for migratory and nonmigratory animals at 1.0 fawn:female in 1995 and 1.25 fawn:female in 1996 (McClure et al. 2005).

Columbian black-tailed deer in Washington produced approximately 1.83 fawns:female and 1.36 fawns:female over a 2-year study period (Bender et al. 2004b). Due to fawn or female mortalities, recruitment rates of black-tailed deer in Washington ranged from 0.90 to 1.33 fawns:female (Bender et al. 2004b). For black-tailed deer in urban Vancouver, Washington, high rates of fawn recruitment compensate for lower survival rates compared to rural populations (Bender et al. 2004b). Productivity in this study was higher than any previously recorded level in the Pacific Northwest (Bender et al. 2004b).

Case Study

The increasing population of deer in Helena, Montana, is a good example of the interaction of biophysical-behavioral, social–structural, valuation, and institution-regulation forces. The ecology of deer is well defined, and a basic principle of any management is knowledge of population levels (Krausman 2002). To supplement baseline data with population estimates for the city, Hickman (2007) conducted a street-by-street survey. He also subjectively documented deer impacts and deer defenses developed by homeowners. Observers (2–4) in a vehicle drove throughout the city at less than 40 km/h counting deer within the city boundaries. Some areas were not included in the survey because of traffic congestion, safety considerations, time considerations, or inaccessibility (e.g., Helena airport). The vehicle stopped if numbers, sex, and age needed to be verified. Deer were recorded by street or block with GPS coordinates as male, female, or fawn. Surveys were conducted two–three times beginning 3 h before sunset or 3 h after sunrise in seven Helena Citizens Council Districts (no. surveys = 20) in December 2006 and January 2007. The maximum number counted for the survey was 289 mule deer along the 280 km of survey routes. There were 36 male:100 females in the observational survey.

Because the deer were within the city and on private- and state-owned land, the social–structural forces were simplified; however, valuation forces were unclear until a series of public comments and surveys were conducted. Public meetings and comments recorded revealed an array of attitudes ranging from complete control of the deer population to no control (Helena Urban Wildlife Task Force 2007). A telephone survey revealed that citizens did not enjoy seeing deer in their neighborhoods (36.4%), deer were a problem (58.6%), and actions were necessary (78.3%). Most supported lethal (54.3%) and nonlethal (82.1%) solutions to control deer, and nearly 90% said the Montana Fish, Wildlife, and Parks should be responsible for deer management along with the City of Helena (72.1%). Only 2.1% of those surveyed were involved in an accident with deer within city limits. Controlling deer would enhance the health of the deer population (48.2%), reduce predator activity in the city (62%), and reduce risks to their pets from urban deer (45.5%) (Helena Urban Wildlife Task Force 2007). Deer were clearly creating concerns in the city (Table 9.7).

Based on public meetings and the survey, biologists, administrators, and managers had a good representation of public valuation forces. In 2003, the Montana Legislature passed HB 249 and SB 410 as amended to read: "7-31-4110. Restriction of wildlife. A city or town may adopt a plan to control, remove, and restrict game animals, as defined in 87-2-101, within the boundaries of the city or town limits for public health and safety purposes. Upon adoption of a plan, the city or town shall notify the department of fish, wildlife, and parks of the plan. If the department of fish, wildlife, and parks approves the plan or approves the plan with conditions, the city or town may implement the plan as approved or as approved with conditions... The plan may allow the hunting of game animals and provide restrictions on the

TABLE 9.7
Reports to Montana Fish, Wildlife and Parks Personnel about Wildlife within the City Limits of Helena, Montana, 2004–2007

		No. Interactions/Year			
Species	**Interaction Report**	**2004**	**2005**	**2006**	**2007**
Mule deer	Injured/sick	22	49	54	60
	Dead	21	26	40	41
	Aggressive	2	52	28	27
	Hanging in fence	3		6	4
	Nuisance	9	22	24	14
	Killed dog				1
	Fawns separated	1	1	7	5
	Poaching				1
	No deer in area				1
White-tailed deer	Injured		1		2
	Dead		3	1	2
Mountain sheep	Dead		1		
Moose	Sighting				1
	Nuisance				
Pronghorn	Sighting				1
American black bear	Getting: garbage				9
	Fruit trees				1
	Grain				1
	Bird feeders				1
	Dog food				1
	Nuisance	2			12
	Dead	1			
	Sightings				12

TABLE 9.7 (continued)
Reports to Montana Fish, Wildlife and Parks Personnel about Wildlife within the City Limits of Helena, Montana, 2004–2007

		No. Interactions/Year			
Species	**Interaction Report**	**2004**	**2005**	**2006**	**2007**
Raccoons	Nuisance				1
Bats	Nuisance			1	1
	Dead	1			
Muskrat	Nuisance	1			
Lagomarph	Nuisance	1			
Porcupine	Nuisance	1	1		
Skunk	Nuisance	1	1		
	Sick/Injured				1
Marmot	Nuisance		1		1
Squirrel	Nuisance	2	1		
	Dead		1		
Fox	Injured/sick	1			
Mountain lion	Sighting	5	5	3	4
	Chasing deer				1
	Nuisance			1	1
	Encounter	1			
Birds:					
Owl	Nuisance				1
	Injured	1	2		1
	Dead			1	
	Injured				1
Nighthawk	Nuisance		1		
Woodpecker	Dead	1			
Finch	Dead	1			
Raven	Aggressive				1
	Injured				1
Duck	Injured			1	
Goose	Nuisance	1			
	Injured			1	
Hawk	Dead	1			
	Dead			1	
Grosbeak	Nuisance			1	
Magpie	Dead	4			
	Dead	1		1	
Flicker	Injured			1	
Falcon	Injured			1	
Seagull	Dead	1			
Crow	Injured		1		
Songbirds			1		
Deformed birds					

feeding of game animals." Once the institutional-regulation forces were in place, the City of Helena began to implement its plan in autumn 2008.

Elk

Exurbanization and its influence on white-tailed deer and mule deer have received more attention than the influences on elk, but elk are clearly affected. Lessons learned from other ungulates can be applied to elk as urbanization spreads into the habitat of elk. Many of the conflicts occur because elk generally spend summer and autumn on high-elevation public land but migrate to lower elevations in winter and spring. Many of these habitats are in private ownership that does not have the same level of protection as public lands (Hayden 1975; Henderson and O'Herren 1992; Haggerty and Travis 2006). All across the west, changes in use of private lands are changing the landscape and wildlife habitat (Henderson and O'Herren 1992; Wait and McNally 2004). These changes include alterations of large agricultural holdings to small residential tracks or shifting priorities of using lands for wildlife instead of livestock (Haggenty and Travis 2006), stream alteration, disturbance of movements, restricting access to lands, nuisance wildlife concerns, and alterations of farmland. Their effects are increased because legislation is not adequate to compensate for the negative influences of urbanization (Henderson and O'Herren 1992).

Numerous studies have evaluated elk use in human-altered landscapes (i.e., logging, roads, fire, livestock), but how they are actually influenced by suburban and exurban development is relatively unpublished. However, these data are not new. In 1967, Klemmendson (1967, p. 268) claimed big game winter range had been declining since the 1930s and that "... while something must be done to counteract the trend of dwindling winter habitat ... it seems unlikely that existing priorities will greatly change." In the eastern United States, big game winter range reduction has led to numerous problems. In the West, with adequate planning and foresight, the massive landscape changes may be planned effectively to benefit wildlife (Berris 1987; Wait and McNally 2004; Haggenty and Travis 2006). However, alterations in policy cannot linger. Elk, like all wildlife, belong to the public but, as large landscapes are purchased in Montana and throughout elk habitat in the West, elk spend more time on private lands and are tolerated by landowners who do not tolerate elk hunting. In essence, this places large areas "out of administrative control" (Haggenty and Travis 2006). Ranch sales appear to be driving elk management in many areas as new owners consider the possession of elk as part of the transaction. "This perspective typically precludes a conceptualization of public access to private land for the purpose of harvesting elk as part of the necessary human ecology of elk management—the ecological commons in which hunters played the role of top predator has dissolved" (Haggerty and Travis 2006, p. 828–829).

The Ungulate–Vehicle Collision Crisis

Deer–vehicle collisions are becoming an increasing problem across the United States and may result in reductions of deer populations, property damage to vehicles, and human injuries or fatalities (Romin and Bissonette 1996; Biggs et al. 2004). Over 1 million DVC occur annually in the United States (with approximately 50% unreported; Conover et al. 1995; Romin and Bissonette 1996). Each year 29,000 human injuries and 211 human fatalities occur as a result of DVC (Conover et al. 1995). The economic losses associated with DVC may be substantial and vary state to state, but can exceed $2 billion annually across the United States in addition to losses associated with human fatalities (Romin and Bissonette 1996; Danielson and Hubbard 1999).

Although most DVC occur on rural, two-lane, high-speed roads (Finder et al. 1999), they are becoming an issue in more urban areas that are experiencing an expansion of human and deer populations. Nielson et al. (2003) characterized DVC in urban areas and found that a higher proportion of DVC occurred in areas with fewer buildings and infrastructure, more patches and higher proportions of forest or shrub cover, more public land patches, and higher landscape diversity. These findings are similar to other studies involving DVC in rural areas (Bashore et al. 1985; Finder et al. 1999; Hubbard et al. 2000).

An overview of DVC management strategies and monitoring efforts was presented by Romin and Bissonette (1996) based on a survey distributed to state wildlife agencies throughout North America. Questionnaires were distributed to agencies regarding the number of deer killed annually on highways from 1982 to1991, how the information was obtained, methods used to reduce deer mortalities on highways, and success of each technique based on personnel reports of scientific evaluation (Romin and Bissonette 1996).

A total of 43 states responded to the survey and 35 reported yearly totals of deer mortality (Romin and Bissonette 1996). Information was obtained for several other western states; Utah reported 1,826–5,502 annual deer mortalities, Wyoming reported 987–1,756, California reported 15,000, and Colorado reported 5,202–7,296 (Romin and Bissonette 1996).

Of the 43 responding states, only Florida has not addressed deer–highway mortality (Romin and Bissonette 1996). Two states (4.7%) used highway lighting, 3 (7%) hazed deer, 6 (14%) modified habitat, 7 (16.3%) reduced speed limits, 7 (16.3%) built highway underpasses or overpasses, 11 (25.6%) used mirrors, 11 (25.6%) built fencing to deter deer from highways, 20 (46.5%) used warning whistles, 22 educated the public (51.1%), 22 installed swareflex reflectors (51.1%), and 40 (93%) used deer-crossing signs (Romin and Bissonette 1996). Between 62% to 95% percent of respondents using deer-crossing signs, public awareness programs, deer warning whistles, and swareflex reflectors reported that they had not conducted any scientific evaluation of the techniques, and evaluation of success was based upon opinion (Romin and Bissonette 1996). Only 13 (30.2%) states have conducted scientific studies of mitigative strategies for reducing DVC, and 9 of these considered only one technique (Romin and Bissonette 1996).

Although relatively few states use highway fencing and deer overpasses or underpasses, 91% of the 11 states using fencing and 63% of the 7 states using overpasses or underpasses reported these as effective strategies, although a majority of the responses were based on opinion (Romin and Bissonette 1996).

Hedlund et al. (2004) reviewed the most commonly used mitigative strategies for reducing DVC, and grouped them into three categories: affecting motorist behavior, affecting deer behavior, and affecting deer populations.

Managing Motorists

Public education about DVC and speed limit reductions have not been sufficiently evaluated but are likely ineffective based on similar results from other campaigns (e.g., impaired driving and other stand-alone education programs; Hedlund et al. 2004). However, a public education campaign can be effective if it is actively enforced (i.e., the seat belt law) or provides information on time and site-specific situations, such as the beginning and location of a mule deer migration (Hedlund et al. 2004).

Passive signs, such as deer-crossing signs, are a commonly used technique; however, they are used so frequently, motorists probably ignore them (Putman 1997). Sullivan et al. (2004) tested temporary warning signs during mule deer migration. Travel speeds during this time dropped and DVC decreased by 50%, although the effect diminished during the second year of the study (Sullivan et al. 2004).

Active signs used to alert motorists, when deer are near the road, were evaluated in a study in Wyoming (Gordon et al. 2001). Vehicle speeds slowed when the sign was lighted, but there is no corresponding DVC data (Gordon et al. 2001). More testing with active signs is needed, and detection technology needs to be improved for these signs to be deemed effective (Hedlund et al. 2004).

Bashore et al. (1985) reported an increase in DVC at the woodland–field interface where deer congregate and motorist visibility decreases. Additionally, the most important topographical feature influencing DVC in Illinois was the distance between the road and forest cover (Finder et al. 1999). Increased highway lighting (Reed and Woodard 1981; Hedlund et al. 2004) has also been ineffective at reducing DVCs. Clearing roadside vegetation increases driver visibility and decreases deer use of the area by reducing habitat quality near the road (Hedlund et al. 2004). Roadside clearing has been recommended in several studies (Putman 1997; Nielson et al. 2003; Hedlund et al. 2004), but stakeholder acceptance is needed as the public may feel habitat modification may affect deer activity or the environment (Nielson et al. 2003).

Public Involvement

There are arrays of techniques and management styles addressing problem ungulates along the rural–urban gradient (e.g., lethal control, nonlethal control). However, nearly everyone agrees that the single most important principle needed to successfully address the problem is community involvement with the planning process. The problems did not arise overnight and were created by cumulative effects of housing, developments, roads, and other habitat alterations. By including municipal, county, state, federal, and other responsible entities into the process, objectives can be broadly established and solutions planned (McAninch and Parker 1991; Kilpatrick and Walter 1997; Lund 1997; Rutberg 1997). Curtis et al. (1993) summarized some of the lessons learned from the public involvement process.

1. Reaching a consensus may not be possible but that does not mean the program failed.
2. Emphasize problem-solving techniques, so mechanisms for strongly held minority opinions are built into the process.

3. Procedures for receiving comments from people in the community should be part of the process.
4. Attempt to involve all interests in the process, particularly those with the ability to block implementation of the recommendations.
5. Use the media aggressively to publicize management planning efforts.
6. Provide ample time and resources for the project to work.
7. Know the timelines and provide timely responses to participants.

Many of the management plans consider reduction by harvesting a certain number of individuals. However, because of the social biology and ecology of white-tailed deer, they form family groups of females. These females are highly philopatric to ancestral group ranges and by removing all deer from 400 to 2,000 ha may create voids in distribution of deer that may not be recolonized for up to 5 years (Porter et al. 1991). The approach described by Porter et al. (1991) is called the rose petal hypothesis. The older females occupy home ranges at the center of the group, and younger individuals occupy overlapping home ranges that extend the periphery. These overlapping home ranges take form roughly analogous to the petals of a rose (Mathews 1989).

If the rose petal hypothesis is applicable, managers may want to consider pulsing controls that are more efficient than steady-state regimes. A pulsing control solution culls the population intermittently and can be effective when animals can be rounded up or baited. Success will also depend on the desired harvest, continual use of the pulsing control, and the acceptability of lethal control. Models for pulsing controls are presented by Rondeau and Conrad (2003).

Desert Bighorn Sheep

The cumulative effects of human activities have altered and eliminated mountain sheep habitat in many areas of the southwestern United States. Mountain sheep populations in North America have declined from >500,000 to approximately 70,000 and are among the rarest ungulates (Valdez and Krausman 1999). As humans continue to use mountain sheep habitat for development, recreation, resource extraction, or other uses, additional declines can be expected. Populations that have expired in Arizona, California, and New Mexico have been associated with human development (Krausman et al. 2001), but cause-and-effect studies were not carried out. However, human activities adjacent to sheep populations have been suggested as significant causes for declines (Gionfriddo and Krausman 1986; Etchberger et al. 1989; Krausman 1993, 1997; Harris et al. 1995; Krausman et al. 1996; Rubin et al. 1998; Turner et al. 2004). Unfortunately, very little research has attempted to alter the problem. This may simply represent a situation that when urbanization advances rapidly it overwhelms natural ecosystems. When development occurs adjacent to and in mountain sheep habitat, sheep often decline and ultimately can become extinct. Society is faced with a difficult choice: restrict suburban expansion and control human activities within sheep habitat, or accept the reality that bighorn sheep and expanding developments are not compatible (Krausman et al. 2001).

General Management of Urban and Exurban Landscapes for Wildlife

Successful management of wildlife along the rural–urban gradient is about managing people as much as managing wildlife, and certainly includes cumulative effects.

> People generally want wildlife in urban and suburban areas, even if they are ambivalent about some of the potential conflicts. Having a rich assemblage of native plants and animals around us is an indication that nature still prospers in the places where we dwell. It is a sign that our habitat still retains some of its ecological integrity. In the long run, the greatest benefit of having a well-connected system of habitats in our cities and suburbs... children as well as adults will have abundant opportunities for contact with wild nature... And with a positive attitude toward nature they will likely be good citizens willing to support strong conservation measures for their broader environment. (Noss 2004:7)

Urban areas will "elicit their greatest loyalty, commitment, and stability when they function as places where people can consistently encounter satisfying connections with natural as well as economic and cultural wealth" (Kellert 2004, p. 14). Furthermore, understanding artificial ecosystems (e.g., urban areas) may yield important insights into the management of natural ecosystems (Savard et al. 2000).

Because people have such a large investment in managing wildlife in urban areas, their attitudes have to be considered when formulating management plans. Human dimensions, which include public attitudes, are also one of the triads of any successful wildlife management plan (Krausman 2000).

Studies have examined attitudes of residents (i.e., newcomers, longtime nonfarm residents, and farm household residents) in exurban areas in Ohio (Smith and Sharp 2005), attitudes of residents in Arizona about elk management in the wildland–urban interface (Lee and Miller 2003; Heydlauff et al. 2006), and a series of studies of public attitudes across the United States (Decker and Gaavin 1987; Baker et al. 2004; Mannan et al. 2004; Siemer et al. 2004) discovering people's attitudes about urban wildlife. Surveys of attitudes are often general but can be specific. Wolch and Lassiter (2004) examined attitudes of African American women in Los Angeles, California, toward wildlife, and Sasidharon and Thapu (2004) examined the ethnicity and variations in concerns for wildlife. In general, society supports urban wildlife even if they do not support the same species assemblages as natural areas (Noss 2004). Urban areas are important in the conservation and management of biodiversity. As such, urban planners increasingly integrate corridor systems to link habitat patches in urban areas with surrounding suburban, rural, and exurban lands (Adams 2005). When planners work with ecologists, connectivity can be maintained between anthropogenic development and wild landscapes to minimize the influence of urbanization on wildlife, sustain native populations, and further conserve biodiversity (Shaw et al. 2009).

Because of the broad jurisdiction over lands, only a few planning efforts have been conducted at sufficiently broad spatial scales to integrate conservation across the urban–rural gradient (Shaw et al. 2004). For example, Montana's Growth Policy Resource Book (2008) makes few statements related to wildlife. A recent effort to integrate urban and exurban planning on a large scale is Pima County's Arizona Sonoran Desert Conservation Plan (SDCP) described by Shaw et al. (2009) and Steidl et al. (2009).

The SDCP is centered around Tucson, Pima County, Arizona, which has been one of the fastest growing communities in the United States for decades. By 2000–2001, approximately 1 ha of natural desert was being lost to urban development every 5 h (Benedict et al. 2005). The population of Pima County is about 1,000,000, and most people are concentrated in the Tucson metropolitan area, leaving most of the county's 14,400 km^2 as wildlands, Indian reservations, rural, or exurban lands. To preserve and maintain the diverse and abundant flora and fauna, the SDCP (http://222.pima.gov/sdcp/) was "guided by five goals: (1) define urban form to prevent urban sprawl and protect natural and cultural resources; (2) provide a natural resource-based framework for making regional land-use decisions; (3) protect habitat for and promote recovery of species listed under ESA; (4) obtain a Section 10 permit under ESA for a multispecies HCP; and (5) develop a system of conservation lands to ensure persistence of the full spectrum of indigenous plants and animals by maintaining or restoring the ecosystems on which they rely, preventing the need for future listings" (Steidl et al. 2009, p. 219). The SDCP began in 1997 with the creation of the Science and Technology Advisory Team to advise the county. More than a decade later, the planning process continues with at least 6 significant accomplishments (Shaw et al. 2009).

1. Unanimous adoption by the County Board of Supervisors of the SDCP and its conservation guidelines as an integral part of the County's comprehensive land-use plan.
2. Passage of a bond initiative in 2004 providing the $174 million for open space including at least $112 million for acquisition of lands and easements to protect land with high biological importance.
3. Purchase of land and easements that by 2008 had already protected over 30,000 ha of high-priority conservation lands.
4. Involvement of hundreds of citizens in educational workshops and public hearings.
5. Involvement of more than 150 scientists as sources of information and as reviewers for the plan.
6. Development of a comprehensive, county-wide geographic database that enables sophisticated environmental modeling.

Important lessons have been learned from this long-term project (Shaw et al. 2009).

1. Conservation planning can provide a reliable basis for well-balanced land-use planning.
2. The goal of a conservation plan should be conservation, not compliance with the bureaucratic requisites of environmental legislation.
3. Science is a process and knowledge in land-use planning science performs rigor, consistency, and replicability; a mechanism to set goals is transparent and accountable should be evaluated and validated by experts and use the peer-review process.
4. Separate scientific and political processes.

5. Urban and exurban landscapes are critical elements in land-use planning for conservation.
6. There are numerous secondary benefits of conservation in urban areas.
7. The scientific process of the SDCP is discussed in more detail by Steidl et al. (2009).

The SDCP is designed to minimize urbanization, suburbanization, and exurban development within Pima County and to maintain and preserve habitats and landscapes where possible through land purchases and trades. The policy is sound; biologists may not know all the life history characteristics of the wildlife they are responsible for but are certain that viable populations will not exist without habitat. This concept is not new. In 1975, Hayden (1975) examined the effects of anthropogenic development on wildlife and their habitat around Lolo, Montana, and made nine management suggestions (most related to habitat).

1. Wildlife habitat can be managed and conserved best when in large undeveloped blocks that are not fragmented with subdivisions.
2. Natural barriers should be maintained between natural and altered landscapes.
3. Connectivity between subdivisions and wildland should be maintained.
4. Wildlife should not be harassed by domestic animals.
5. County commissioners need to be part of the planning process, so fragile and critical habitats are not negatively influenced.
6. Education is warranted to develop a land ethic for those purchasing lands along the rural–urban gradient.
7. The use of conservation easements should be encouraged.
8. Additional tax breaks should be encouraged, so large private landscapes are not fragmented further.
9. Zoning restrictions should be enacted and enforced to prevent development of riparian forests, sloughs, marshes, and other important valley bottom areas.

Each of these recommendations benefits habitats by minimizing disturbance and maintaining and enhancing wildlife habitat. To maintain large blocks of land, conservation development (i.e., cluster building with open space) has been in use for decades (Zipperer et al. 2000; Compas 2007; Milder et al. 2008) but rarely evaluated. Land trusts, landowners, and developers in the eastern United States used revenue from limited development to finance the protection of land and natural resources. Ten of these cluster developments (0.07–1.79 dwellings/ha) were randomly selected from 101 that had been identified. Each unit was designed to emphasize conservation. To evaluate their effectiveness as "conservation units," land alteration, edge effect, spatial configuration, and connectivity, impervious surface, riparian buffers, impacts to site construction targets, restoration, and land management were measured and contrasted with traditional subdivisions. The conservation and limited development projects protected sensitive conservation resources and resulted in significantly more conservation benefits than traditional subdivisions. More than 85% of the conservation developments were protected as interior habitat and the layout of development addressed conservation, restoration, and stewardship needs of the site-specific conservation targets (Milder et al. 2008).

In Gallatin County, Montana, planners and ecologists were also sensitive toward minimizing the impacts of exurban growth on habitat loss and fragmentation (Compus 2007). By contrasting development patterns from 1973 to 2004, Compus (2007, p. 58) found that "(1) major subdivisions became more 'clustered' and less land consumptive, (2) minor sub-divisions revealed the opposite trend and are recently consuming more land, (3) distances from existing development decreased for major subdivisions, and (4) increasing numbers of parcels were near or within riparian areas. These findings indicated a differential impact of planning across scales and types of subdivision and a mixed success of planning in mitigating the environmental impacts of rural residential development."

Maintaining connectivity is a relatively new approach in urban areas. As part of an overall biodiversity conservation strategy for Coquitlam, British Columbia, Canada, Rudd et al. (2002) performed a connectivity analysis to determine the numbers and patterns of corridors required to connect urban green spaces. For an urban area of 2,600 ha, with 54 green spaces (636.5 ha), Rudd et al. (2002) determined that 325 linkages were necessary to connect 50% of the green spaces. Within an urban zone that would demand backyard habitat, roads, and right-of-ways be used for connectivity. More research will be required in this arena.

Large-scale approaches to maintaining and preserving wildlands in Montana are under way. Acquisition and leasing important habitats by state wildlife agencies, federal natural resource agencies, and NGO (e.g., Montana Fish, Wildlife and Parks, U.S. Fish and Wildlife Service, Forest Service, Bureau of Land Management, National Park Services, the Nature Conservancy, The Trust for Public Land) provide for successful programs (Henderson and O'Herren 1992). Some examples include Montana's acquisition and lease of thousands of hectares of winter range in the Blackfoot-Clearwater Wildlife Management area, acquisition of the Lee Metcalf National Wildlife Refuge, streamside Rattlesnake Greenway with Land and Water Conservation Funds administered by the U.S. Department of Agriculture, Forest Service, and the Montana Legacy Project. The Montana Legacy Project has plans to conserve important forest habitats in northwestern Montana owned by Plum Creek Timber Company. The goals are to preserve vital wildlife habitat and water resources, conserve traditional access for hunting, fishing, and outdoor recreation, and to keep sustainable harvesting operations in the forests and timber in local mills. This project has partnered with the Nature Conservancy and The Trust for Public Land, and will purchase 126,467 ha of timberland for >$500 million. Details are being worked out, but most lands will eventually be absorbed by the U.S. Forest Service and the State of Montana.

Large (i.e., The Montana Legacy Project) and smaller acquisitions or conservation easements are beneficial toward land preservation for outdoor recreation, education, protection of natural habitats and open space, and the preservation of historical areas and structures (Henderson and O'Herren 1992). Because of the positive benefits to wildlife and wildlife habitat, land trusts are becoming more popular and active throughout the state (i.e., The Montana Legacy Project, Montana Land Reliance, The Nature Conservancy, Five Valleys Land Trust,

Gallatin Valley Land Trust, Flathead Land Trust, and Vital Grounds [Henderson and O'Herren 1992]).

Land-use planners have to be part of the solution of maintaining biodiversity in the face of the rapid land-use changes. Eight suggestions for land use planners and developers to consider how they impact natural landscapes were provided by Dale et al. (2000) and the Environment Law Institute (2003).

1. Examine the impacts of local decisions in a regional context.
2. Plan for long-term change and unexpected events.
3. Preserve rare landscape elements and associated species.
4. Avoid land uses that deplete natural resources over a broad area.
5. Retain large contiguous or connected areas that contain critical habitats.
6. Minimize the introduction and spread of nonnative species.
7. Avoid or compensate for effects of development on ecological processes.
8. Implement land-use and land-management practices that are compatible with the natural potential of the area.

Research Needs

More than 90% of Americans work or live in metropolitan areas causing problems for wildlife. Simply developing impervious surface cover ≥10% can be detrimental to sensitive aquatic species. When cover levels reach 25%, populations are impaired, resulting in ecosystems that are nonsupporting of native species (Center for Watershed Protection 2003). Because urbanization is a growing issue in North America, many of the problems aquatic species face have already occurred (e.g., containments, fish passage, impervious surface cover, altered fish assemblages, physical alteration of riparian areas). To enhance riparian habitats and species, habitat restoration of streams in urban areas will be necessary (Ehrenfeld 2000). In exurban landscapes, managers, planners, and biologists should work together to meet their common goals and monitoring should become a standard practice, so we can learn from the examples provided in rural and suburban landscapes.

The habitat for amphibians and reptiles altered by development will also need to be restored if the maintenance of native species is an important concern. In addition, roads need to be carefully planned as they fragment habitats (Carr and Fahrig 2001) and cut off connectivity to important habitat patches (Noël et al. 2007).

Avifauna is similarly negatively influenced by anthropogenic influences. Overall, species richness and diversity decline as urbanization increases. As such, numerous suggestions have been made to enhance and maintain avifauna. Unfortunately, the issues have not been tested, and research will be necessary to determine their success in enhancing populations. Common research needs identified include educating residents about their negative impacts on wildlife, developing buffer zones around sensitive areas, maintaining native plant communities that are more than 1 ha, maintaining and enhancing riparian corridors, determining how urbanization influences insects, and determining if nonnative vegetation can be used to restore habitats for avifauna. Cluster development of homes has also been suggested as a way to enhance wildlife habitats, but additional research is necessary to determine the ecological role of cluster housing.

As humans infiltrate relatively undisturbed landscapes, they bring the problems society is now trying to deal with: domestic animals, anthropogenic food that is available to wildlife, use of key areas precluding use by wildlife, poor or no management, or management without enforcement. To overcome these problems if humans and wildlife are to coexist, we need to recognize the negative impacts caused by domestic animals and pass and enforce regulations to minimize them. Numerous problems are caused by uncontrolled anthropogenic food available to wildlife, and some habitats are being altered to the extent that natural forage is no longer available. Each of these are serious problems that will cause challenges to those charged with managing wildlife resources, and it will be an expensive, time consuming, and sometimes controversial process if habitats for wildlife along the urban–exurban gradient are to be maintained and enhanced. Planning with enforcement has to be a key ingredient, or the unplanned, random, and chaotic urban development scheme will continue to alter habitats.

There are various calls for research to address the serious problems of urbanization and wildlife. The U.S. Environmental Protection Agency advocates four key areas of research (Munns 2006).

1. We need mechanistically based extrapolation research to improve the basis for predicting the response of wildlife from existing information.
2. Researchers should coordinate efforts to determine how toxicology influences population biology across heterogeneous landscapes.
3. Researchers should identify techniques that adequately measure the numerous stressors on wildlife.
4. Researchers need to identify the spatial and temporal scales appropriate for wildlife risk assessments.

There are numerous stressors to terrestrial and aquatic resources along the urban rural gradient: invasive and exotic species, nutrient enrichment, direct human disturbance, toxic chemicals, plus others (Munns 2006). To scientifically evaluate any stressors, we need better "... development of population dynamics models to evaluate the effects of multiple stressors at varying spatial scales, methods for extrapolating across endpoints and species with reasonable confidence, stressor-response relations and methods for combining them in predictive and diagnostic assessments, and accessible data sets describing the ecology of terrestrial and aquatic species" (Munns 2006:23).

In developing research to address the complex issue of altered wildlife habitat, better descriptions of the urban–rural gradient are needed and the selection of research sites should be representative. Researchers should also ensure that the metrics used to determine influences are adequate, so research can identify the mechanism affected by human activity that links population processes to community level patterns (Donnelly and Marzluff 2004).

The conflicts created for wildlife along the rural–urban gradient are human caused and will only be addressed properly (if they can) by collaborative cooperation. The first step in any program is to establish the objectives. What is it society wants: what species, how many, where, and when? Urbanization clearly has negative consequences to most wildlife, and only with careful planning and monitoring will the full effects be known or successful programs documented. There is abundant information along the rural–urban gradient that can be used in managing the influence of exurban development on wildlife. It all begins with open communication, clear statement of objectives, and appropriate use of biophysical-behavioral, social–structural, valuation, and institutional-regulatory forces to achieve the objectives desired.

CONCLUSION

Urbanization is the primary cause of species endangerment and a leading threat to biodiversity in the contiguous United States. Unfortunately, urban sprawl has only recently been addressed as a serious issue. Even states like Montana, which is geographically the fourth largest state in the United States and has <1,000,000 people, will not be immune to the negative issues that are facing wildlife due to development along the rural–urban gradient. Our objective was to review relevant literature that addresses how the cumulative effects of suburban and exurban growth influence fauna.

We cited >350 references in this review: 53% were related to game animals, 21% to nongame, and 16% to management and other topics. Most (96%) were peer-reviewed and either descriptive (67%) or gradient studies (17%). Only 7% used treatment and controls. Most study periods were short: <1 year (25%), 2–5 years (38%), and >5 years (8%). Over 25% of studies did not specify the length of their study. Our review concentrated on reptiles and amphibians, avifauna, and mammals.

Lizards, snakes, turtles, salamanders, and anurans have all decreased in the face of anthropogenic activities. Biologists and managers concerned with amphibian and reptile conservation need to be aware that relatively limited development of watersheds alone, not necessarily in the riparian area itself, or even directly upstream, can influence reptiles and amphibians.

Many of the avifauna studies examined were designed to measure various characteristics associated with avifauna communities and altered landscapes from urbanization and most covered large areas. In all cases, native avifauna benefited from native vegetation in undeveloped land or surrounded by undeveloped land. The problems with urbanization and avifauna have been clearly established, but few recommendations have been made to mitigate for urban, suburban, and exurban centers.

Most mammals do not benefit from urbanization (but some do), and serious problems arise due to larger species (e.g., coyotes, American black bears, ungulates) using urban areas. There is an abundant source of information about the biology of these species, documentation of the conflicts when associations with humans occur, and recommendations that have been suggested to minimize wildlife–human conflicts. Management plans and mitigation for conflict resolution between humans and American black bears and coyotes have been successful in some areas because information about the animal, habitat, and human attitudes were considered. The public involvement associated with human–wildlife conflicts is a critical component of successful management.

As with predators, the biggest concern of humans and ungulates revolves around damage and human safety along rural–urban gradients. Some authors have indicated the problems with human land uses and ungulate habitat, but research explaining how to enhance and maintain habitats for ungulates is not well developed. However, it is critical that the public be involved in all decisions. Because people have such a large investment in managing wildlife in urban areas, their attitudes have to be considered when formulating management plans.

Purchasing wildlife habitat for conservation is likely the most important way to minimize the anthropogenic effects of urbanization on populations. The arena of understanding wildlife and fish among the rural–urban gradient is relatively new to research. The conflicts created for wildlife along the gradient are human caused and will only be addressed properly by collaborative cooperation that leads to clear objectives and mechanisms to meet them. The problem has to be addressed, or the chaotic and planned consumption of habitat for wildlife will continue decreasing the quality of life for all organisms as cumulative effects accumulate.

10 Cumulative Effects on Freshwater Fishes

Scott A. Bonar and William J. Matter

CONTENTS

INTRODUCTION

Consider this scenario. You are a biologist conducting feeding experiments with fish in large tanks. Bacterial infections occur in your fish, and you treat them frequently with medicines designed to control the infections. The health of your fish improves for a few days, and then they fall ill again. Considerable time is lost caring for fish and waiting for their health to improve so they can be used in experiments.

Eventually, several fish leap out of their tanks—at a most inconvenient time when the director is touring the facility! A talented aquaculturist, the director reminds you to check the water quality in tanks. You find that before-dawn dissolved oxygen levels in tanks fell to levels which, although not lethal to the fish, contributed to their stress. This stress enabled ever-present bacteria to gain a foothold and ensured that fish were frequently sick. You increase oxygen to the tanks, and the disease problems stop. One factor alone could not have caused the problem, but the cumulative effect of poor water quality and presence of bacteria resulted in poor health of the fish.

The above scenario, which happened to one of the authors, illustrates that early in their careers, most fisheries biologists learn how critical the accumulation of effects on fish can be. Organisms in aquatic systems are probably far more likely to experience effects of multiple and cumulative abiotic and biotic forces than single, noninteracting forces. However, management agencies traditionally address one effect at a time: overfishing, habitat destruction, introduced species, disease, or pollution—and hope the problem disappears. Unfortunately, that does not always work and the importance of cumulative effects on fish populations is becoming increasingly recognized. Our objectives are to discuss how cumulative effects influence freshwater fishes, provide some of the more common examples of how effects accumulate in fish populations, and show how to assess and manage them.

HOW EFFECTS ACCUMULATE

The most devastating impacts to fish populations may not result from a single event, but a combination of individual minor actions occurring over time or space (Council on Environmental Quality 1997). These combinations of effects, or "cumulative effects," can occur in several ways.

Single Source, Additive Effect

An effect from a single source may occur at multiple times and have an additive effect. For example, an industrial or sewage treatment plant might release a pollutant at multiple times in a water body. Continuous, cumulative discharge of nitrogen pollutants was associated with loss of aquatic life below a chemical plant on the River Petite Paise in France (Dauba et al. 1997). Cleanup of the sewage outfall over 20 years allowed fish to recolonize the river.

Single Source, Multiplicative Effect

The effect from a single source may interact with the fish community, creating a multiplicative effect. The biomagnification of heavy metals in Columbia River fishes from the lead smelter in Trail, British Columbia, Canada, is an example of a multiplicative effect from a single source (Schmitt et al. 2002). These effects can biomagnify throughout the biota, affecting larger predators, such as largemouth bass (*Micropterus salmoides*), more than fishes occupying lower trophic levels such as bluegill (*Lepomis machrochirus*).

Multiple Sources, Additive Effect

Effects can result from multiple actions in an additive manner. Examples might include the additive effects of multiple water wells depleting an aquifer, or multiple commercial fishing operations harvesting a fish population.

Multiple Sources, Multiplicative Effect

Multiple actions that have multiplicative effects are some of the most complex to understand, but are typical, especially in larger water bodies, or waters in proximity to people. Managers and biologists spend millions of dollars and thousands of hours trying to understand the effects of multiple actions on Pacific salmon (*Oncorhynchus* spp.) populations migrating through the Columbia River. Dams pool water and slow river flow, forcing outmigrating salmon fry to expend more energy to swim downstream than they would have used if pushed along by downstream current. In addition, reservoirs behind dams support numerous piscivorous fishes that prey on young migrating salmon. Diversions from rivers for agriculture further reduce river flow, and development in the watershed can cause changes in water quality and sedimentation that reduce survival of salmon eggs.

CUMULATIVE FACTORS THAT AFFECT FISH POPULATIONS

In the Mekong or Mississippi rivers, the Great Lakes, small streams in Norway, or ponds in Alabama, major factors affecting freshwater fishes are similar. These include overharvest, introduced species, habitat destruction (that can encompass degradation in water quality and in physical and biotic habitat conditions), and disease and parasites. These forces can occur singly or in combination; however, they rarely occur alone, even though this is how they are usually studied and managed.

Cumulative Effects of Water Removal

In 1928, near the dusty town of Parker, Arizona, a war was about to start. This was not between two nations, or between Native Americans and European settlers, but between two states: Arizona and California. The governor of Arizona, B. B. Moeur, declared martial law on the Arizona side of the river because of California's attempts to appropriate more water from the Colorado River than Arizona thought legal. A total of 100 Arizona National Guard troops in 18 trucks arrived, some armed with mounted machine guns, to prevent California from constructing any structure on the Arizona side of the river that might help it pump water. Fortunately, the issue was settled by the federal courts before bloodshed occurred (Reisner 1986).

"Whiskey is for drinking, water is for fighting" was a common mantra across the American West. Waterways in arid regions have been diverted, dammed, diked, pumped, and evaporated (Figure 10.1). Walk suburban streets in Phoenix, Arizona, and Las Vegas, Nevada, and you will see few signs they are surrounded by desert. Phoenix contains verdant green expansive lawns, lakes within subdivisions, world-class golf courses, and swimming pools that dot the backyards of a third of the homes (City of Phoenix 2010). Travel to Las Vegas and see pirate ships conducting mock battles on artificial lakes, ride in a gondola down a simulated Venetian canal, and see a huge fountain keep time to a medley of tunes. The irony is that this water is sucked from the driest land of North America. As may be expected, providing water for the desert's fish and other aquatic species is usually a lesser priority.

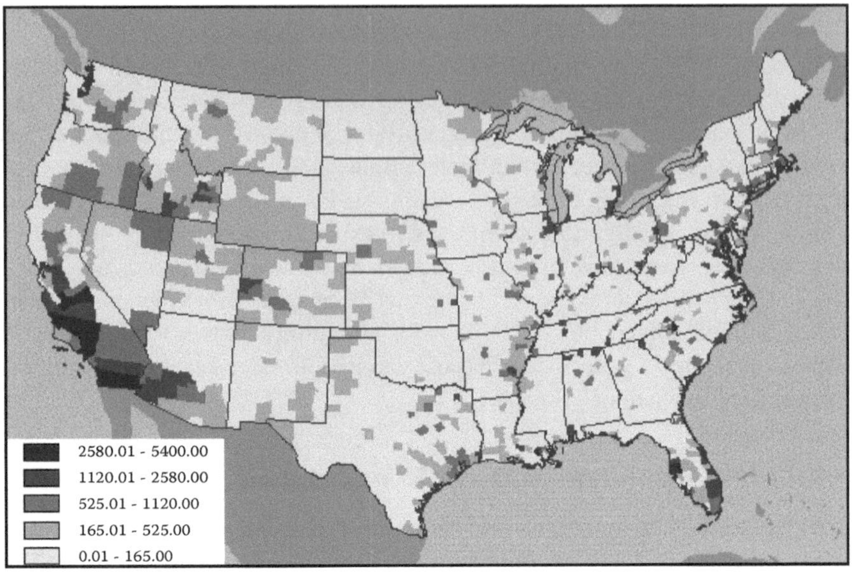

FIGURE 10.1 Total ground and surface water withdrawals in millions of gallons per day in 2000. (From National Atlas of the United States; available from http://www.nationalatlas.gov/natlas/Natlasstart.asp.)

Freshwater is in short supply throughout the world. Under current climate change projections, almost half the population of the world will be living in areas of high water stress by 2030 (United Nations 2006). Most of the fishes that live in arid climates of the world have been affected in some manner by water removal from surface waters and aquifers. Water removal is most often a multiple-source additive effect. Most commonly, water is removed through diversion of surface flow and by pumping of ground water. Individual effects can be very small, from a small well supplying water to a rural residence to huge diversions from the Colorado River to quench the thirst of cities such as Los Angeles and Phoenix. The cumulative impacts of hundreds of diversions and wells over time have substantially altered stream flow.

Water diverted directly from surface waters has the most direct effect on fish habitat. The amount of water affects carrying capacity and the amount of fish biomass it can sustain. Lowering water levels decreases potential fish biomass and can affect water quality and water temperature. Streams subject to reduced discharge are more subject to daily water temperature fluctuation, and increased warming during the day. Warmer water holds less oxygen than cool water, further increasing the potential for stress in fish. In September 2002, low flow in the Klamath River, California, was blamed for higher water temperatures, which in turn contributed to an infection of salmonids by the ubiquitous pathogens *Ichthyopthirius* and *Columnaris*. This resulted in the deaths of over 33,000 salmonids (California Department of Fish and Game 2004). Lower flow can force small fish out of drying stream margins and make them more available to predators (Riley et al. 2009). Low flow can result in reduced connectivity of stream sections. Riffles may dry and force some fish into pools that may not be able to support them. Increasingly, the genetic implications of flow reductions are being considered (Sato and Harada 2008, 2010). Lack of flow between sections of once perennial rivers fragments formerly connected fish populations. Habitat fragmentation may lead to inbreeding depression and declines in fish abundance (Fagan et al 2002; Sato and Harado 2008, 2010).

Water does not have to be diverted from surface waters to affect fish habitat. In many regions, water is obtained mainly from underground, and cumulative depletion of ground water aquifers can have considerable effects (Bartoloino and Cunningham 2003). The additive effect of individual water wells on the aquifer below the San Pedro River, Arizona, provides an example of profound effects on surface water flow (Arizona Department of Water Resources 2009). The human population in the Upper San Pedro River Basin increased approximately 1,200 people/year between 1980 and 2000. As of 2005, there were 6,127 registered water wells in the Upper San Pedro Basin. The annual stream flow near the settlement of Charleston on the San Pedro River decreased by about 66% from 1913 to 2002 (Thomas 2006). In July 2005, stream flow at the Charleston gauge went to zero for the first time in more than a century of keeping records. Many sections of the river are now dry and flow, even in the deepest sections, has greatly diminished. The additive effects of water removal, resulting in loss of flow is characteristic of most Southwest rivers.

Stringent laws are used to allocate surface and groundwater. Originally, regulation of water distribution concentrated primarily on surface waters. Early water laws were derived from Spanish and English laws, and were designed to allocate water among people. Spanish laws allow an individual to take water once the community needs are met. English law, which originated in a humid climate, allows users to take water from water flowing through their property with few restrictions. Water law in the western United States is more restrictive than that of more humid climates (Pearce 2003; Water Education Foundation and University of Arizona Water Resource Research Center 2007). For example, in Arizona, the doctrine of prior appropriation is employed. "First in time, first in right" means that water cannot be removed if it will affect flows allocated previously. Groundwater in Arizona is regulated by the 1980 Groundwater Management Act. Land developers in Arizona must demonstrate an "assured water supply" that will be physically, legally, and continuously available for the next 100 years before the developer can record plats or sell parcels (Arizona Administrative Code, Title 12, Chapter 15). Only recently have environmental uses of water been emphasized (Water Education Foundation and University of Arizona Water Resource Research Center 2007). For example, in 1990 the Arizona Department of Water Resources (ADWR) upheld instream flow "rights" to support fish and wildlife populations by The Nature Conservancy in two southeastern Arizona streams. The right of the ADWR to withhold water for instream uses was later upheld by the Arizona Court of Appeals.

Perhaps one of the most famous cases where law was used to protect water rights for a fish population occurred in the 1970s. Close to Death Valley National Park lies a small area with more endemic species in a small locale that any other place on the North American continent (Minckley and Deacon 1990). Ash Meadows is an aquatic "island" in a vast sea of desert, containing numerous springs, fed by water that slowly percolates from nearby mountains. Adjacent to Ash Meadows is a cleft in the rock, containing a water-filled cavern. This cavern, Devil's Hole, contains a fish called the Devils Hole pupfish (*Cyprinodon diabolis*). The Devils Hole pupfish depends on a single rock shelf for spawning, located immediately under the water's surface. Because of its unique nature, Devil's Hole was made part of Death Valley National Monument by President Harry S. Truman. However, even though Devil's Hole is protected, the spawning shelf started to dry up in 1968, due to nearby water well pumping to provide water for alfalfa production. Court battles and rapid attention by conservationists during the late 1960s and early 1970s protected the water level in Devil's Hole. The U.S. Supreme Court ruled that actions outside of a national monument or park (e.g., groundwater pumping) could not be allowed to affect the survival of a species in a national park. This landmark decision was important during the formation of the U.S. Endangered Species Act.

Cumulative Watershed Development Effects

"The most beautiful river on earth. Its current gentle, waters clear, and bosom smooth and unbroken by rocks and rapids, a single instance only excepted." This quote does not describe the Yellowstone, the Yukon, or the River of No Return, but the Ohio River, as Thomas Jefferson saw it, in the 18th century (Jefferson 1781–82, p. 14). The view of the river as a pristine resource changed soon after settlement. Agricultural lands were developed along the river, and as early as 1876, erosion was of such a concern that Ohio required landowners to give a day of labor each year to plant willows along the riverbanks. This may be the first law of its kind to promote river bank stabilization (Frost 1976). By 1994, a U.S. Environmental Protection Agency (EPA) report stated that siltation and excess organic matter were the major pollutants in streams and rivers in the Ohio River Valley. It was determined that 57% of streams were impaired by siltation, and 32% of the waters were impaired by organic enrichment. These pollutants were due primarily to cumulative effects of watershed development, principally resource extraction and agriculture (Environmental Protection Agency 1994). In fact, a later report to Congress listed agriculture as the primary source of pollutants to the nation's rivers and streams (Environmental Protection Agency 2009).

Urbanization and resource extraction (e.g., timber harvest, mining, agriculture) in a watershed can have significant effects on fish communities. The impacts to aquatic communities are almost never due to a single source of watershed disturbance. Urbanization affects flooding and lowers base flows in streams, chiefly by increasing the amount of impervious surfaces such as parking lots, roads, and roof-tops. Wang et al. (2001) reported that connected impervious surfaces of 8%–12% in the watershed were a threshold above which major changes in conditions of Wisconsin streams could occur. Just over 10% impervious surface cover has had major affects on fish diversity, abundance, and recruitment in many systems (Schueler 2000). Brabec et al. (2002) concluded that 3.6%–12% cover by impervious surfaces could have significant effects on fish populations, and that a simple threshold of impervious cover was not appropriate because the types of pervious cover and riparian buffers might mitigate runoff from impervious surfaces. Typically the "flashy" stream conditions caused by rapid runoff from impervious surfaces result in replacement of species that are unable to tolerate these conditions. In the King County, Washington, metropolitan area the more "flashy" stream environments favored cutthroat trout (*Oncorhychus clarki*), which were able to tolerate the more variable flow conditions, over sockeye salmon (*Oncorhychus nerka*; Lucchetti and Fuerstenberg 1992).

Removal of riparian vegetation by logging, grazing, and urbanization can cause stream warming and increased erosion. Increased flow can also result in incised channels, or down cutting (Figure 10.2), where the erosive power of the stream exceeds the capability of the streambed to resist erosion. Down cutting generally starts with the removal or degradation of riparian vegetation, which results in rapid water flow into the stream. The stream bed is eroded by high-velocity water, and a deep channel is cut. The groundwater level is lowered and much preexisting riparian vegetation dies. At the upper end of such channels, water may spill over a waterfall, or into a plunge pool, which is the site of active down cutting (i.e., nick point). This nick point moves upstream. Below the nick point, high earth banks become unstable and slough off into the stream channel. The channel widens and shallows because of this sloughing. Downstream, the channel may gradually reestablish itself as sediment is deposited and riparian vegetation is reestablished. Down cutting is a major feature of many streams where riparian communities have been removed.

How runoff enters streams can also affect flow, habitat, and fish populations. Small fishes depend on shallow, slow-flow side channels and marshes as rearing habitat. More

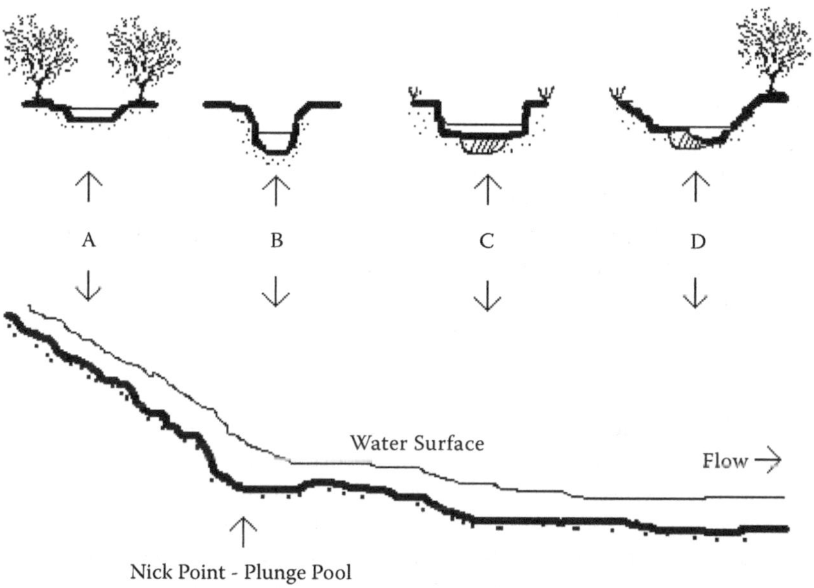

FIGURE 10.2 Downcutting process in streams. Upstream is the original channel (A) with riparian vegetation intact. Downstream is an area of active down cutting (B), usually started when riparian vegetation is removed, and the streambed is no longer resilient to the higher flows previously held in check by the vegetation. This site of down cutting is moving upstream. Below the down cutting site, sides of the deeper channel slough into the stream bed, resulting in a wide shallow channel (C). Farther downstream, sediments aggregate, riparian vegetation regrows, and the channel gradually restores (D). (Adapted from Hunter 1991.)

than 98% of the American prairie and vast tracts of woodland have been replaced with agricultural lands (Blann et al. 2009). Runoff can enter streams directly from fields or through channels draining farm fields. Channels draining fields often transform mosaic marshlands providing fish-rearing habitat and a buffer from high runoff flows into linear systems such as ditches or subsurface piping (Blann et al. 2009).

Pollution from developed watersheds can affect fish communities independently of stream flow. Sediment can smother fish eggs, eliminate interstitial spaces used by invertebrates, and provide nutrients that increase biological oxygen demand, resulting in lowered oxygen. Acid mine drainage can lower the pH of receiving waterways so that they are uninhabitable to fish and wildlife populations. Heavy metals from mining operations can bioaccumulate in fish and wildlife tissues.

Even slight physical modifications in watersheds can affect predation among fishes. Increased ambient light levels resulted in slowing or stopping of the outmigration of sockeye salmon fry, resulting in substantially increased predation by cottids in a Washington stream (Tabor et al. 2004). This work suggests that even urban lighting may affect nearby streams and lakes

Reversing negative effects of watershed development involves corrective land-use practices over the entire watershed, such as increasing the amount of perennial riparian and upland vegetation (Vondracek et al. 2005) or reducing the percent of impervious surfaces. Management practices have focused on slowing and treating runoff before it reaches streams or water bodies, often by directing water through settling ponds, constructed wetlands, and buffer strips. Settling ponds collect water runoff and allow pollutants to sink to the bottom of the pond before the water is returned to the water body. Constructed wetlands can collect and filter sediment and wastes. Settling ponds and constructed wetlands help reduce the adverse affects of mine waste (Chadwick and Canton 1982; Moshiri 1993), stormwater (Moshiri 1993; Wanielista and Yousef 1993; Bernhardt and Palmer 2007), various industrial pollutants (Moshiri 1993), and agriculture runoff (Moshiri 1993; Bouchard et al. 1995). Buffer strips are areas of riparian vegetation, along the margins of a stream, lake, or river that are left in place to stop and hold runoff. Width of effective buffer strips has been studied in different regions; in Australian logging operations, no significant effects were noticed when buffer strip width exceeded 30 m (Davies and Nelson 1994). Osborne and Kovacic (1993) reviewed studies of riparian buffers in U.S. agriculture and logging operations and reported that buffers between 10–30 m effectively maintained stream temperature, but effective width for sediment trapping was less clear and depended on vegetation type, soil characteristics, slope, and other factors. However, even riparian buffers cannot completely mitigate the effects of watershed development in agricultural or logged areas (Nerbonne and Vondracek 2001; Pollock et al. 2009). In an analysis of 40 previously logged watersheds on the Olympic Peninsula, stream temperatures in logged watersheds were significantly higher than those that were not logged, even after canopies had grown back over streams (Pollock et al. 2009). These effects may be the result of microclimate effects of the logging operations, increases in temperature of the shallow groundwater flowing into streams from logged areas, more light entering stream corridors through riparian canopies in logged watersheds than those that are not logged, or other factors away from the stream that still affect stream temperatures (Brosofske et al. 1997; Pollock et al. 2009).

Cumulative Impacts of Instream Physical Habitat Changes

Some have argued that dams, specifically a dam, won World War II. The construction of Grand Coulee Dam was one of the pinnacles of the Roosevelt dam building years. Constructed on the Columbia River where it flowed through the middle of a vast, windswept, isolated, sagebrush (*Artemisia tridentate*)-filled steppe, Grand Coulee Dam was the largest concrete structure in the world, and a power-producer extraordinaire. Massive electric power was needed to produce aluminum, and aluminum was required to build the great air armadas the Allies needed to defeat the Axis. The aluminum produced by power generated from Grand Coulee Dam was used by aircraft manufacturers such as Boeing to build bombers, fighters, and transports. Electricity produced by the dam was also used to power plutonium production reactors and reprocessing facilities at the Hanford Site, a key part of the Manhattan Project that was responsible for the atomic bomb dropped on Japan (Reisner 1986). However, salmon were not a priority for those building Grand Coulee Dam, and no fish passage facilities were built. A total of 1,600 km of salmon spawning habitat was lost in the Upper Columbia River Basin.

Cumulative physical changes, such as dams, made at multiple sites within river or lake systems, impact fish populations. Natural flooding patterns are important in the reproduction of many species, but patterns of flooding were altered with the damming of rivers and regulation of water releases to match demands for power and water supply. The Colorado River historically had huge floods, with at least 15 floods with a peak discharge greater than 5,500 m^3/s (O'Connor et al. 1994). However, the frequency and magnitude of floods are greatly reduced because of the damming of the Colorado. Physical habitat changes caused by reduced flooding and cold water released from the hypolimnion of reservoirs have changed the environment available for many warm-water native Colorado River fishes such as bonytail chub (*Gila elegans*), razorback sucker (*Xyrauchen texanus*), humpback chub (*Gila cypha*), and Colorado pikeminnow (*Ptychocheilus lucius*). For riverine species to persist, they must be able to survive and reproduce within the unique environments below large dams or in the lake-like environments of reservoirs created upstream.

Channelization, culverts, diking, removal of vegetation, and addition of bank protection are other instream physical changes that have impacted fishes. One of these by itself may not be significant, but the cumulative effects of many have considerable impact. Along a new section of highway on Canada's Labrador coast, 53% of the culverts posed problems to fish passage because of poor design or construction (Gibson et al. 2005). Channelization of rivers can result in lower fish density and diversity than comparable nonchannelized areas (Oscoz et al. 2005).

In-stream and in-lake habitat changes can also affect water quality. When rivers are dammed, water flow is slowed, and flocculent material can sink to the bottom. When the material contains pollutants, it can be trapped in the sediments of the bottom of the reservoir. This sediment is typically unavailable to fish unless through turnover, or reservoir discharge, the sediments are resuspended. Debol'skii (1996) reports the presence of large gel-like concentrations in stagnant areas of the bottom of Volga River reservoirs in Russia. These formations range in size from hundreds of square meters to tens of square kilometers and up to 2 m thick. Rapid drawdown of these reservoirs to allow discharge of these pollutants into the water column could have catastrophic effects on the fisheries resources of the Caspian Sea and Volga River.

Remediation of in-stream effects usually means removal or altering of the physical structure that is causing the impact, or intensively managing or mitigating fish and wildlife impacts. Dam removal is receiving increased emphasis as a management tool (Shuman 1995; Bednarek 2001) and has been employed in selected areas. In-stream and in-lake habitat improvement projects have been employed frequently. In-stream habitat improvement projects include the addition of timber, boulders, or nutrients (Roni et al. 2002), whereas in-lake habitat improvement can consist of aquatic plant control, dredging of accumulated sediments, or bank stabilization (Kohler and Hubert 1999). Aquatic sites that have been isolated through road building, culvert construction, and other means can be reconnected, often adding large tracts of fish habitat. Reconnection can be an effective method where invasion of nonnative species is not a concern. The sequence of restoration activities can be prioritized by first protecting existing high-quality habitat, next restoring habitat connectivity and natural watershed processes to the degree possible, and finally concentrating on in-stream or in-lake habitat enhancement (Roni et al. 2002).

Cumulative Effects of Multiwater Quality Stressors

Water pollution can cause stress to fish, resulting in the release of corticosteroids and increase in immunosuppression (Donaldson et al. 1984; Kime 1998). This increase in immunosuppresion can result in a lessened ability of fishes to fight diseases. The increasing water temperature stressed the fish in the Klamath River mentioned above, resulting in a kill of salmonids by their increased susceptibility to diseases.

Pollutants can enter water bodies through point sources (e.g., discharge pipes, factories, canals) or nonpoint sources (i.e., runoff from agricultural fields, lawns, timbered areas, failing septic tanks). Nonpoint source pollution is currently the leading cause of water pollution in the United States, and can consist of nutrients, sediments, and other materials. For most freshwaters, phosphorus is a limiting nutrient to fish production. Phosphorus additions up to a certain level can increase fish production. However, excess phosphorus or other nutrients introduced into a water body can encourage too much primary production, resulting in depleted oxygen concentrations, especially at night when plants are respiring, and from biological oxygen demand from decaying matter in the water body. When dissolved oxygen falls below 5 ppm, fish kills can result. Some species are highly susceptible to oxygen below the 5 ppm level (e.g., trout, salmon), while others can tolerate lower amounts of dissolved oxygen in the 2–3 ppm range (e.g., catfishes, pupfishes, carps). Usually a single source of nutrients is not responsible for excess primary production. It is the cumulative effects of many sources of nutrients to a water body. Agriculture, lawn fertilizers, failing septic tanks, and nutrient runoff from parking lots all need to be accounted for. Quantification of sources of this pollution can be conducted and the contributors to lake eutrophication can be identified and prioritized (Guo et al. 2004).

Humans flush a broad variety of compounds into wastewater systems, many of which persist through treatment processes and are discharged into rivers, reservoirs, and lakes. Some of these organic wastewater compounds (OWCs) mimic normal hormones, especially sex hormones, or disrupt normal hormonal processes in aquatic animals, especially fish (Kime 1998). Known and suspected endocrine disrupting compounds come from detergents and their metabolites, fire and flame retardants, fragrance and flavoring compounds, fuels, herbicides and insecticides, plasticizers and antioxidants, steroids, and prescription and nonprescription drugs, including human hormones. Even at concentrations far below lethal levels, some of these endocrine disruptors can cause feminization of male fish or masculinization of females or disrupt the intricate, species-specific endocrine processes that govern maturation, production of eggs and sperm, and reproductive behaviors such as territoriality, mate selection, and spawning. The chronic effects of low doses of mixtures of OWCs are poorly known, but they may have long-term harmful effects on individuals and populations through alteration of fertility, fecundity, and population viability. As sources of freshwater become scarce, especially in arid regions, water reuse will likely increase the probability of finding harmful OWCs in water bodies.

Cumulative Effects of Multiple Sources of Harvest

Overharvest has been common in sport and commercial freshwater fisheries (Kohler and Hubert 1999). Overfishing can put entire populations or species at risk, such as the overfishing of beluga sturgeon (*Huso huso*) for caviar (Billard and Lecointre 2001), or it can suppress specific size groups (Noble and Jones 1999), such as large black basses (*Micropterus* spp.) or walleye (*Sander vitreus*). Overharvest is a cumulative effect, occurring over a short period of time by multiple sources (e.g., different anglers and methods of harvest), over longer periods of time with fewer anglers, or any permutation in between. While overharvest is most commonly discussed in the context of marine fish populations, it can be significant in freshwaters also. The effects in freshwater systems are through commercial and recreational fisheries. Lewin et al. (2006) reported that exploitation rates in recreational angling fisheries (i.e., the proportion of fish from a population that can be removed over a specific time interval, typically a year) have varied from less than 10% to greater than 80%, depending on the effort. Fish harvest can have a variety of effects. Population structure can be truncated to smaller, younger individuals because larger individuals are more attractive to anglers (Olson and Cunningham 1989; Beard and Kampa 1999). Some species are favored over others by fishers, causing a decline in favored species and an increase in nonfavored fish. Small population sizes of the overharvest species may lead to loss of genetic diversity, evolutionary responses, and dispensatory responses (Lewin et al. 2006).

The full extent of cumulative overharvest on freshwater fish populations may be underestimated (Humphries and Winemiller 2009). North America and Australia, continents that have exhibited a relatively recent explosion in human population, provide the most dramatic examples of how substantial the effects of cumulative overfishing may have been (Humphries and Winemiller 2009). Freshwater fish populations on both continents were large and diverse before European settlement. Fish in the Ohio River valley were described as being so numerous that a spear thrown into the water rarely missed one (Trautman 1981). Lewis and Clark reported superabundance of salmon in the Pacific Northwest region of the United States, citing how this abundance contributed to large populations of indigenous people. Nineteenth century explorers and ranchers in Australia were astonished at the numbers of fishes that could be easily caught. Winemiller et al. (2008) and Humphries and Winemiller (2009) report examples of huge fish populations in areas that received almost no fishing, such as the isolated Casiquiare River basin in southern Venezuela. They argue for more protected areas in freshwaters.

Cumulative effects of overfishing are corrected through a variety of regulations (Noble and Jones 1999). Closed waters (i.e., protected areas), protected species, and protected populations, where no harvest can be tolerated, may be necessary, especially for species that are threatened or of special management significance. Minimum size, maximum size, and slot limit regulations are all used to allow some harvest, but reduce pressure on some part of the population to protect spawning fish, or specific size groups that are being overharvested. Creel or bag limits are used primarily to distribute the catch among anglers and do not reduce total harvest until the creel limit is reached. Gear restrictions are also used to make it more difficult for anglers and commercial fishers to remove too many fish.

Cumulative Effects of Multiple Biological Stressors

Movement of fishes to the western United States was a priority of the U.S. Fish Commission in the late 19th and early 20th centuries. Fish from the Mississippi River drainage and Europe were moved, primarily by rail and ship, to Western drainages to provide angling opportunities and food for immigrants and pioneers who wanted familiar fish (Dextrase and Coscarelli 1999). The scope of this movement was immense. From 1903 to 1923, over 72 billion fish were distributed by fish cars and detached motor vehicles traveling over 16 million km. During a 2-month period, fish were sent to almost all states, and cars were operating 24 h/day (Leonard 1979). These introductions often were successful in establishing nonnative fishes in Western rivers, and today 52% of the length of Western streams now contains nonnative fishes (Schade and Bonar 2005).

Exotic species have been introduced to freshwaters with considerable frequency throughout the world. These introductions were made to provide food or sport, and some introductions were accidental. Introduced species often have impacted native fish populations through predation, competition, hybridization, and modification of the environment (Courtenay and Stauffer 1984; Ogutu-Ohwayo 1990; Li and Moyle 1999; Wydoski and Wiley 1999). All aquatic systems have a limited capacity for providing the space and resources required by native species. As introduced species become established and proliferate, they nearly always use some resources used previously by native species (Courtenay and Stauffer 1984; Chick and Pegg 2001). Introduced aquatic species can also impact habitat conditions for aquatic species, such as cover and water quality (Claudi and Leach 1999).

In some instances, introduced species can hybridize with native fishes, producing genetic mixtures or sterile hybrids (Carmichael et al. 1993). Novel diseases introduced into aquatic systems are an unintended consequence of introducing nonnative organisms. Recently, viral hemorrhagic septicemia, a fish disease originally from Europe that has substantial effects on a variety of fishes and listed as reportable with the World Organization for Animal Health, was discovered in the Great Lakes (Groocock et al. 2007). Asian fish tapeworm (*Bothriocephalus acanthodii*) has spread throughout waterways of the United States because of transfer of cyprinids for bait and aquatic plant control. Whirling disease and largemouth bass virus are other examples of diseases brought into new areas with stocked fish. Notable examples of introduced species affecting ecosystems include introductions of largemouth bass in lakes of Central America, colonization of the U.S. Great Lakes by sea lamprey (*Petromyzon marinus*) and zebra mussels (*Dreissena polymorpha*), stocking and proliferation of Nile perch (*Lates niloticus*) in African rift lakes, expansion of hydrilla (*Hydrilla verticillata*) across the southern United States, the introduction of common carp (*Cyprinus carpio*) to Australia, and hybridization between nonnative rainbow trout (*Oncorhynchus mykiss*) and Apache trout (*O. apache*; Courtenay and Stauffer 1984; Ogutu-Ohwayo 1990; Carmichael et al. 1993; Langeland 1996; Claudi and Leach 1999; Wydoski and Wiley 1999; Helfman 2007).

It is instructive to reflect on just how large a problem invasive species pose. As of 2000, the costs to control, manage, and prevent invasive species in the United States were $137 billion each year (Pimentel et al. 2000). In comparison, the first Gulf War cost approximately $61 billion (U.S. Department of Defense 1992), the revenue generated by all U.S. recreational fishing in 2001 was $36 billion (U.S. Department of the Interior, Fish and Wildlife Service and U.S. Census Bureau 2001), and the cost of all the World's Natural Disasters in 2002 (United Nations Environment Programme News Release 2002/78) was $70 billion.

Cumulative impacts of invasive species are particularly difficult to predict. Many species are intercepted at a nation's border by agricultural inspectors. Most of those that get through will not become established, and of those that do, only a proportion (around 15% in the United States) cause severe harm (U.S. Congress, Office of Technology Assessment 1993). However, the effects of many introduced species can accumulate on an ecosystem.

What allows an introduced species to flourish in a new environment? Certainly the natural enemies of an organism in a new environment do not exist as they did in the native range. Torchin et al. (2003) examined 26 host species of mollusks, crustaceans, fishes, birds, mammals, amphibians, and reptiles, and found that introduced populations contained half the number of parasites found in the same species in their native range. Release of organisms from their natural enemies is one reason why introduced organisms may flourish in their new environments (Keane and Crawley 2002; Shea and Chesson 2002).

Some of the most famous water bodies where multiple exotic species have had devastating cumulative impacts are the Great Lakes of North America. Since the 1800s, over 140 species have been introduced into the Great Lakes (Mills et al. 1993; Mills et al. 1994; Glassner-Shwayder 2000). For the last 40 years, the rate of introductions has averaged 1.2/year. Nonindigenous species have entered the Great Lakes through many pathways. Construction of the Erie Canal and the Saint Lawrence Seaway removed barriers to colonization of the lakes by exotics such as sea lamprey. When the first barge traveled from Lake Erie to New York Harbor, it brought a keg of Lake Erie water to be symbolically dumped into New York harbor. Water from the Thames, Seine, Elbe, Rhine, Orinoco, Ganges, and Nile rivers were also dumped into the harbor to symbolize the event (Mills et al. 1994). No doubt little attention was paid to any organisms that entered the harbor through this ceremony, which foreshadowed the large-scale release of ballast water into the Great Lakes from ships entering the newly opened water bodies. Ballast water from ships traveling from around the globe has been released into the lakes, polluting them with organisms moved from other continents. Some nonnative fishes were stocked or appeared by accident in other ways.

Some of the major exotic species that proliferated in the Great Lakes were the sea lamprey, a parasite that uses a rasping tongue to cut holes in fish and suck out body fluids; alewives (*Alosa pseudoharengus*) and rainbow smelt (*Osmerus mordax*), which compete with the young of other fishes for zooplankton; zebra and quagga mussels (*D. bugensis*) that clog water intake structures and compete for food of native organisms, especially filter feeders; and round (*Neogobius melanostomus*) and tubenose (*Proterorhinus marmoratus*) gobies that displace small-bodied benthic fishes such as mottled sculpin (*Cottus bairdi*); and Eurasian ruffe (*Gymnocephalus cernuus*) that displace and compete for food with native fishes also (Jude and Leach 1999; Bauer et al. 2007). Pacific salmon were introduced to the Great Lakes to provide important opportunities for recreation in a put-grow-and-take fishery (Jude and Leach 1999).

Introductions of fishes and invertebrates and other factors such as habitat degradation and overfishing resulted in the declines of native fish stocks in the Great Lakes, most notably lake trout (*Salvelinus namaycush*), lake whitefish (*Coregonus clupeaformis*), cisco (*Coregonus artedi*), yellow perch (*Perca flavescens*), and walleye. Recent emphasis of management agencies has been in rebuilding stocks of native fishes (Jude and Leach 1999).

Prevention and quarantine is a primary method of protection against invasive exotic species. For the Great Lakes and large river shipping, programs are in place to require ships to dump ballast water offshore in marine waters so potentially invasive freshwater organisms cannot survive. Programs such as the 100 Meridian initiative were developed to inform those transporting boats, scuba equipment, fishing tackle, or other gear across geographic regions that unwanted invaders might be inadvertently transferred unless precautions are taken (Mangin 2001). Biologists sampling different water bodies are informed how their practices can be altered to minimize the transfer of invasive species (Jacks et al. 2009). Once species have invaded, they are much more difficult to control. Chemical, mechanical, and biological control methods are all employed (Kohler and Hubert 1999; Claudi and Leach 1999), but are usually only effective in removing or suppressing the abundance of introduced species in a small area.

Physical Habitat/Biotic Interactions

The sea lamprey would not have invaded the Great Lakes successfully unless a few physical habitat changes were shifted in its favor. First, a physical corridor was constructed that allowed its invasion from the Atlantic Ocean—the Erie Canal and the Saint Lawrence Seaway (Mills et al. 1994). Lamprey then became established in lakes Huron, Michigan, and Superior. Effects of lamprey in Lake Ontario would not have been as great without concomitant deforestation of the watershed. The newly silted tributaries to Lake Ontario provided ideal habitat for larval sea lampreys, and allowed their populations to explode (Jude and Leach 1999). Thus, the cumulative effects of physical (i.e., corridors and watershed construction) and biological (i.e., sea lamprey) changes combined to help lamprey abundance grow and stocks of native lake trout to collapse.

Modifications in habitat have often led to conditions that favor one fish species over another. Moyle and Light (1996) report that the most important factor determining the success of an invading fish species is the match between the invader and hydrologic conditions. Although their work was in California streams, it applies to other areas. Furthermore, they conclude that in aquatic systems with high levels of human disturbance, a much wider range of species can invade than in systems with low levels of human disturbance.

Dams are a prime example where physical habitat alterations interact with the biota to produce cumulative effects on fishes and other native species. The Colorado River in the western United States was historically a turbid, warm, highly fluctuating river. Before dams were built on the river, flood flows of the Colorado River scoured the basin and created significant habitat modifications, building sandbars and even creating the Salton Sea. The reddish muddy flow of the Colorado River moved so thickly into the upper Gulf of California that it may have led to the name given it by the early Spanish explorers, the Vermillion Sea (Adler 2007). The fishes that evolved to survive under Colorado River conditions are like no others in the world. They had dorsal humps for orientation in the fast current; small eyes because good eyesight was of little use in the dark, murky waters; strong caudal muscles to fight the powerful current; and adaptations to spawn and rear young in warm waters. When dams were built to tame the Colorado River, the heavy load of sediment sank to the bottom of reservoirs. Water discharges from dams came from the hypolimnion of reservoirs and were significantly cooler than in pre-dam conditions, making much of the river more suitable for salmonids and cool-water fishes than the native warm water species. Water clarity increased, making it easier for sight feeding predators to eat the young of the native fishes. Fishes from the eastern United States and Europe were introduced into the river and the newly created reservoirs. In many instances the changes in habitat combined with increased predation from and competition with nonnative fish species, worked in a cumulative manner to impact native fishes such as razorback sucker, humpback chub, flannelmouth sucker, and Colorado pikeminnow. If exotic fishes had not been introduced, the effects of the reservoir pooling caused by the dams could have been much less. If the exotic species had been stocked but the dams never built, the impact of nonnative species may have been less because they would not have survived as well as the native fishes that evolved in the fast-flowing, turbid conditions (Meffe 1984). The cumulative effects of dams and exotic species had a huge impact on the fish populations.

Dams can combine with preexisting native fishes to change community dynamics as well. The riverine environment of the Columbia River system used to carry salmon fry downstream to the ocean where they grew to adults. However, the pooling of the Columbia River, caused by multiple dams, slowed the salmon fry from moving downstream. Also salmon fry passing through the dams were concentrated in the forebays, resulting in heavy predation by native northern pikeminnow (*Ptychocheilus oregonensis*). Northern pikeminnow predation was so great that bounty programs were put into effect to lower this source of predation to the juvenile salmon (Friesen and Ward 1999).

Physical structure can concentrate fishes to be eaten by other predators besides fishes. Gull predation was responsible for an estimated loss of about 10% of juvenile salmonids near a dam and water diversion structure on the Yakima River, Washington (Major et al. 2005). Approximately 2% of the spring migration of salmonids on the Columbia River was estimated to be eaten by gulls below Wanapum Dam, and predation rate was significantly correlated with passage rate through the turbines and spillways (Ruggerone 1986). Pinnipeds and other mammals such as river otter (*Lutra canadensis*) all can feed significantly on migrating fishes concentrated below dams and at water intakes.

Physical removal of riparian vegetation from stream banks can warm stream water, enabling fishes that prefer warmer waters to move into the affected area. Down cutting of streams, resulting also from destabilization on stream banks, can result in a shallower, wider channel, more susceptible to the effects of warm (or very cold) temperatures that can favor other fish species. Carveth et al. (2006) found that many native desert fishes had lower temperature tolerances than some of the most prevalent introduced fishes. Reduction of the cooling cover of riparian vegetation, coupled with climate change, would result in warmer water temperatures preferred by some of the introduced fish species.

ANALYZING CUMULATIVE EFFECTS ON FISHES

There are a multitude of ways effects can accumulate. Fortunately, examples and techniques are available to demonstrate how cumulative effects can be identified, analyzed, and corrected.

Examining Cumulative Effects in the NEPA Process

When development in the watershed is being considered, especially by an agency, it is often necessary to submit to the National Environmental Policy Act (NEPA) process. This process was placed into law to require those proposing an action to consider all possible effects of that action, including cumulative effects that could occur over time or over space. Chapter 2 describes a cumulative effects analysis in NEPA in more detail. The Council on Environmental Quality (1997) provides a useful overview for analyzing cumulative effects (Chapter 4).

A first step in a cumulative effects analysis is to identify the geographic scope and the time scale of interest. In aquatic systems, the geographic scope of interest is typically the watershed. Next, list all possible factors affecting the resource and prioritize which cumulative effects may be significant. Narrative, qualitative, and quantitative means are all available to prioritize the effects. In this manner, one can usually come up with some of the most important effects, either singly or in combination.

Using Structured Decision-Making to Choose among Alternative Management Actions to Address Cumulative Effects

When numerous effects accumulate, trying to determine which effects should be addressed first or what management alternatives should be used to address the effects is a daunting challenge for natural resource managers. The inclusion of many different effects and many different alternatives can make the decision process complex. Fortunately, decision science has developed strategies so biologists can make decisions about how to manage cumulative effects and a variety of alternatives to address the effects in an organized, structured manner (Hammond et al. 1999; Runge et al. 2009). This process is called structured decision making (Runge et al. 2009).

We will use an example of management of bull trout (*Salvelinus confluentus*) populations to demonstrate how structured decision making works. Assume a section of stream in the Pacific Northwest contains an imperiled bull trout population. This population has been affected by a series of cumulative anthropogenic effects. The watershed was logged, and because of the poor logging practices used, the timber was cut close to the water's edge. The logging activity resulted in sediment running off from the watershed into the stream and in warmer stream temperatures because of loss of riparian shade. Brook trout have invaded the watershed, competing and hybridizing with the bull trout. The watershed is isolated, but three wells were created to provide water to three houses located near the stream. Because bull trout require colder water than many other trout as part of their life history, global climate change is thought to further imperil the fish.

Now the challenge is to choose a set of management strategies to address the cumulative effects in the most effective manner possible. A simple method to carry out structured decision-making problems is called PrOACT. This acronym refers to the steps of structured decision making: **Pr**oblem definition, **o**bjectives, **a**ctions or alternatives, **c**onsequences, **t**radeoffs and optimization. Problem definition involves defining a problem in such a way that it is made into a question requiring a decision. For example, the problem statement "Bull trout numbers are declining" is first changed into "How do we increase numbers of bull trout?" Framing the problem appropriately is arguably the most important step of the process. Next, objectives are defined. In a cumulative effects problem, objectives can address each of the effects and are included with some indication of direction (maximize, minimize). The bull trout example objectives might be: maximize spawning habitat, minimize predation, minimize hybridization with brook trout, minimize angling pressure, maximize survival of juveniles, and maximize survival of adults. Alternatives are developed that might address each of the effects. Thinking creatively when developing different alternatives and creating groups of alternatives (portfolios) that might address more than one objective is beneficial at this stage. Once objectives and alternatives are refined, they are put into a consequences table that gives the consequences of each alternative for each objective (Table 10.1).

The consequence table presents all of the objectives, alternatives, and consequences in a structured manner. The next task is to reduce the set of alternatives and objectives to a manageable set, and decide which alternative should be selected. Alternatives that score lower than other alternatives on every objective, called

TABLE 10.1
A Structured Decision-Making (SDM) Consequences Table for an Imperiled Population of Bull Trout[a]

			Alternatives			
Objective	**Weight**	**Units**	**Status Quo**	**Restrict Logging Only**	**Restrict Logging and Electrofish Removal of Brook Trout**	**Electrofish Removal of Brook Trout and Limit Water Removal from Watershed**
Maximize spawning habitat	80	Stream km of spawning habitat	4	1 (tie)	1 (tie)	3
Minimize water temperature	80	Water temperature (°C)	4	1 (tie)	1 (tie)	3
Maximize riparian cover	70	Percent stream covered	4	1 (tie)	1 (tie)	3
Minimize water withdrawal from the watershed	50	Acre or feet/day removed	2 (tie)	2 (tie)	2 (tie)	1
Minimize stream turbidity	70	NTUs	2 (tie)	1 (tie)	1 (tie)	2 (tie)
Minimize cost	30	$	1	3	4	2
Maximize constituent "buy-in"	60	1–5 scale, 5 happiest with alternative	4	2	1	3
Minimize brook trout population	65	Population size of brook trout	2 (tie)	2 (tie)	1 (tie)	1 (tie)

[a] The problem statement would be "How do we increase numbers of bull trout?" Alternatives are ranked as to how well each would meet each objective, with "1" being the highest rated. Weights rank the importance of each objective. Weights can be assigned by thorough consensus of stakeholders, managers, or by some other means. In this table, the "restrict logging and electrofish removal of brook trout" alternative best meets the highest priority objectives.

dominated alternatives, can be eliminated. If the various alternatives do not have different consequences for an objective, that objective can be eliminated, not because it is unimportant, but because choosing different alternatives would have no effect on the consequences for that objective.

Once the simplified consequences table is constructed, the final alternatives can be selected using a variety of methods. How willing is the manager to accept the risks associated with each of the alternatives? Is the decision to choose a particular alternative linked to important decisions that will need to be made in the future? Do the gains in one objective using one alternative counterbalance the losses in another objective using another alternative? Can objectives be weighted through input from experts or participants to identify which are the most important, and thus ascertain which are the most important to address? Detailed discussions on how all these techniques can be used to select the best alternatives is beyond the scope of this chapter, but is explained in Hammond et al. (1999), Runge et al. (2009), and a variety of other literature associated with decision analysis.

CONCLUSION

Most effects on fish populations are cumulative. Traditionally, additive effects have been most studied: for example, effects of many agricultural diversions on stream flow and the amount of impervious surfaces in urban areas. However, the role of more complex cumulative effects resulting from multiple sources with multiplicative effects, such as how watershed practices impact biological effects, are becoming increasingly considered. New sophisticated analyses techniques that allow consideration of these effects will be within reach of a larger number of biologists in the future.

11 Sage-Grouse and Cumulative Impacts of Energy Development

David E. Naugle, Kevin E. Doherty, Brett L. Walker, Holly E. Copeland, and Jason D. Tack

CONTENTS

INTRODUCTION

World demand for energy has increased by more than 50% since 1950, and a similar increase is projected between 2010 and 2030 (National Petroleum Council 2007). Fossil fuels will remain the largest source of energy worldwide, with oil, natural gas, and coal accounting for more than 80% of world demand. Projected growth in U.S. energy demand is 0.5%–1.3% annually (National Petroleum Council 2007), and development of domestic reserves will expand through the first half of the 21st century. Western states and provinces will continue to play a major role in providing additional domestic energy resources to the United States and Canada, which is anticipated to place unprecedented pressure on the conservation of wildlife populations throughout the West.

The sagebrush (*Artemisia* spp.) ecosystem is representative of the struggle to maintain biodiversity in a landscape that bears the debt of the ever-increasing demand for natural resources. One species impacted by domestic energy production is the greater sage-grouse (*Centrocercus urophasianus*), a game bird endemic to western semiarid sagebrush landscapes in western North America (Schroeder et al. 1999). Previously widespread, the sage-grouse has been extirpated from approximately half of its historic range (Schroeder et al. 2004), and populations have declined by 1.8% to 11.6% annually since 1970 in about half of the populations studied (Garton et al. 2010). Energy development has emerged as a major issue in conservation because areas currently under development contain some of the highest densities of sage-grouse (Connelly et al. 2004) and other sagebrush obligate species (Knick et al. 2003) in western North America.

The sage-grouse is considered a "landscape" species because it requires large, intact habitats with sagebrush to maintain robust populations (Connelly et al. 2010). As a result, the size of sage-grouse breeding populations is often used as an indicator of the overall health of the sagebrush ecosystem (Hanser and Knick 2010). There are few early studies evaluating impacts of energy development to sage-grouse populations (Naugle et al. 2010), but research has increased rapidly in concert with the pace and extent of development. The goal of this chapter is to provide a scientific understanding of impacts of energy development on sage-grouse and to recommend biologically based solutions. Objectives are to synthesize current data regarding the biological response of sage-grouse to energy development, identify ecological and behavioral mechanisms causing population-level impacts, evaluate empirically the extent to which current and anticipated development impacts populations, and outline a strategy for landscape conservation analogous in scale to that of ongoing and anticipated impacts of development.

BIOLOGICAL RESPONSE OF SAGE-GROUSE TO ENERGY DEVELOPMENT

We searched the literature for studies that investigated relationships between sage-grouse and energy development using methods described in Naugle et al. (2010). We included theses and dissertations but excluded from review documents that included only cautionary statements regarding potential impacts or anecdotal data. Eleven studies reported negative impacts of energy development to sage-grouse, whereas no studies reported a positive influence of development on populations or habitats (Table 11.1). Breeding populations were severely impacted at well densities (8 pads/2.6 km^2) commonly permitted in conventional oil and gas fields (Holloran 2005, Walker et al. 2007a). Furthermore, analyses showed no difference in lek persistence at 1 well pad/2.6 km^2 (Doherty 2008), but declines in males at large leks (more than 25 males) were apparent at approximately 1 well pad/2.6 km^2 (Tack 2009, Figure 9).

Negative impacts are known for four different sage-grouse populations in three different types of development including shallow coal-bed natural gas in the Powder River Basin of northeastern Wyoming and extreme southeastern Montana (Walker et al. 2007a; Doherty et al. 2008), deep gas in the Pinedale Anticline Project Area in southwestern Wyoming (Lyon and Anderson 2003; Holloran 2005; Holloran et al. 2010), oil extraction in the Manyberries Oil Field in southeast Alberta (Aldridge and Boyce 2007), and oil and gas development in the Cedar Creek Anticline near the state boundaries between southeastern Montana, western North Dakota, and northwestern South Dakota (Tack 2009). Population trends in the Powder River Basin indicated that from 2001 to 2005, lek-count indices inside gas fields declined by 82%, whereas indices outside development declined by 12% (Figure 11.1). Of leks active in 1997, only 38% inside gas fields remained active as of 2004–2005 compared to 84% outside development (Walker et al. 2007a). Male lek attendance in the Pinedale Anticline decreased

TABLE 11.1
Research Studies on Effects of Oil and Gas Development on Greater Sage-Grouse

Citation and Study Location	Research Outlet	Pretreatment Design	Pretreatment or Control	Years	Sample Size
Lyon and Anderson (2003), Pinedale Mesa, SW Wyoming	Scientific journal	Observational	N/Y	2	48 females from 6 leks
Walker et al. (2007a), Powder River Basin, NE Wyoming and SE Montana	Scientific journal	Correlative	Y/Y	8	97–154 lek complexes/year for trends, 276 lek complexes in status analysis
Aldridge and Boyce (2007), SE Alberta, Canada	Scientific journal	Correlative	N/N	4	113 nests, 669 locations on 35 broods, 41 chicks from 22 broods
Doherty et al. (2008), Powder River Basin, NE Wyoming and SE Montana	Scientific journal	Correlative	N/N	4	435 locations to build model, 74 new locations from different years to test it
Holloran et al. (2010), Pinedale Anticline Project Area and Jonah II gas field, SW Wyoming	Scientific journal	Correlative and observational	N/Y	6	135 yearling females, and 34 yearling males, from 17 leks, and 62 yearling sage-grouse nests
Johnson et al. (2010), range-wide	Scientific journal	Correlative	N/N	11	3,679 leks with at least 4 counts from 1997–2007
Holloran et al. (2007), Pinedale Anticline Project Area and Jonah II gas field, SW Wyoming	U.S. Geological Survey	Correlative and observational	N/Y	2	86 yearlings (52 females), 23 yearlings (17 females) with known maternity
Holloran (2005), Pinedale Anticline Project Area and Jonah II gas field, SW Wyoming	Ph.D. dissertation	Correlative and observational	N/Y	7	Counts of 21 leks, 209 females from 14 leks, 162 nests
Doherty (2008), Powder River Basin, NE Wyoming and SE Montana	Ph.D. dissertation	Correlative and observational	N/Y	10	1190 active leks and 154 inactive leks
Kaiser (2006), Pinedale Anticline Project Area and Jonah II gas field, SW Wyoming	M.S. thesis	Correlative and observational	N/Y	1	18 leks, 83 females (23 yearlings), plus 20 yearling males
Tack (2009), Montana, SE Alberta, SW Saskatchewan, SW North Dakota, and NW South Dakota	M.S. thesis	Correlative and observational	N/N	10	802 active and 297 inactive lek complexes and

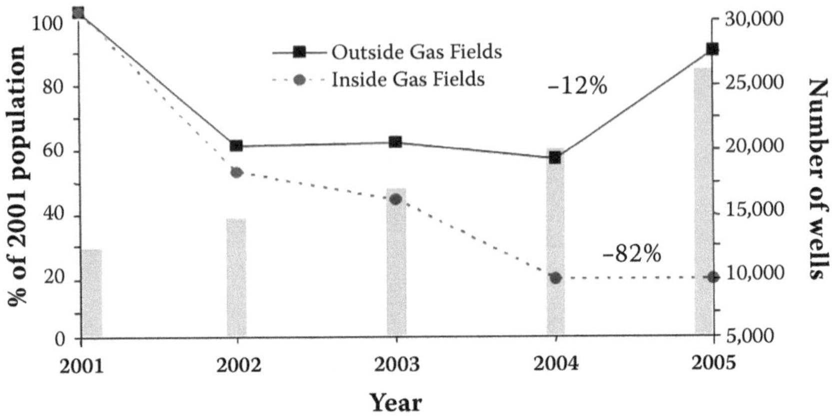

FIGURE 11.1 Population indices based on male lek attendance for sage-grouse in the Powder River Basin (PRB), Montana and Wyoming, 2001–2005 for leks categorized as inside or outside of coal-bed natural gas fields on a year-by-year basis. Leks in gas fields had ≥40% energy development within 3.2 km or >25% development within 3.2 km and ≥1 well within 350 m of the lek center. Number of producing gas wells in the PRB shows the overall increase in development coincident with declines in sage-grouse population indices. (Modified from Walker, B.L. et al. 2007a. *Journal of Wildlife Management* 71:2644–2654.)

with distance to the nearest active drilling rig (Figure 11.2), producing gas well, and main haul road, and declines were most severe (40%–100%) at breeding sites within 5 km of an active drilling rig or within 3 km of a producing gas well or main haul road (Holloran 2005). In an endangered population in Alberta, Canada, where low chick survival (12% to 56 days) limits population growth, risk of chick mortality in the Manyberries Oil Field was 1.5 times higher for each additional well site visible within 1 km of a brood location (Aldridge and Boyce 2007). At the Cedar Creek Anticline in southeastern Montana, male abundance at leks decreased by 52% at 16 leks with more than 1 well pad/2.6 km^2 within 3.2 km, and no males were counted in 2009 at 4 of the 16 impacted leks that had multiple displaying males in 2008 (Tack 2009).

Studies also have quantified the distance from leks at which impacts of development become negligible and have assessed the efficacy of the stipulation by the U.S. Bureau of Land Management (BLM) of no surface disturbance within 0.4 km of a lek. Impacts to leks from energy development were most severe near the lek, remained discernable out to distances of more than 6 km (Holloran 2005; Walker et al. 2007a; Tack 2009; Johnson et al. 2010), and often resulted in extirpation of leks within gas fields (Holloran 2005; Walker et al. 2007a). Curvilinear relationships showed that lek counts decreased with distance to the nearest active drilling rig, producing well, or main haul road, and that development within 4.7 to 6.2 km of leks influenced counts of displaying males (Holloran 2005; Figure 11.2). All well-supported models in Walker et al. (2007a) indicated a strong negative effect of energy development, estimated as proportion of development within either 0.8 km or 3.2 km, on lek persistence. A model with development at 6.4 km had considerably less support (5–7 $\Delta AICc$ units lower), but the regression coefficient ($\beta = -5.11$, SE = 2.04) indicated that negative

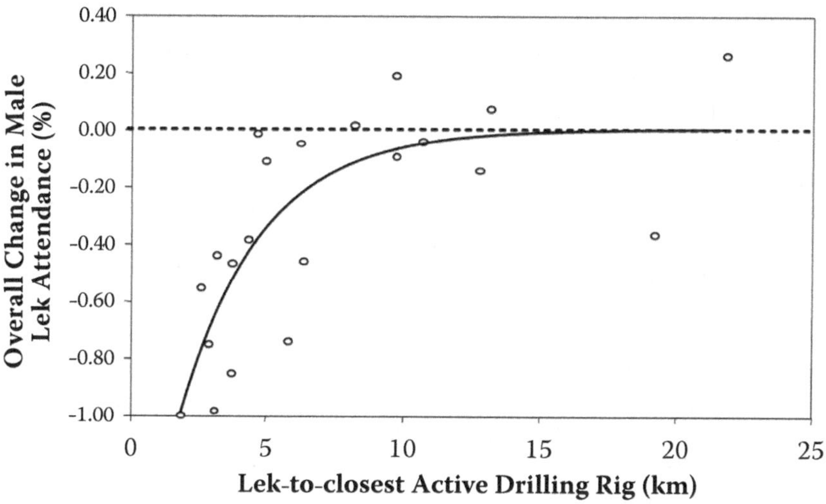

FIGURE 11.2 Relationship between number of sage-grouse males attending leks and average distance from leks to closest active gas drilling rig, Pinedale Anticline Project Area, southwestern Wyoming, 1998–2004. Each point along the regression line represents one lek ($N = 21$). (Modified from Holloran, M.J. 2005. Greater sage-grouse (*Centrocercus urophasianus*) population response to gas field development in western Wyoming. Ph.D. dissertation, University of Wyoming, Laramie, Wyoming.)

impacts of development within 6.4 km were still apparent. Walker et al. (2007a) used the resulting model to demonstrate the 0.4-km lease stipulation used by the BLM was insufficient to conserve breeding sage-grouse populations in fully developed gas fields. A 0.4-km buffer leaves 98% of the landscape within 3.2 km of a lek open to full-scale development. Full-field development of 98% of the landscape within 3.2 km of leks in a typical landscape in the Powder River Basin reduced the average probability of lek persistence from 87% to 5% (Walker et al. 2007a). Two recent studies reported negative impacts apparent out to 12.3 km on large lek occurrence (>25 males; Tack 2009) and out to 18 km on lek trends (Johnson et al. 2010), the largest scales that have yet been evaluated.

Negative responses of sage-grouse to energy development were consistent among studies regardless of whether they examined lek dynamics or demographic rates of specific cohorts within populations. Sage-grouse populations decline when birds avoid infrastructure in ≥1 seasons (Doherty et al. 2008), when cumulative impacts of roads, power lines, and other aspects of the energy footprint negatively affect reproduction or survival (Aldridge and Boyce 2007), or both demographic rates (Lyon and Anderson 2003; Holloran 2005; Holloran et al. 2010). Avoidance of energy development at the scale of entire oil and gas fields should not be considered a simple shift in habitat use, but rather a reduction in the distribution of sage-grouse (Walker et al. 2007a). Avoidance is likely to result in true population declines if density-dependence, competition, or displacement of birds into poorer-quality adjacent habitats lowers survival or reproduction (Holloran and Anderson 2005; Aldridge and Boyce 2007; Holloran et al. 2010). High site fidelity in sage-grouse also suggests that unfamiliarity with new habitats may also reduce survival, as in other grouse species (Yoder et al. 2004). Sage-grouse in the Powder River Basin were 1.3 times less likely to use otherwise suitable winter habitats that had been developed from energy (12 wells/4 km^2), and avoidance of developed areas was most pronounced when it occurred in high-quality winter habitat with abundant sagebrush (Doherty et al. 2008).

Long-term studies in the Pinedale Anticline Project Area in southwestern Wyoming present the most complete picture of cumulative impacts, and provide a mechanistic explanation for declines to populations. Early in development, nest sites were farther from disturbed than undisturbed leks, rate of nest initiation from disturbed leks was 24% lower than for birds breeding on undisturbed leks, and 26% fewer females from disturbed leks initiated nests in consecutive years (Lyon and Anderson 2003). As development progressed, adult females remained in traditional nesting areas regardless of increasing levels of development, but yearlings that had not yet imprinted on habitats inside the gas field avoided development by nesting farther from roads (Holloran 2005). The most recent study confirmed that yearling females avoided infrastructure when selecting nest sites, and yearling males avoided leks inside of development and were displaced to the periphery of the gas field (Holloran et al. 2010). Recruitment of males to leks also declined as distance within the external limit of development increased, indicating a high likelihood of lek loss near the center of developed oil and gas fields (Kaiser 2006).

The most important finding from studies in Pinedale was that sage-grouse declines are in part explained by lower annual survival of female sage-grouse, and that the impact to survival resulted in a population-level decline (Holloran 2005). However, we are still lacking a clear picture of long-term effects of behavioral avoidance coupled with decreased survival. The population decline observed in sage-grouse is similar to that observed in Kansas for the lesser prairie-chicken (*Tympanuchus pallidicinctus*; Hagen 2003), a federally threatened species that also avoided otherwise suitable sand-sagebrush (*Artemisia filifolia*) habitats proximal to oil and gas development (Pitman et al. 2005; Johnson et al. 2006). High site fidelity but low survival of adult sage-grouse combined with lek avoidance by younger birds (Holloran et al. 2010) resulted in a time lag of 3–4 years between the onset of development activities and lek loss (Holloran 2005). The time lag observed by Holloran (2005) in the Anticline matched that for leks that became inactive 3–4 years following natural gas development in the Powder River Basin (Walker et al. 2007a). This knowledge of time lags suggests that ongoing development in the Cedar Creek Anticline will result in additional impacts to fringe populations in eastern Montana and western North and South Dakota (Tack 2009).

Mechanisms that attribute to avoidance and decreased fitness have not been empirically tested, but rather suggested from observational studies. For example, lek abandonment may increase if repeatedly disturbed by raptors perching on power lines near leks (Ellis 1984), by vehicle traffic on nearby roads (Lyon and Anderson 2003), or by noise and human activity associated with energy development during the breeding season (Holloran 2005; Kaiser 2006). Collisions with nearby power lines and vehicles and increased predation by raptors may also increase mortality of birds at leks (Connelly et al. 2000a,b). Alternatively, roads and power lines may indirectly affect lek persistence by altering productivity of local populations or survival at other times of the year. For example, sage-grouse mortality associated with power lines and roads occurs year-round (Beck et al. 2006, Aldridge and Boyce 2007), and ponds created by coal-bed natural gas development may increase risk of West Nile virus (WNv) mortality in late summer (Walker et al. 2004; Zou et al. 2006; Walker et al. 2007b). Loss and degradation of sagebrush habitat can also reduce carrying capacity of local breeding populations (Swenson et al. 1987; Braun 1998; Connelly et al. 2000b; Crawford et al. 2004). Alternatively, birds may simply avoid otherwise suitable habitat as the density of roads, power lines, or energy development increases (Lyon and Anderson 2003; Holloran 2005; Kaiser 2006; Doherty et al. 2008).

A SHIFT IN PARADIGM: BUSINESS AS USUAL VERSUS LANDSCAPE CONSERVATION

The unequivocal answer to energy development in the West is not a "No!" but rather a "Where?" to reduce environmental impacts while still extracting resources to meet increasing domestic demand for energy. The U.S. government has already leased more than 7,000,000 ha of the federal mineral estate and the number of producing wells tripled from 11,000 in the 1980s to more than 33,000 in 2007 (Naugle et al. 2010). Managers struggling to keep pace with development have implemented reactive measures in hopes of mitigating disturbance around leks. Protective measures such as not allowing energy infrastructure within varied distances around leks and timing restrictions on drilling failed to maintain populations, and it has become apparent that sage-grouse conservation and energy development are incompatible in the same landscapes.

Budgetary constraints to research and maintain wildlife populations means that "conservation triage" (i.e., prioritization of limited resources to maximize biological return on investment; Bottrill et al. 2008, 2009) is unavoidable to meet the high social and economic costs to maintain sage-grouse populations. The focus for conservation should be to prioritize and conserve remaining intact landscapes, rather than trying to maintain small declining populations at the cost of further loss in the best remaining areas. The challenge will be to implement conservation at a scale that matches energy development to offset the spatial extent of anticipated impacts. Scientists need to work with managers to develop proactive decision-support tools that identify priority landscapes that will maintain large populations, develop management prescriptions that increase populations in priority landscapes and offset losses in developed landscapes, and identify ecological corridors among priority populations to maintain connectivity. Despite ongoing development, no comprehensive range-wide plan is in place to conserve large and functioning landscapes to maintain sage-grouse populations.

CONSERVATION PLANNING USING CORE REGIONS TO REDUCE IMPACTS

Analytical frameworks are available to evaluate options for reducing impacts to sage-grouse populations at highest risk of oil, gas, and wind development. For example Doherty et al. (2011) used lek-count data (n = 2,336 leks) to delineate high abundance population centers, or "core-regions," that contained 25%, 50%, 75%, and 100% of the known breeding populations in Wyoming, Montana, Colorado, Utah, and North and South Dakota, respectively (Figure 11.3). Core regions can be overlaid spatially with authorized oil and gas leases and the potential for commercial development of wind energy (Figure 11.4), and the resulting output can then be used to identify the least-at-risk core populations to energy development to prioritize for immediate conservation. Areas that share high energy potential and high sage-grouse density

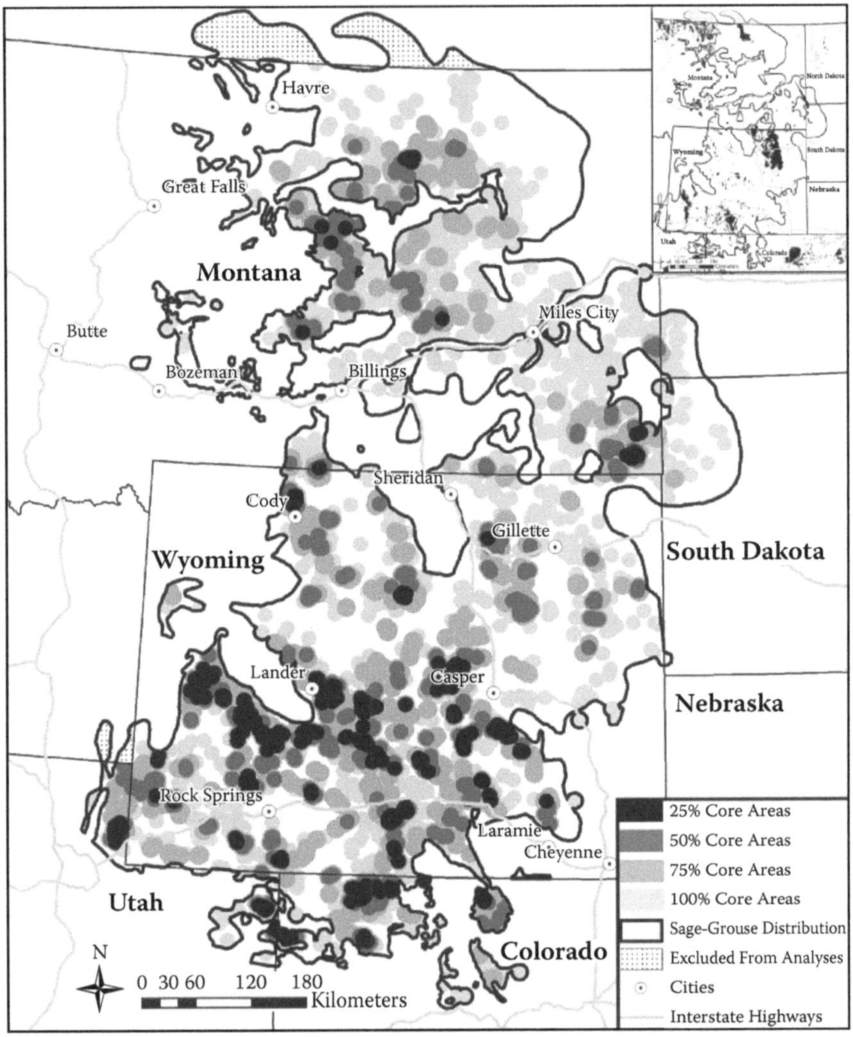

FIGURE 11.3 Core areas that contain 25%, 50%, 75%, and 100% of the known breeding population of greater sage-grouse in their eastern range (Modified from Doherty, K. et al. 2011. *Studies in Avian Biology,* University of Colorado Press, Boulder.) Distribution boundaries are the combined areas of sage-grouse management zones I and II (Connelly, J.W. et al. 2004. Conservation assessment of greater sage-grouse and sagebrush habitats. Unpublished report, Western Association of Fish and Wildlife Agencies, Cheyenne, Wyoming). Inset depicts locations of producing oil and gas wells (black triangles) as of September 2007.

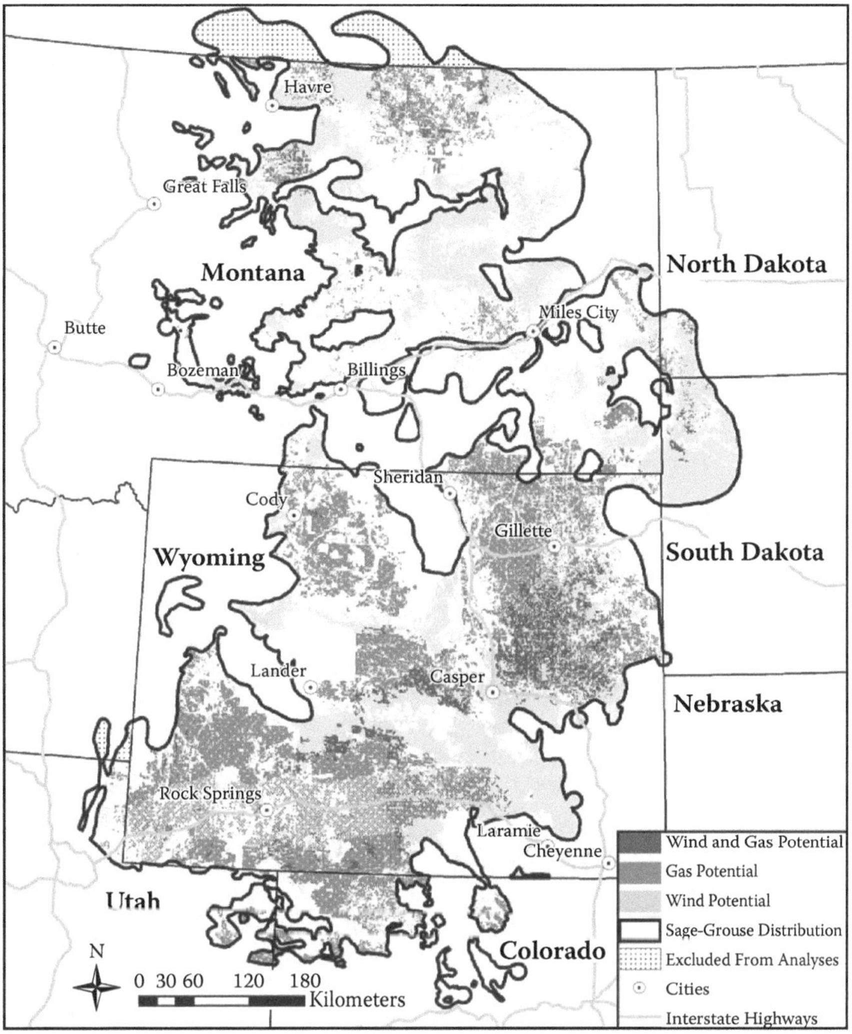

FIGURE 11.4 Potential for oil and gas and wind development in the eastern range of greater sage-grouse (management zones I and II). (Modified from Doherty, K. et al. 2011. *Studies in Avian Biology*, University of Colorado Press, Boulder.)

will require policy reform to reduce threats, while areas with high energy potential but low biological value can act as areas to "trade" development for conservation (Figure 11.5).

Clumped distributions of populations suggest that a disproportionately large proportion of breeding birds can be conserved within core regions. For example, 75% of the breeding population in the eastern range of sage-grouse was captured within only 30% of the area (Doherty et al. 2011). Wyoming is a key to conservation of the species because it contains 64% of the known eastern breeding population and is at greatest combined risk from wind energy and oil and gas development (Doherty et al. 2011,

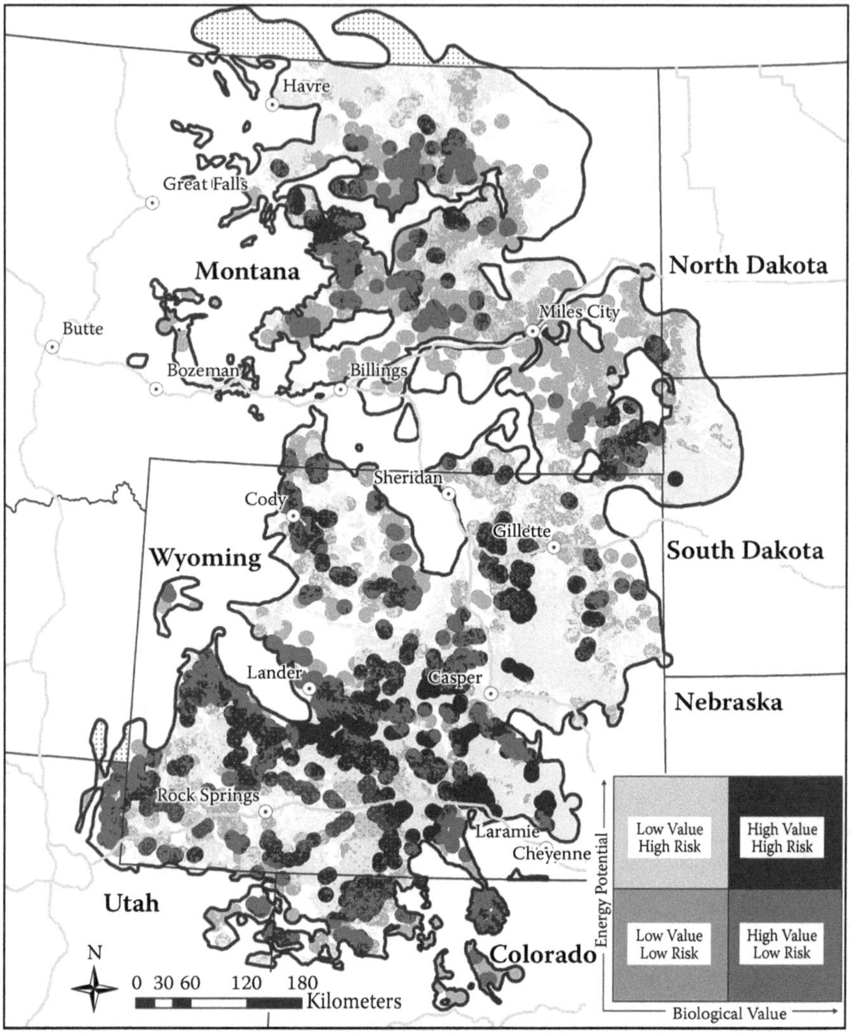

FIGURE 11.5 Overlay of biological value (25% to 75% core regions = high value) with energy potential for oil and gas or wind development to assess risk of development to greater sage-grouse core areas.

Tables 1 and 3). Risks to core regions vary dramatically, and each state and province will need to do its part to ameliorate these risks to maintain sage-grouse distribution and abundance. Successful implementation of landscape conservation in one state is insufficient to compensate for losses in others.

Core regions provide a vision for decision makers to spatially prioritize conservation targets. Several western states adopted the initial concept and have subsequently refined core regions by linking them with the best available habitat maps and expert knowledge of seasonal habitat needs outside the breeding season. Core regions have been heralded as a way of partnering with industry to fund conservation in priority landscapes and as a basis for forecasting development scenarios to aid in conservation design. Identification of core regions provides a biological foundation for implementing community-based landscape conservation. Landscape-scale conservation in priority areas is the most defensible and realistic solution to the dilemma between energy development and sage-grouse conservation in the West. Maintaining large landscapes with minimum disturbance is paramount to sage-grouse conservation and will require collaborative efforts from a diverse group of stakeholders.

A pressing need in conservation plans is a better understanding of connectivity among sage-grouse populations (Oyler-McCance and Quinn 2010). Our understanding of sage-grouse movements, dispersal, and connectivity is limited because telemetry studies have not been conducted to document how individual populations move during dispersal or seasonal migration (Knick and Hanser 2010). Analytic advances in landscape genetics (Murphy et al. 2010) and noninvasive sampling of genetic material from feathers collected off leks may provide efficient means of quantifying connectivity among sage-grouse populations (Storfer et al. 2007). However, genetic samples alone can obscure emerging or disrupted patterns of animal movements (Fedy et al. 2008). Using GPS technology with sage-grouse is a promising technique that allows researchers to identify movements previously missed by VHF telemetry equipment that are critical to population persistence and that may not be detected by genetic data (i.e., migration).

Lastly, researchers should focus on finding links between prescriptive management actions and sage-grouse productivity. The initial requirement is an understanding of how different vital rates influence overall population growth, as well as the plasticity and ability to manage influential vital rates. Once key vital rates are identified, management practices that bolster vital rates should be implemented to help maintain and enhance populations. Tools to manage sage-grouse populations will vary across the species range with biotic and abiotic characteristics of different landscapes and local constraints to populations. Some populations may benefit from changing grazing regimes, removing conifers, or managing invasive species, yet these will ultimately depend on the site. The ultimate measure of our management success will be the biological return on investment as measured in number of birds.

OFFSETS TO REDUCE IMPACTS RESULTING FROM DEVELOPMENT

Mandatory off-site mitigation for sage grouse beyond that of voluntary compliance and the corporate mantra of sustainability may someday become a reality (Kiesecker et al. 2009). If and when it does, biodiversity offsets could provide a mechanism to compensate for unavoidable damages from new energy development as the United States increases domestic production. To date, proponents argue that offsets provide a partial solution for funding conservation while opponents contend the practice is flawed because offsets are negotiated without the science necessary to back up resulting decisions. Missing in negotiations is a biologically based currency for estimating sufficiency of offsets and a framework for applying proceeds to maximize conservation benefits.

One new study provides a common currency for offsets for sage-grouse by estimating number of impacted birds at levels of oil and gas development commonly permitted (Doherty et al. 2010). Analyses used lek count data from across Wyoming (n = 1,344) to test for differences in rates of lek inactivity and changes in bird abundance at five intensities of energy development including "control" leks with no development. Impacts are indiscernible at 12 wells per 32.2 km^2. Above this threshold lek losses were 2–5 times greater inside than outside of development and bird abundance at remaining leks declined by –32% to –77% (Doherty et al. 2010). Documented impacts relative to development intensity can be used to forecast biological trade-offs of newly proposed or ongoing developments, and when drilling is approved, anticipated bird declines form the biological currency for negotiating offsets. Implications suggest that offsetting risks using birds as the currency can be implemented immediately; monetary costs for offsets will be determined by true conservation cost to mitigate risks to other populations of equal or greater number. If this information is blended with landscape level conservation planning, the mitigation hierarchy can be improved by steering planned developments away from conservation priorities, ensuring that compensatory mitigation projects deliver a higher return for conservation that equates to an equal number of birds in the highest priority areas, provide on-site mitigation recommendations, and provide a biologically based cost for mitigating unavoidable impacts.

CONCLUSION

The severity of impacts to sage-grouse populations from various types of energy development dictate the need to shift from local to landscape conservation. This shift should transcend state and other political boundaries to develop and implement a plan for conservation of sage-grouse populations across the western United States and Canada. Planning tools are available that overlay the best-known areas for sage-grouse with the extent of current and projected development for all of Sage-Grouse Management Zones I and II (Doherty et al. 2011). Maps depicting locations of the largest remaining sage-grouse populations and their relative risk of loss provide decision-makers with the information they need to implement community-based landscape conservation. Ultimately, sources of all cumulative impacts—not just energy development—must be managed collectively to maintain populations over time in priority landscapes.

Cumulative effects from energy development to sage-grouse populations are a primary reason we need a broad-scale and long-term strategy for planning and implementing conservation. A scientifically defensible strategy can be constructed, and the most reliable measure of success will be long-term maintenance of robust sage-grouse populations in their natural habitats. Foregoing development in priority landscapes is the obvious approach necessary to conserve large populations. The challenge will be for governments, industries, and communities to implement solutions at a sufficiently large scale across multiple jurisdictions to meet the biological requirements of sage-grouse. New best-management practices can be applied and rigorously tested in landscapes less critical to conservation. We have the capability and opportunity to reduce future losses of sage-grouse to energy development, yet populations continue to decline as energy production increases; thus, the need for inter-jurisdictional cooperation is paramount. Political wrangling, lawsuits, regulatory uncertainty, and repeated attempts to list the species as federally threatened or endangered will continue until we demonstrate success in collaborative landscape planning and on-the-ground actions that benefit sage-grouse populations.

References

Adamczewski, J. Z., C. C. Gates, B. M. Soutar, and R. J. Hudson. 1988. Limiting effects of snow on seasonal habitat use and diets of caribou (*Rangifer tarandus groenlandicus*) on Coats Island, Northwest Territories, Canada. *Canadian Journal of Zoology* 66:1986–1996.

Adams, L. W. 2005. Urban wildlife ecology and conservation: A brief history of the discipline. *Urban Ecosystems* 8:139–156.

Adler, R. W. 2007. *Restoring Colorado River Ecosystems: A Troubled Sense of Immensity*. Island Press, Washington, D.C.

Aldridge, C. L., and M. S. Boyce. 2007. Linking occurrence and fitness to persistence: Habitat-based approach for endangered greater sage-grouse. *Ecological Applications* 17:508–526.

Alexis, A., and P. Cox. 2005. The California Almanac of Emissions and Air Quality. California Air Resources Board, Sacramento, California.

Andelt, W. F., and B. R. Mahan. 1980. Behavior of an urban coyote. *American Midland Naturalist* 103:399–400.

Andelt, W. F., and J. S. Gipson. 1979. Home range, activity, and daily movements of coyotes. *Journal of Wildlife Management* 43:944–951.

Anderson, D. R., K. P. Burnham, and W. L. Thompson. 2000. Null hypothesis testing: Problems, prevalence, and an alternative. *Journal of Wildlife Management* 64:912–923.

Anderson, S. H., and K. J. Gutzwiller. 1994. Habitat evaluation methods. Pages 592–622 in Bookhout, T.A., editor. *Research and Management Techniques for Wildlife and Habitats*. The Wildlife Society, Bethesda, Maryland.

Arizona Department of Water Resources. 2009. Arizona Water Atlas, Volume 3. Southeastern Arizona Planning Area, Phoenix, Arizona.

Arizona Game and Fish Department. 2010. Heritage Data Management System. Special Status Species by County, Taxon, Scientific Name. Phoenix, Arizona. <http://www.azgfd.gov/w_c/edits/documents/ssspecies_bycounty_002.pdf> Accessed March 2010.

Arizona Land and Water Trust. 2010. The Sonoran Desert web page. http://www.alwt.org/wherewework/sonorandesert.shtml. Accessed June 15, 2010.

Arizona State Parks. 2008. Arizona statewide comprehensive outdoor recreation plan. <http://azstateparks.com/publications/#SCORP>. Accessed April 29, 2010.

Armitage, D. 2005. Collaborative environmental assessment in the Northwest Territories, Canada. *Environmental Impact Assessment Review* 25:239–258.

Athabasca Landscape Team. 2009. Athabasca caribou landscape management options report. Unpublished report submitted to the Alberta Caribou Committee Governance Board, Edmonton, Alberta, Canada.

Atkinson, K. T., and D. M. Shackleton. 1991. Coyote, *Canis latrans*, ecology in a rural-urban environment. *Canadian Field-Naturalist* 105:49–54.

Atwood, T. C., H. P. Weeks, and T. M. Gehring. 2004. Spatial ecology of coyotes along a suburban-to-rural gradient. *Journal of Wildlife Management* 68:1000–1009.

Aumann, C., D. R. Farr, and S. Boutin. 2007. Multiple use, overlapping tenures, and the challenge of sustainable forestry in Alberta. *Forestry Chronicle* 83:642–650.

Austin, M.A., D. A. Buffett, D. J. Nicolson, G. G. E. Scudder, and V. Stevens (editors). 2008. Taking nature's pulse: The status of biodiversity in British Columbia. Biodiversity BC, Victoria, British Columbia, Canada.

AXYS Environmental Consulting. 2001. Thresholds for addressing cumulative effects on terrestrial and avian wildlife in the Yukon. Report prepared for Department of Indian and Northern Affairs, Environmental Directorate and Environment Canada, Whitehorse, Yukon, Canada.

AXYS Environmental Consulting. 2003a. A cumulative effects assessment and management framework (CEAMF) for Northeastern British Columbia. Volume 1. Prepared for the BC Oil and Gas Commission and the Muskwa-Kechika Advisory Board, Fort St. John, British Columbia, Canada.

AXYS Environmental Consulting. 2003b. Cumulative effects indicators, thresholds, and case studies. Volume 2. Prepared for the BC Oil and Gas Commission and the Muskwa-Kechika Advisory Board, Fort St. John, British Columbia, Canada.

AXYS Environmental Consulting and Penner and Associates. 1998. Environmental effects report: Wildlife. Prepared for Diavik Diamonds Project, Yellowknife, Northwest Territories, Canada.

Aycrigg, J. L., and W. F. Porter. 1997. Sociospatial dynamics of white-tailed deer in the central Adirondack Mountains, New York. *Journal of Mammalogy* 78:468–482.

Ayres, L., L. Chow, and D. Graber. 1986. Black bear activity patterns and human induced modifications in Sequoia National Park. *International Conference on Bear Research and Management* 6:151–154.

Baker, R. O., and R. M. Timm. 1998. Management of conflicts between urban coyotes and humans in southern California. *Proceedings of the Vertebrate Pest Conference* 18:299–312.

Baker, P. S. Funk, S. Harris, T. Newman, G. Saunders, and P. White. 2004. The impact of human attitudes on the social and spatial organization of urban foxes (*Vulpes vulpes*). Pages 153–163 in W. W. Shaw, L. K. Harris, and L. VanDruff, editors. *Fourth International Symposium on Urban Wildlife Conservation*, Tucson, Arizona.

Banta, B. H., and D. Morfka. 1966. An annotated check list of the recent amphibians and reptiles inhabiting the city and county of San Francisco, California. *Wasmann Journal of Biology* 24:223–238.

Bardecki, M. J. 1990. Coping with cumulative impacts: An assessment of legislative and administrative mechanisms. *Impact Assessment Bulletin* 8:319–344.

Barnes, J. L., L. Matthews, A. Griffiths, and C. L. Horvath. 2001. Addressing cumulative environmental effects: Determining significance. *Proceedings of Cumulative Environmental Effects Management, Tools and Approaches*. Alberta Society of Professional Biologists, Alberta Institute of Agrologists, and Association of Professional Biologists of British Columbia, Calgary, Alberta, Canada.

Bartoloino, J. R., and W. L. Cunningham. 2003. Ground-water depletion across the nation. U.S. Geological Survey Fact Sheet 103-03.

Bashore, T. L., W. M. Tzikowski, and E. D. Bellis. 1985. Analysis of deer-vehicle collision sites in Pennsylvania. *Journal of Wildlife Management* 49:769–774.

Batten, E. 1972. Breeding bird species diversity in relation to increasing urbanization. *Bird Study* 19:157–166.

Bauer, C. R., A. M. Bobeldyk, and G. A. Lamberti. 2007. Predicting habitat use and trophic interactions of Eurasian ruffe, round gobies, and zebra mussels in nearshore areas of the Great Lakes. *Biological Invasions* 9:667–678.

Baumol, W. J. and W. E. Oates. 1988. *The Theory of Environmental Policy*, 2nd edition. Cambridge University Press, New York.

Baxter, W., W. A. Ross, and H. Spaling. 2001. Improving the practice of cumulative effects assessment in Canada. *Impact Assessment and Project Appraisal* 19:253–262.

Beard, T. D., and J. M. Kampa. 1999. Changes in bluegill, black crappie, and yellow perch populations in Wisconsin during 1967–1991. *North American Journal of Fisheries Management* 19:1037–1043.

Beck, J. L., K. P. Reese, J. W. Connelly, and M. B. Lucia. 2006. Movements and survival of juvenile greater sage-grouse in southeastern Idaho. *Wildlife Society Bulletin* 34:1070–1078.

Beckmann, J. P., and J. Berger. 2003a. Using black bears to test ideal-free distribution models experimentally. *Journal of Mammalogy* 84:594–606.

Beckmann, J. P., and J. Berger. 2003b. Rapid ecological behavioral changes in carnivores: The responses of black bears (*Ursus americanus*) to altered food. *Journal of the Zoological Society of London* 261:207–212.

Beckmann, J. P., C. W. Lackey, and J. Berger. 2004. Evaluation of deterrent techniques and dogs to alter behavior of "nuisance" black bears. *Wildlife Society Bulletin* 32:1141–1146.

Beckmann, J. P., and C. W. Lackey. 2008. Carnivores, urban landscapes, and longitudinal studies: A case history of black bears. *Human-Wildlife Conflicts* 2:168–174.

Bedford, B. L., and E. M. Preston. 1988. Developing the scientific basis for assessing cumulative effects of wetland loss and degradation on landscape functions: Status, perspectives, and prospects. *Environmental Management* 12:755.

Bednarek, A. T. 2001. Undamming rivers: A review of the ecological impacts of dam removal. *Environmental Management* 27:803–814.

Beier, P. 1995. Dispersal of juvenile cougars in fragmented habitat. *Journal of Wildlife Management* 59:228–237.

Beissinger, S. R., and D. R. Osborne. 1982. Effects of urbanization on avian community organization. *Condor* 84:75–83.

Bellamy, J. A., D. H. Walker, G. T. McDonald, and G. J. Syme. 2001. A systems approach to evaluation of natural resource management initiatives. *Journal of Environmental Management* 63:407–423.

Bender, L. C., J. C. Lewis, and D. P. Anderson. 2004. Population ecology of Columbian black-tailed deer in urban Vancouver, Washington. *Northwestern Naturalist* 85:53–59.

Bender, L. C., G. A. Schirato, R. D. Spencer, K. R. McCallister, and B. L. Murphy. 2004c. Survival, cause-specific mortality, and harvesting of male black-tailed deer in Washington. *Journal of Wildlife Management* 68:870–878.

Benedict, M., J. Drohan, and J. Gavely. 2005. Sonoran Desert Conservation Plan, Pima County Arizona, Green Infrastructure—Linking Lands for Nature and People: Case study series 6. The Conservation Fund, Arlington, Virginia.

Bennett, V. J., M. Beard, P. A. Zollner, E. Fernandez-Juricic, L. Westphal, and C. L. LeBlanc. 2009. Understanding wildlife responses to human disturbance through simulation modeling: A management tool. *Ecological Complexity* 6:113–134.

Berger, T. R. 1977. Northern frontier, northern homeland. The report of the Mackenzie Valley pipeline inquiry, Vol. I. Ministry of Supply and Services, Ottawa, Ontario, Canada.

Berger, J. 2004. The last mile: How to sustain long-distance migration in mammals. *Conservation Biology* 18:320–331.

Bergerud, A. T., R. D. Jakimchuk, and D. R. Carruthers. 1984. The buffalo of the north: Caribou (*Rangifer tarandus*) and human developments. *Arctic* 37:7–22.

Bergerud, A. T., S. N. Luttich, and L. Camps. 2008. *The Return of Caribou to Ungava*. McGill-Queen's University Press, Montreal, Quebec, Canada.

Beringer, J., L. P. Hansen, J. A. Demand, J. Sartwell, M. Wallendorf, and R. Mange. 2002. Efficacy of translocation to control urban deer in Missouri: Costs, efficiency, and outcome. *Wildlife Society Bulletin* 30:767–774.

Berkes, F., J. Colding, and C. Folke. 2003. *Navigating Social-Ecological Systems: Building Resilience for Complexity and Change*. Cambridge University Press, Cambridge, United Kingdom.

Berkes F., G. P. Kofinas, F. S. Chapin. 2009. Conservation, community and livelihoods: Sustaining, renewing, and adapting cultural connections to the land. Pages 129–147 in F. S. I. Chapin, G. P. Kofinas, and C. Folke, editors. *Principles of Ecosystem Stewardship: Resilience-Based Natural Resource Management in a Changing World*. Springer Science + Business Media, New York.

Bernhardt, E. S., and M. A. Palmer. 2007. Restoring streams in an urbanizing world. *Freshwater Biology* 53:738–751.

Berris, C. R. 1987. Interactions of elk and residential development: Planning, design, and attitudinal considerations. *Landscape Journal* 6:31–41.

Berry, M. E., C. E. Bock, and S. L. Haire. 1998. Abundance of diurnal raptors on open space grasslands in an urbanized landscape. *Condor* 100:601–608.

Bezzel, E. 1984. Birdlife in intensively used rural and urban environments. *Ornins Fennica* 62:90–95

Biggs, J., S. Sherwoood, S. Michalak, L. Hansen, and C. Bare. 2004. Animal-related accidents at the Los Alamos Laboratory, New Mexico. *The Southwestern Naturalist* 49:384–394.

Billard, R., and G. Lecointre. 2001. Biology and conservation of sturgeon and paddlefish. *Reviews in Fish Biology and Fisheries* 10:355–392.

Blair, R. 2004. The effects of urban sprawl on birds at multiple levels of biological organization. *Ecology and Society* 9:2 (online: http://www.ecologyandsociety.org/vol9/iss5/art2.). Accessed August 2, 2010.

Blair, R. B. 1996. Land use and avian species diversity along an urban gradient. *Ecological Applications* 6:506–519.

Blann, K. L., J. L. Anderson, G. R. Sands, and B. Vondracek. 2009. Effects of agricultural drainage on aquatic ecosystems: A review. *Critical Reviews in Environmental Science and Technology* 39:909–1001.

Bliss, L. C., G. M. Courtin, D. L. Pattie, R. R. Riewe, D. W. A. Whitfield, and P. Widden. 1973. Arctic tundra ecosystems. *Annual Review of Ecology and Systematics* 4:359–399.

Boal, C. W., T. S. Estabrook, and A. E. Duerr. 2002. Productivity of loggerhead shrikes nesting in an urban interface. Pages 104–109 in W. W. Shaw, L. K. Harris, and L. VanDruff, editors. *Fourth International Symposium on Urban Wildlife Conservation*, Tucson, Arizona.

Bock, C. E., K. T. Vierling, S. L. Haire, J. D. Boone, and W. W. Merkle. 2002. Patterns of rodent abundance on open-space grasslands in relation to suburban edges. *Conservation Biology* 16:1653–1658.

Bock, C. E., Z. F. Jones, and J. H. Bock. 2006a. Abundance of cottontails (*Sylvilagus*) in an exurbanizing southwestern savanna. *Southwestern Naturalist* 51:352–357.

Bock, C. E., Z. F. Jones, and J. H. Bock. 2006b. Rodent communities in an exurbanizing southwestern landscape (USA). *Conservation Biology* 20:1442–1250.

Bonnell, S., and K. Storey. 2000. Addressing cumulative effects through strategic environmental assessment: A case study of small hydro development in Newfoundland, Canada. *Journal of Environmental Assessment Policy and Management* 2:477–499.

Bonnell, S., C. Leeder, and R. Pottle. 2001. Assessing the cumulative environmental effects of the Trans Labrador Highway (Red Bay to Cartwright): Methods, challenges and lessons. Proceedings of Cumulative Environmental Effects Management, Tools and Approaches. Alberta Society of Professional Biologists, Alberta Institute of Agrologists, and Association of Professional Biologists of British Columbia, Calgary, Alberta, Canada.

Boren, J. C., D. M. Engle, M. W. Palmer, R. E. Masters, and T. Criner. 1999. Land use change effects on breeding bird community composition. *Journal of Range Management* 52:420–430.

Bottrill, M. C., L. N. Joseph, J. Carwardine, M. Bode, C. Cook, E. T. Game, H. Grantham, S. Kark, S. Linke, E. McDonald-Madden, R. L. Pressey, S. Walker, K. A. Wilson, and H. P. Possingham. 2008. Is conservation triage just smart decision making? *Trends in Ecology and Evolution* 23:649–654.

Bottrill, M. C., L. N. Joseph, J. Carwardine, M. Bode, C. Cook, E. T. Game, H. Grantham, S. Kark, S. Linke, E. McDonald-Madden, R. L. Pressey, S. Walker, K. A. Wilson, and H. P. Possingham. 2009. Finite conservation funds mean triage is unavoidable. *Trends in Ecology and Evolution* 24:183–184.

Bouchard, R., M. Higgins, and C. Rock. 1995. Using constructed wetland-pond systems to treat agricultural runoff: A watershed perspective. *Lake and Reservoir Management* 11:29–36.

Bounds, D. L., and W. W. Shaw. 1997. Movements of suburban and rural coyotes at Saguaro National Park, Arizona. *Southwestern Naturalist* 42:94–121.

Bowers, M.A., and B. Breland. 1996. Foraging of gray squirrels on the urban-rural gradient: Use of the GUD to assess anthropogenic impact. *Ecological Society of America* 64:1135–1142.

Boyd, D. R. 2001. Canada vs. the OECD: An environmental comparison. Eco-Research Chair of Environmental Law and Policy, University of Victoria, British Columbia, Canada.

Brabec, E., S. Schulte, and P. L. Richards. 2002. Impervious surfaces and water quality: A review of current literature and its implications for watershed planning. *Journal of Planning Literature* 16:499–514.

Braun, C. E. 1998. Sage grouse declines in western North America: What are the problems? *Proceedings of the Western Association of State Fish and Wildlife Agencies* 78:139–156.

Breck, S. W., N. Lance, and P. Callahan. 2006. A shocking device for protection of concentrated food sources from black bears. *Wildlife Society Bulletin* 34:23–26.

Brinkman, T. J., J. A. Jenks, C. S. DePerno, B. S. Haroldson, and R. G. Osborn. 2004. Survival of white-tailed deer in an intensively farmed region of Minnesota. *Wildlife Society Bulletin* 32:726–731.

British Columbia Environmental Assessment Office. 2003. Guide to the British Columbia Environmental Assessment Process. Victoria, British Columbia, Canada.

British Columbia Government. 2004. Order Establishing Provincial Non-spatial Old Growth Objectives. Available from http://ilmbwww.gov.bc.ca/slrp/lrmp/policiesguidelinesand assessements/oldgrowth/pdf/Old_Growth_Order_May18th_FINAL.pdf. Accessed August 3, 2010.

British Columbia Ministry of Forests. 1995. Forest practices code of British Columbia: Biodiversity guidebook. Queens Printer, Victoria, British Columbia. Available from www.for.gov.bc.ca/tasb/legsregs/fpc/fpcguide/biodiv/biotoc.htm. Accessed August 3, 2010.

British Columbia Oil and Gas Commission. 2004. Geophysical Guidelines for the Muskwa-Kechika Management Area. Fort St. John, British Columbia, Canada.

Brosofske, K. D., J. Chen, R. J. Naiman, and J. F. Franklin. 1997. Harvesting effects on microclimatic gradients from small streams to uplands in western Washington. *Ecological Applications* 7:1188–1200.

Brotton, J., and G. Wall. 1997. Climate change and the Bathurst caribou herd in the Northwest Territories, Canada. *Climatic Change* 35:35–52.

Brown, D. E. Editor. 1994. *Biotic Communities: Southwestern United States and Northwestern Mexico*. University of Utah Press, Logan.

Brown A. L, and R. Therivel. 2000. Principles to guide the development of strategic environmental assessment methodology. *Impact Assessment and Project Appraisal* 18:183–189.

Brundige, G. C. 1993. Predation ecology of the eastern coyote (*Canis latrans)* in the Adirondacks, New York. Dissertation, State University of New York, Syracuse.

Bureau of Economic Analysis. 2009. Industry Economic Accounts. <http://bea.gov/industry/index.htm>. Accessed January 6, 2010.

Bureau of Land Management. 2006. Southern Arizona project to mitigate environmental damages resulting from illegal immigration; A summary of 2003–2005 accomplishments. http://www.blm.gov/pgdata/etc/medialib/blm/az/pdfs/undoc_aliens.Par.62736.File.dat/complete_summary_03-05.pdf. Accessed March 16, 2010.

Burris, R. K., and L. W. Canter. 1997. Cumulative impacts are not properly addressed in environmental assessments. *Environmental Impact Assessment Review* 17:5–18.

Burton, D. L., and K. A. Doblar. 2004. Morbidity and mortality of urban wildlife in the midwestern United States. Pages 171–181 in W. W. Shaw, L. K. Harris, and L. VanDruff, editors. *Fourth International Symposium on Urban Wildlife Conservation*, Tucson, Arizona.

California Department of Fish and Game. 2004. September 2002 Klamath River fish-kill: Final analysis of contributing factors and impacts. California Department of Fish and Game, Northern California-North Coast Region. The Resources Agency, Sacramento, California.

Callicott, J. B. 2002. From the balance of nature to the flux of nature. Pages 90–105 in R. L. Knight and S. Riedel, editors. *Aldo Leopold and the Ecological Conscience*. Oxford University Press, New York.

Calver, M., S. Thomas, S. Bradley, and H. McCutcheon. 2007. Reducing the rate of predation on wildlife by pet cats: The efficacy and practicability of collar-mounted pounce protectors. *Biological Conservation* 137:341–348.

Cam, E., J. D. Nichols, J. R. Sauer, J. E. Hines, and C. H. Flather. 2000. Relative species richness and community completeness: Birds and urbanization in the mid-Atlantic states. *Ecological Applications* 10:1196–1210.

Cameron, R. D., and J. M. Ver Hoef. 1994. Predicting parturition rate of caribou from autumn body mass. *Journal of Wildlife Management* 58:674–679.

Cameron, R. D., W. T. Smith, R. G. White, and B. Griffith. 2005. Central Arctic caribou and petroleum development: Distributional, nutritional, and reproductive implications. *Arctic* 58:1–9.

Campbell, C. A., and A. I. Dagg. 1976. Bird populations in downtown and suburban Kitchener-Waterloo, Ontario. *Ontario Field Biologist* 30:1–22.

Canadian Arctic Resources Committee. 2007. Industry and caribou: Can they coexist? *Northern Perspectives* 31:22–23.

Canadian Environmental Assessment Agency. 1999a. Comprehensive study report for the Diavik diamonds project. The Canadian Environmental Assessment Agency, Ottawa, Ontario, Canada.

Canadian Environmental Assessment Agency. 1999b. The 1999 Cabinet directive on the environmental assessment of policy, plan and program proposals. Public Works and Government Services Canada, Ottawa, Ontario, Canada.

Canadian Environmental Assessment Agency. 2007. Operational policy statement—Addressing cumulative environmental effects under the Canadian Environmental Assessment Act. Ottawa, Ontario, Canada.

Canter, L.W. 1996. *Environmental Impact Assessment*, second edition. McGraw-Hill, New York.

Carlson, M., E. Bayne, and B. Stelfox. 2007. Seeking a balance: Assessing the future impacts of conservation and development in the Mackenzie watershed. Canadian Boreal Initiative. Ottawa, Ontario, Canada.

Carmichael, G. J., J. N. Hanson, M. E. Schmidt, and D. C. Morizot. 1993. Introgression among Apache, cutthroat, and rainbow trout in Arizona. *Transactions of the American Fisheries Society* 122:121–130.

Carr, L. W., and L. Fahrig. 2001. Effect of road traffic on two amphibian species of differing vagility. *Conservation Biology* 15:1071–1078.

Carroll, C., R. F. Noss, and P. C. Paquet. 2001. Carnivores as focal species for conservation planning in the Rocky Mountain region. *Ecological Applications* 11:961–980.

Carroll, C., M. K. Phillips, N. H. Schumaker, and D. W. Smith. 2003. Impacts of landscape change on wolf restoration success; planning a reintroduction program on static and dynamic spatial models. *Conservation Biology* 17:536–548.

Carver, E., and J. Caudill. 2007. Banking on nature: The economic benefits to local communities of national wildlife refuge visitation. U.S. Fish and Wildlife Service, Washington, D.C.

Carveth, C. J., A. M. Widmer, and S. A. Bonar. 2006. Comparison of upper thermal tolerances of native and nonnative fish species in Arizona. *Transactions of the American Fisheries Society* 135:1433–1440.

Cashore, B., G. Hoberg, M. Howlett, J. Rayner, and J. Wilson. 2001. *In Search of Sustainability: British Columbia Forest Policy in the 1990s*. University of British Columbia Press, Vancouver, British Columbia, Canada.

Center for Watershed Protection. 2003. Impacts of impervious cover on aquatic systems. *Watershed Protection Research Monograph* 1:1–142.

Central Intelligence Agency. 2009. The world factbook: Canada. <https://www.cia.gov/library/publications/the-world-factbook/>. Accessed July 15, 2009.

Chadwick, J. W., and S. P. Canton. 1982. Coal mine drainage effects on a lotic ecosystem in northwest Colorado, USA. *Hydrobiologia* 107:25–33.

Chick, J. H., and M. A. Pegg. 2001. Invasive carp in the Mississippi River basin. *Science* 292: 2250–2251.

Christensen, J., and M. Grant. 2007. How political change paved the way for indigenous knowledge: The Mackenzie Valley resource management act. *Arctic* 60:115–123.

City of Phoenix. 2010. <http://phoenix.gov/waterservices/wrc/home/outdoor/pool.html>.

Claudi, R., and J. H. Leach. 1999. *Nonindigenous Freshwater Organisms. Vectors, Biology and Impacts*. Lewis Publishers, New York.

Clergeau, P. 2008. Can biodiversity be urban? *Biofuture* 285:1.

Clergeau, P., J. P. L. Savard, G. Mennechez, and G. Falardeau. 1998. Bird abundance and diversity along an urban-rural gradient: A comparative study between two cities on different continents. *Condor* 100:413–425.

Cocklin, C. 1993. What does cumulative effects analysis have to do with sustainable development? *Canadian Journal of Regional Science* XVI: 453–479.

Coleman, J. E., B. W. Jacobsen, and E. Reimers. 2001. Summer response distances of Svalbard reindeer *Rangifer rangifer plathyrhynchus* to provocation by humans on foot. *Wildlife Biology* 7:275–283.

Compas, E. 2007. Measuring exurban change in the American west: A case study in Gallatin County, Montana, 1973–2004. *Landscape and Urban Planning* 82:56–65.

Conant, R. 1951. The reptiles of Ohio, second edition. *American Midland Naturalist* 20:1–284.

Conacher, A. J. 1994. The integration of land-use planning and management with environmental impact assessment: Some Australian and Canadian perspectives. *Impact Assessment* 4:347–373.

Connelly, J. W., A. D. Apa, R. B. Smith, and K. P. Reese. 2000a. Effects of predation and hunting on adult sage grouse *Centrocercus urophasianus* in Idaho. *Wildlife Biology* 6:227–232.

Connelly, J. W., M. A. Schroeder, A. R. Sands, and C. E. Braun. 2000b. Guidelines to manage sage grouse populations and their habitats. *Wildlife Society Bulletin* 28:967–985.

Connelly, J. W., S. T. Knick, M. A. Schroeder, and S. J. Stiver. 2004. Conservation assessment of greater sage-grouse and sagebrush habitats. Unpublished report, Western Association of Fish and Wildlife Agencies, Cheyenne, Wyoming.

Connelly, J. W., E. T. Rinkes, and C. E. Braun. 2010. Characteristics of greater sage-grouse habitats: A landscape species at micro and macro scales. *Studies in Avian Biology*: in press.

Conover, M. R. 1995. What is the urban deer problem and where did it come from? Pages 11–18 in J. B. McAninch, editor. *Proceedings of the 1993 Symposium on the North Central Section*. The Wildlife Society, Bethesda, Maryland.

Cooper, L., and W. Sheate. 2002. Cumulative effects assessment: A review of UK environmental impact statements. *Environmental Impact Assessment Review* 22:415–439.

Cornicelli, L. A. 1992. White-tailed deer use of a suburban area in southern Illinois. Thesis, Southern Illinois University, Carbondale.

Cornicelli, L, A. Woolf, and J. L. Roseberry. 1996. White-tailed deer use of a suburban environment in southern Illinois. *Transactions of the Illinois Academy of Science* 89:93–103.

Costanza, R., editor. 1991. *Ecological Economics: The Science and Management of Sustainability*. Columbia University Press, New York.

Costanza, R., d'Arge, R., de Groot, R. Farber, S., Grasso, M., Hannon, B., Limburg, K., Naeem, S., O'Neill, R. V., Paruelo, J., Raskin, R. G., Sutton, P., van den Belt, M., 1997. The value of the world's ecosystem services and natural capital. *Nature* 387: 253–260.

Council on Environmental Quality. 1986. Regulations for implementing the procedural provisions of the National Environmental Policy Act, Washington, D.C.

Council on Environmental Quality. 1997. Considering cumulative effects under the National Environmental Policy Act, Washington, D.C.

Court, J., C. Wright, and A. Guthrie. 1994. Assessment of cumulative impact and strategic assessment in environmental impact assessment. Commonwealth Environment Protection Agency, Canberra, Australia.

Courtenay, W. R., and J. R. Stauffer. 1984. *Distribution, Biology and Management of Exotic Fishes*. Johns Hopkins University Press, Baltimore, Maryland.

Courtois, R., J.-P. Ouellet, L. Breton, A. Gingras, and C. Dussault. 2007. Effects of forest disturbance on density, space use, and mortality of woodland caribou. *Ecoscience* 14:491–498.

Crain, C. M., K. Kroeker, and B. S. Halpern. 2008. Interactive and cumulative effects of multiple human stressors in marine systems. *Ecology Letters* 11:1304–1315.

Crawford, J. A., R. A. Olson, N. E. West, J. C. Mosley, M. A. Schroeder, T. D. Whitson, R. F. Miller, M. A. Gregg, and C. S. Boyd. 2004. Ecology and management of sage-grouse and sage-grouse habitat. *Journal of Range Management* 57:2–19.

Creasey, J. R. 1998. Cumulative effects and the wellsite approval process. Thesis, University of Calgary, Calgary, Alberta, Canada.

Cronin, M. A., H. A. Whitlaw, and W. B. Ballard. 2000. Northern Alaska oil field and caribou. *Wildlife Society Bulletin* 28:919–922.

Crooks, K. R. 2002. Relative sensitivities of mammalian carnivores to habitat fragmentation. *Conservation Biology* 16:488–502.

Crowe, D. M. 1983. Comprehensive planning for wildlife resources. Wyoming Game and Fish Department, Cheyenne, Wyoming.

Curtis, P. D., R. J. Stout, B. A. Knuth, L. A. Myers, and T. M. Rockwell. 1993. Selecting deer management options in a suburban environment: A case study from Rochester, New York. *Transactions of the North American Wildlife and Natural Resources Conference* 57:102–116.

Customs and Border Protection 2008. Environmental waiver/environmental stewardship plan information (cbp.gov 12/09/2008). Accessed March 2010.

Custred, G. 2000. Alien crossings. *The American Spectator* 33:38–43.

Czech, B. 2000a. Economic growth as the limiting factor for wildlife conservation. *Wildlife Society Bulletin* 28:4–15.

Czech, B. 2000b. Shoveling fuel for a runaway train: Errant economists, shameful spenders, and plan to stop them all. University of California Press, Berkeley.

Czech, B. 2008. Prospects for reconciling the conflict between economic growth and biodiversity conservation with technological progress. *Conservation Biology* 22:1389–1398.

Czech, B., P. R. Krausman, and P. K. Devers. 2000. Economic associations among causes of species endangerment in the United States. *Bioscience* 50:593–601.

Dale, V., S. Brown, R. Haeuber, N. Hobbs, N. Huntly, R. Naiman, W. Reisbsame, M. Turner, and T. Valone. 2000. Ecological Society of America report: Ecological principles and guidelines for managing the use of land. *Ecological Applications* 10:639–670.

Danielson, B. J., and M. W. Hubbard. 1999. A literature review for assessing the status of current methods of reducing deer-vehicle collisions. Iowa Department of Transportation, Ames.

Daniels, G. D., and J. B. Kirkpatrick. 2006. Does variation in garden characteristics influence the conservation of birds in suburbia? *Biological Conservation* 133:326–335.

Dasmann, R. F. 1966. Wildlife and the new conservation. *The Wildlife Society News* 105:48–49.

Dauba, F., S. Lek, S. Mastrorillo, and G. H. Copp. 1997. Long-term recovery of macrobenthos and fish assemblages after water pollution abatement measures in the River Petite Baise (France). *Archives of Contamination and Toxicology* 33:277–285.

Davey, L. H., J. L. Barnes, C. L. Horvath, and A. Griffiths. 2001. Addressing cumulative environmental effects: Sectoral and regional environmental assessment. In A. J. Kennedy, editor. *Proceedings: Cumulative Environmental Effects Management, Tools and Approaches*. Alberta Society of Professional Biologists, Alberta Institute of Agrologists, and Association of Professional Biologists of British Columbia, Calgary, Alberta, Canada.

Davies, P. E., and M. Nelson. 1994. Relationships between riparian buffer widths and the effects of logging on stream habitat, invertebrate community composition and fish abundance. *Australian Journal of Marine and Freshwater Research* 45:1289–1305.

Debol'skii, V. K. 1996. Problem of preventing secondary pollution of reservoirs. *Hydrotechnical Construction* 30: 691–692. (Translated from *Gidrotekhnicheskoe Stroitel'stvo* 11:46–47, November 1996; Plenum Publishing Corporation.)

Decker, D. J., and T. A. Gavin. 1987. Public attitudes toward a suburban deer herd. *Wildlife Society Bulletin* 15:173–180.

Decker, D. J., T. L. Brown, and W. F. Siemer. 2001. *Human Dimensions of Wildlife Management in North America*. The Wildlife Society, Bethesda, Maryland.

DeGraaf, R. M. 1991. Winter foraging guild structure and habitat associations in suburban bird communities. *Landscape and Urban Planning* 21:173–180.

DeNicola, A. J., K. C. VerCauteren, P. D. Curtis, and S. E. Hygnstrom. 2000. Managing white-tailed deer in suburban environments: A technical guide. Cornell Cooperative Extension, Ithaca, New York.

Department of Homeland Security, Office of Immigration Statistics. 2008 annual report. Immigration enforcement actions 2006. <http://www.dhs.gov/files/statistics/publications/yearbook.shtm>. Accessed April 29, 2010.

DeStefano, S., and R. M. DeGraaf. 2003. Exploring the ecology of suburban wildlife. *Frontiers in Ecology and the Environment* 1:95–101.

DeSteiguer, J. E. 1995. Three theories from economics about the environment. *BioScience* 45:552–557.

Dextrase, A. J., and M. A. Coscarelli. 1999. Intentional introductions of nonindigenous freshwater organisms in North America. Pages 61–98 in R. Claudi and J. H. Leach, editors. *Nonindigenous Freshwater Organisms. Vectors, Biology, and Impacts*. Lewis Publishers, New York.

Dixon, J., and B. E. Montz. 1995. From concept to practice: implementing cumulative impact assessment in New Zealand. *Environmental Management* 19:445–456.

Doherty, K. E. 2008. Sage-grouse and energy development: Integrating science with conservation planning to reduce impacts. Dissertation, University of Montana, Missoula.

Doherty, K. E., D. E. Naugle, B. L. Walker, and J. M. Graham. 2008. Greater sage-grouse winter habitat selection and energy development. *Journal of Wildlife Management* 72:187–195.

Doherty, K. E., D. E. Naugle, and J. S. Evans. 2010. A currency for offsetting energy development impacts: Horse-trading sage-grouse on the open market. *PLoS One* 5. e10339.

Doherty, K., D. E. Naugle, H. Copeland, A. Pocewicz, and J. M. Kiesecker. 2011. Energy development and conservation tradeoffs: Systematic planning for sage-grouse in their eastern range. *Studies in Avian Biology*. University of Colorado Press, Boulder.

Donaldson, E. M., U. H. M. Fagerlund, and J. R. McBride. 1984. Aspects of the endocrine stress response to pollutants in salmonids. Pages 213–221 in V. W. Cairns, P. V. Hodson, and J. O. Nriagu, editors. *Contaminants Effects on Fisheries*. Wiley-Interscience, New York.

Donihee, J. 1999. Resource development and the Mackenzie Valley Resource Management Act. *Resources* 66:1–5.

Donnelly, R. E., and J. M. Marzluff. 2004. Designing research to advance the management of birds in urbanizing areas. Page 114 in W. W. Shaw, L. K. Harris, and L. VanDruff, editors. *Fourth International Symposium on Urban Wildlife Conservation*, Tucson, Arizona.

Dowlatabadi, H., M. Boyle, S. Rolwley, and M. Kandlikar. 2004. Bridging the gap between project-level assessments and regional development dynamics: A methodology for estimating cumulative effects. Canadian Environmental Assessment Agency, Ottawa, Ontario, Canada.

Dubé, M. 2003. Cumulative effect assessment in Canada: A regional framework for aquatic ecosystems. *Environmental Impact Assessment Review* 23:723–745.

Dubé, M., and K. Munkittrick. 2001. Integration of effects-based and stressor-based approaches into a holistic framework for cumulative effects assessment in aquatic ecosystems. *Human and Ecological Risk Assessment* 7:247–248.

Dubé, M., B. Johnson, G. Dunn, J. Culp, K. Cash, K. Munkittrick, I. Wong, K. Hedley, W. Booty, D. Lam, O. Resler, and A. Storey. 2006. Development of a new approach to cumulative effects assessment: A northern river ecosystem example. *Environmental Monitoring and Assessment* 113:87–115.

Duchamp, J. E., D. W. Sparks, and J. O. Whitaker. 2004. Foraging-habitat selection by bats at an urban-rural interface: Comparison between a successful and a less successful species. *Canadian Journal of Zoology-Revue Canadienne de Zoologie* 82:1157–1164.

Duinker, P. N., and L. A. Greig. 2006. The impotence of cumulative effects assessment in Canada: Ailments and ideas for redeployment. *Environmental Management* 37:153–161.

Duinker, P. N., and L. A. Greig. 2007. Scenario analysis in environmental impact assessment: Improving explorations of the future. *Environmental Impact Assessment Review* 27:206–219.

Dusek, G. L. 1987. Ecology of white-tailed deer in upland ponderosa pine habitat is southeastern Montana. *Prairie Naturalist* 19:1–17.

Dusek, G. L., R. J. Mackie, J. D. Herriges, and B. B. Compton. 1989. Population ecology of white-tailed deer along the lower Yellowstone River. *Wildlife Monographs* 104:1–68.

Ehrenfeld, J. G. 2000. Evaluating wetlands within an urban context. *Ecological Engineering* 15:253–265.

Elith, J., and J. Leathwick. 2009. Species distribution models: Ecological explanation and prediction across space and time. *Annual Reviews for Ecology, Evolution and Systematics* 40:677–697.

Ellis, K. L. 1984. Behavior of lekking sage grouse in response to a perched golden eagle. *Western Birds* 15:37–38.

Ellis, S. C. 2005. Meaningful consideration? A review of traditional knowledge in environmental decision making. *Arctic* 58:66–77.

Emlen, J. T. 1974. An urban bird community in Tucson, Arizona; derivation, structure, regulation. *Condor* 76:184–197.

Environmental Consultants. 2009. Environmental assessment DWG decorative rock quarry. AZ-420.USDI, Bureau of Land Management, Tucson, Arizona.

Environmental Protection Agency. 1994. National water quality inventory: 1994 report to Congress. Environmental Protection Agency, Washington, D.C.

Environmental Protection Agency. 2009. National water quality inventory: Report to Congress 2004 reporting cycle. EPA 841-R-08-001, Environmental Protection Agency, Washington, D.C.

Erickson, A. W., J. E. Nellor, and G. A. Petrides. 1964. The black bear in Michigan. Michigan State University, Agriculture Experiment Station Research Bulletin 4.

Erickson, J.D. 2000. Endangering the economics of extinction. *Wildlife Society Bulletin* 28:34–41.

Esty, D. C., M. Levy, T. Srebotnjak, and A. de Sherbinin. 2005. Environmental sustainability index: Benchmarking national environmental stewardship. Yale Center for Environmental Law & Policy, New Haven, Connecticut.

Etchberger, R. C., P. R. Krausman, and R. Mazaika. 1989. Mountain sheep habitat characteristics in the Pusch Ridge Wilderness, Arizona. *Journal of Wildlife Management* 53:902–907.

Etter, D. R., K. M. Hollis, T. R. Van Deelen, D. R. Ludwig, J. E. Chelsvig, C. L. Anchor, and R. E. Warner. 2002. Survival and movements of white-tailed deer in suburban Chicago, Illinois. *Journal of Wildlife Management* 66:500–510.

European Commission. 1999. Guidelines for the assessment of indirect and cumulative impacts as well as impact interactions. Brussels, Belgium.

Evelyn, M. J., D. A. Stiles, and R. A. Young. 2004. Conservation of bats in suburban landscapes: Roost selection by *Myotis yumanensis* in a residential area in California. *Biological Conservation* 115:463–473.

Fagan, W. F., P. J. Unmack, C. Burgess, and W. L. Minckely. 2002. Rarity, fragmentation, and extinction risk in desert fishes. *Ecology* 83:3250–3256.

Fahrig L., and T. Rytwinski. 2009. Effects of roads on animal abundance: An empirical review and synthesis. *Ecology and Society* 14:21.

Federal Aviation Association and National Park Service Technical Team. 2009. Assessment of the proposed noise metrics and impact intensities for the Grand Canyon National Park over flights EIS (status and next steps). U.S. National Park Service, Washington, D.C.

Federal Energy and Regulatory Commission. 1987. Salmon River Basin, fifteen hydroelectric projects, Idaho. Federal Energy Regulatory Commission, FERC/FEIS-0044, Washington, D.C.

Federal Environmental Assessment Review Office. 1994. A Reference Guide for the Canadian Environmental Assessment Act—Addressing Cumulative Environmental Effects. Federal Environmental Assessment Review Office, Hull, Quebec, Canada.

Fedriani, J. M., T. K. Fuller, and R. M. Sauvajot. 2001. Does availability of anthropogenic food enhance densities of omnivorous mammals? An example with coyotes in southern California. *Ecography* 24:325–331.

Fedy, B. C., K. Martin, C. Ritland, and J. Young. 2008. Genetic and ecological data provide incongruent interpretations of population structure and dispersal in naturally subdivided populations of white-tailed ptarmigan (*Lagopus leucura*). *Molecular Ecology* 17:1905–1917.

Ferguson, H. L. 2004. Winter raptor composition, abundance and distribution around urban Spokane, eastern Washington. Pages 123–134 in W. W. Shaw, L. K. Harris, and L. VanDruff, editors. *Fourth International Symposium on Urban Wildlife Conservation*, Tucson, Arizona.

Ficetola, G. F., and F. De Bernardi. 2004. Amphibians in a human-dominated landscape: The community structure is related to habitat features and isolation. *Biological Conservation* 119:219–230.

Finder, R. A., J. L. Roseberry, and A. Woolf. 1999. Site and landscape conditions at white-tailed deer/vehicle collision locations in Illinois. *Landscape and Urban Planning* 44:77–85.

Fitter, R. S. R. 1945. *London's Natural History*. Bloomsbury Books, London.

Fletcher, R. J., and R. R. Koford. 2002. Habitat and landscape associations of breeding birds in native and restored grasslands. *Journal of Wildlife Management* 66:1011–1022.

Folke, C., T. Hahn, P. Olsson, and J. Norberg. 2005. Adaptive governance of social-ecological systems. *Annual Review of Environment and Resources* 30:441–473.

Forchhammer, M. C., E. Post, N. C. Stenseth, and D. M. Boertmann. 2002. Long-term responses in arctic ungulate dynamics to changes in climatic and trophic processes. *Population Ecology* 44:113–120.

Fox, A. D., M. Desholm, J. Kahlert, T. K. Christensen, and I. K. Petersen. 2006. Information needs to support environmental impact assessment of the effects of European marine offshore wind farms on birds. *IBIS* 148:129–144.

Frame, T. M., T. Gunton, and J. C. Day. 2004. The role of collaboration in environmental management: An evaluation of land and resource planning in British Columbia. *Journal of Environmental Planning and Management* 47:59–82.

Franklin, J. F. 1994. Developing information essential to policy, planning, and management decision-making: The promise of GIS. Pages 18–24 in V. A. Sample, editor. *Remote Sensing and GIS in Ecosystem Management.* Island Press, Washington, D.C.

Fraterrigo, J. M., and J. A. Wiens. 2005. Bird communities of the Colorado Rocky Mountains along a gradient of exurban development. *Landscape and Urban Planning* 71:263–275.

Freedman, A. H., K. M. Portier, and M. E. Sunquist. 2003. Life history analysis for black bears (*Ursus americanus*) in a changing demographic landscape. *Ecological Modeling* 167:47–64.

Frid, A., and L. Dill. 2002. Human-caused disturbance stimuli as a form of predation risk. *Conservation Ecology* 6:11.

Friesen, T. A., and D. L. Ward. 1999. Management of northern pikeminnow and implications for juvenile salmonid survival in the lower Columbia and Snake Rivers. *North American Journal of Fisheries Management* 19:406–420.

Frost, S. L. 1976. The steam of history—An Ohio water resources chronology. Department of Natural Resources, Columbus, Ohio. Cited in Frost, S. L., and W. L. Mitsch. 1989. Resource development and conservation history along the Ohio River. *Ohio Journal of Science* 89:143–152.

Gagné, S. A., and L. Fahrig. 2007. Effect of landscape context on anuran communities in breeding ponds in the National Capital Region, Canada. *Landscape Ecology* 22:205–215.

Gaines, W., P. Singleton, and R. Ross. 2003. Assessing the cumulative effects of linear recreation routes on wildlife habitats on the Okanogan and Wenatchee National Forests. USDA Forest Service General Technical Report PNW-GTR-586. U.S. Department of Agriculture, Forest Service, Pacific Northwest Research Station, Portland, Oregon.

Ganster, Paul. 2009. [Letter to President Barack Obama, 2009]. Good neighbor environmental board, presidential advisory committee on environmental and infrastructure issues along the U.S. border with Mexico. Found at <http://www.enviropic.org/Enviropic.org/Project_files/GNEB%20letter%20to%20president.pdf>.

Garton, E. O., J. W. Connelly, C. A. Hagen, J. S. Horne, A. Moser, and M. A. Schroeder. 2010. Greater sage-grouse population dynamics and probability of persistence. *Studies in Avian Biology*: in press.

Gauderman, W. J., R. McConnell, F. Gilliland, S. London, D. Thomas, E. Avol, H. Vora, K. Berhane, E. B. Rappaport, F. Lurmann, H. G. Margolis, and J. Peters. 2000. Association between air pollution and lung function growth in southern California children. *American Journal of Respiratory and Critical Care Medicine*, 162:1383–1390.

Gehrt, S. D., and J. E. Chelsvig. 2004. Species-specific patterns of bat activity in an urban landscape. *Ecological Applications* 14:625–635.

Geist, V. 1971. A behavioral approach to the management of wild ungulates. Pages 413–424 in E. Duffy and A. S. Watt, editors. *The Scientific Management of Animal and Plant Communities for Conservation*. Blackwell Science Publishing, Oxford, United Kingdom.

Geist, V. 1988. How markets in wildlife meats and parts, and the sale of hunting privileges, jeopardize wildlife conservation. *Conservation Biology*. 2:15–26.

Geist, V. 2008. Large predators: Them and us! *Fair Chase* Fall:14–19.

Germaine, S. S., and B. F. Wakeling. 2001. Lizard species distributions and habitat occupation along an urban gradient in Tucson, Arizona, USA. *Biological Conservation* 97:229–237.

Germaine, S. S., S. S. Rosenstock, R. E. Schweinsburg, and W. S. Richardson. 1998. Relationships among breeding birds, habitat, and residential development in greater Tucson, Arizona. *Ecological Applications* 8:680–691.

Gibbs, K. E., R. L. Mackey, and D. J. Currie. 2009. Human land use, agriculture, pesticides and losses of imperiled species. *Diversity and Distributions* 15:242–253.

Gibeau, M. L. 1998. Use of urban habitats by coyotes in the vicinity of Banff, Alberta. *Urban Ecosystems* 2:129–139.

Gibson, R. J., R. L. Haedrich, and C. Michael Wernerheim. 2005. Loss of fish habitat as a consequence of inappropriately constructed stream crossings. *Fisheries* 30:10–17.

Gill, D. A. 1965. Coyote and urban man: A geographical analysis of the relationship between coyote and man in Los Angeles. Thesis, University of California, Los Angeles.

Gionfriddo, J. P., and P. R. Krausman. 1986. Summer habitat use by mountain sheep. *Journal of Wildlife Management* 50:331–336.

Gipson, P. S., and J. A. Sealander. 1972. Home range and activity of the coyote (*Canis latrans frustor*) in Arkansas. *Proceedings of the Southeastern Association of Game and Fish Commissioners* 26:82–95.

Glassner-Shwayder, K. 2000. Briefing paper: Great Lakes nonindigenous invasive species. A product of the Great Lakes Nonindigenous Invasive Species Workshop. October 20–21, 1999. Chicago, Illinois. U.S. Environmental Protection Agency, Chicago, Illinois.

Goodrich, J. M., and S. W. Buskirk. 1995. Control of abundant native vertebrates for conservation of endangered species. *Conservation Biology* 9:1357–1364.

Gordon, K. M., S. H. Anderson, B. Gribble, and M. Johnson. 2001. Evaluation of the FLASH (Flashing Light Animal Sensing Host) system in Nugget Canyon, Wyoming, Report FHWA-WY-01/03F, University of Wyoming, Laramie.

Government of Northwest Territories. 2009. Approach to regulatory improvement. Government of Northwest Territories, Yellowknife, Northwest Territories, Canada.

Government of Northwest Territories, Environment and Natural Resources. 2009. Bathurst caribou: Population. http://www.enr.gov.nt.ca/_live/pages/wpPages/Bathurst_Caribou_Herd.aspx. Accessed December 6, 2009.

Gowdy, J. M. 2000. Terms and concepts in ecological economics. *Wildlife Society Bulletin* 28:26–33.

Granholm, S. L., E. Gerstler, R. R. Everitt, D. P. Bernard, and E. C. Vlachos. 1987. Issues, methods, and institutional processes for assessing cumulative biological impacts. Report 009.5-87.5. Pacific Gas and Electric Company, San Ramon, California.

Green, R. J. 1984. Native and exotic birds in a suburban habitat. *Australian Wildlife Research* 11:181–190.

Griffith, B., D. C. Douglas, N. E. Walsh, D. D. Young, T. R. McCabe, D. E. Russell, R. G. White, R. D. Cameron, and K. R. Whitten. 2002. The Porcupine caribou herd. Pages 8–37 in D. C. Douglas, P. E. Reynolds, and E. B. Rhode, editors. *Arctic Refuge Coastal Plain Terrestrial Wildlife Research Summaries*. U.S. Geological Survey, Biological Resources Division, Biological Science Report USGS/BRD/BSR-2002-0001.

Grijalva, R. M. 2007. [Letter to Secretary of Department of Homeland Security Michael Chertoff. July 18. <http://www.grijalva.house.gov/index.cfm?sectionid=13&parentid=5&sectiontree=5,13&itemid=72> Accessed August 2010.

Grimmond, C. S., C. Souch, M. Hubble. 1996. Influence of tree cover on summertime surface energy balance fluxes, San Gabriel Valley, Los Angeles. *Climate Research* 6:45–57.

Grinder, M. I., and P. R. Krausman. 1998. Ecology and management of coyotes in Tucson, Arizona. *Proceedings of the Vertebrate Pest Conference* 18:293–298.

Grinder, M. I., and P. R. Krausman. 2001. Home range, habitat use, and nocturnal activity of coyotes in an urban environment. *Journal of Wildlife Management* 65:887–898.

Groocock, G. H., R. G. Getchell, G. A. Wooster, K. L. Britt, W. N. Batts, J. R. Winton, R. N. Casey, J. W. Casey, and P. R. Bowser. 2007. Detection of viral hemorrhagic septicemia in round gobies in New York State (USA) waters of Lake Ontario and the St. Lawrence River. *Diseases of Aquatic Organisms* 76:187–192.

Grubbs, S. E., and P. R. Krausman. 2009a. Use of the urban landscape by coyotes. *Southwestern Naturalist* 54:1–12.

Grubbs, S. E., and P. R. Krausman. 2009b. Observations of coyote-cat interactions. *Journal of Wildlife Management* 73:683–685.

Grund, M. D., J. B. McAninch, and E. P. Wiggers. 2002. Seasonal movements and habitat use of female white-tailed deer associated with an urban park. *Journal of Wildlife Management* 66:123–130.

Grzybowski, A., and Associates. 2001. Regional environmental effects assessment and strategic land use planning in British Columbia. Canadian Environmental Assessment Agency Research and Development Monograph Series, http://www.ceaa.gc.ca/015/001/010/index_e.htm. Accessed August 3, 2010.

Guisan, A., and W. Thuiller. 2005. Predicting species distribution: Offering more than simple habitat models. *Ecology Letters* 8:993–1009.

Gunderson, L. H. 2000. Ecological resilience—In theory and application. *Annual Review in Ecology and Systematics* 31:425–439.

Gunn, A. 2003. Voles, lemmings and caribou—Population cycles revisited? *Rangifer Special Issue No.* 14:105–111.

Gunn, A., J. Dragon, and J. Boulanger. 2002. Seasonal movements of satellite-collared caribou from the Bathurst herd. Final Report to the West Kitikmeot/Slave Study Society, Yellowknife, Northwest Territories, Canada.

Gunn, A., B. Griffith, G. Kofinas, and D. Russell. 2001. Cumulative impacts and the Bathurst caribou herd: Proposed tools for assessing the effects. Unpublished Report. Government of the Northwest Territories, Department of Resources, Wildlife and Economic Development, Yellowknife, Northwest Territories, Canada.

Gunn, A., D. Russell, R. G. White and G. Kofinas. 2009. Facing a future of change: Wild migratory caribou and reindeer. *Arctic* 62:iii–vi.

Gunn, J. H. 2009. Integrating strategic environmental assessment and cumulative effects assessment in Canada. Dissertation, University of Saskatchewan, Saskatoon, Saskatchewan, Canada.

Guo, H. Y., X. R. Wang, and J. G. Zhu. 2004. Quantification and index of non-point source pollution in Taihu Lake region with GIS. *Environmental Geochemistry and Health* 26:147–156.

Gustafson, E. J., D. E. Lytle, R. Swaty, and C. Loehle. 2007. Simulating the cumulative effects of multiple forest management strategies on landscape measures of forest sustainability. *Landscape Ecology* 22:141–156.

Gustine, D. D., K. L. Parker, R. J. Lay, M. P. Gillingham, and D. C. Heard. 2006. Calving strategies and calf survival of woodland caribou in a multi-predator ecosystem in northern British Columbia. *Wildlife Monographs* 165:1–32.

Hadidian, S., and J. Smith. 2001. *Urban Animals.* The Humane Society of the United States, New York.

Hagemoen, R. I. M., and E. Reimers. 2002. Reindeer summer activity pattern in relation to weather and insect harassment. *Journal of Animal Ecology* 71:883–892.

Hagen, C. A. 2003. A demographic analysis of lesser prairie-chicken populations in southwestern Kansas: Survival, population viability, and habitat use. Dissertation, Kansas State University, Manhattan.

Haggerty, J. H. and W. R. Travis. 2006. Out of administrative control: Absentee owners, resident elk and the shifting nature of wildlife management in southwestern Montana. *Geoforum* 37:816–830.

Hall, J. V., V. Brajer, and F. W. Lurmann. 2006. The health and related economic benefits of attaining healthful air in the San Joaquin Valley. Working paper. California State University, Institute for Economic and Environmental Studies. Fullerton. <http://cbeweb-1.fullerton.edu/Centers/iees>. Accessed February 21, 2007.

Hall, C. A. S., P. W. Jones, T. M. Donovan, and J. P. Gibbs. 2000. The implications of mainstream economics for wildlife conservation. *Wildlife Society Bulletin* 28:16–25.

Halpern, B. S., K. L. McLeod, A. A. Rosenberg, and L. B. Crowder. 2008. Managing for cumulative impacts in ecosystem-based management through ocean zoning. *Ocean and Coastal Management* 51:203–211.

Hammond, J. S., R. L. Kenney, and H. Raiffa. 1999. *Smart Choices: A Practical Guide to Making Better Life Decisions*. Broadway Books, New York.

Hansen, A. J., R. L. Knight, J. M. Marzluff, S. Powell, K. Brown, P. H. Gude, and A. Jones. 2005. Effects of exurban development on biodiversity: Patterns, mechanisms, and research needs. *Ecological Applications* 15:1893–1905.

Hanser, S. E., and S. T. Knick. 2010. Greater sage-grouse as an umbrella species for shrubland passerine birds: A multiscale assessment. *Studies in Avian Biology*: in press.

Harriman, J. A., and B. F. Noble. 2008. Characterizing project and strategic approaches to regional cumulative effects assessment in Canada. *Journal of Environmental Assessment Policy and Management* 10:25–50.

Harris, H. J., R. D. Wegner, V. A. Harris, and D. S. DeValut. 1994. A method for assessing environmental risks: A case study of Green Bay, Lake Michigan. *Environmental Management* 18:295–306.

Harris, L. K., P. R. Krausman, and W. W. Shaw. 1995. Human attitudes and mountain sheep in a wilderness setting. *Wildlife Society Bulletin* 23:66–72.

Harrison, D. J., and T. G. Chapin. 1998. Extent and connectivity of habitat for wolves in eastern North America. *Wildlife Society Bulletin* 26:767–775.

Harrod, R. J., and S. Reichard. 2001. Fire and invasive species within the temperate and boreal coniferous forests of western North America. Pages 95–101 in K.E.M. Galley and T.P. Wilson, editors. *Proceedings of the Invasive Species Workshop: The Role of Fire in the Control and Spread of Invasive Species*. Fire Conference 2000: the First National Congress on Fire Ecology, Prevention, and Management. Miscellaneous Publication No. 11, Tall Timbers Research Station, Tallahassee, Florida.

Haskell, S. P., and W. B. Ballard. 2008. Annual re-habituation of calving caribou to oilfields in northern Alaska: Implications for expanding development. *Canadian Journal of Zoology* 86:627–637.

Hastings, J. R., and R. M. Turner. 1965. *The Changing Mile: An Ecological Study of Vegetation Change with Time in the Lower Mile of an Arid and Semiarid Region*. University of Arizona Press, Tucson.

Havlick, D. G. 2002. *No Place Distant*. Island Press, Washington, D.C.

Hawbaker, T. J., V. C. Radeloff, M. K. Clayton, R. B. Hammer, and C. E. Gonzalez-Abraham. 2006. Road development, housing growth, and landscape fragmentation in northern Wisconsin: 1937–1999. *Ecological Applications* 16:1222–1237.

Hawkins, C. C., Grant, W. E., and M. T. Longnecker. 2004. Effect of house cats, being fed in parks, on California birds and rodents. Pages 164–170 in W. W. Shaw, L. K. Harris, and L. VanDruff, editors. *Fourth International Symposium on Urban Wildlife Conservation*, Tucson, Arizona.

Hawkins, R. E., and W. D. Klimstra. 1970. A preliminary study of the social organization of white-tailed deer. *Journal of Wildlife Management* 66:123–130.

Hayden, B. 1975. Some effects of rural subdivision on wildlife and wildlife habitat around Lolo, Montana. Thesis, University of Montana, Missoula.

Hazell S., and H. Benevides. 2000. Toward a legal framework for SEA in Canada. Pages 47–68 in M. R. Partidario and C. Clark, editors. *Perspectives on Strategic Environmental Assessment*. Lewis, New York.

Hedlund, J. H., P. D. Curtis, G. Curtis, and A. F. Williams. 2004. Methods to reduce traffic crashes involving deer: What works and what does not. *Traffic Injury Prevention* 5:122–131.

Hegmann, G., C. Cocklin, R. Creasey, S. Dupuis, A. Kennedy, L. Kingsley, W. Ross, H. Spaling, and D. Stalker. 1999. Cumulative effects assessment practitioners guide. Prepared by AXYS Environmental Consulting Ltd. and the CEA Working Group for the Canadian Environmental Assessment Agency, Hull, Quebec, Canada.

Heilbroner, R. L. 1992. *The Worldly Philosophers: The Lives, Times, and Ideas of the Great Economic Thinkers*. Sixth edition. Simon & Schuster, New York.

Helena Urban Wildlife Task Force. 2007. Findings and recommendations of the Helena Urban Wildlife Task Force. Montana Department of Fish, Wildlife and Parks, Helena.

Helfman, G. 2007. *Fish Conservation. A Guide to Understanding and Restoring Global Aquatic Biodiversity and Fishery Resources*. Island Press, Washington, D.C.

Henderson, R. E., and A. O'Harren. 1992. Winter ranges for elk and deer: Victims of uncontrolled subdivisions? *Western Wildlife* 18:20–25.

Henderson, D. W., R. J. Warren, J. A. Cromwell, and R. Joseph Hamilton. 2000. Responses of urban deer to a 50% reduction in local herd density. *Wildlife Society Bulletin* 28:902–910.

Hennings, L. A., and W. D. Edge. 2003. Riparian bird community structure in Portland, Oregon: Habitat, urbanization, and spatial scale patterns. *Condor* 105:288–302.

Heydlauff, A., P. R. Krausman, W. W. Shaw, and S. Marsh. 2006. Perceptions of elk and elk management in Arizona. *Wildlife Society Bulletin* 33:27–35.

Hickman, G. 2007. Helena urban deer inventory 2006–2007. Pages J5–J58 in *Findings and Recommendations of the Helena Urban Wildlife Task Force*. Helena, Montana.

Holling, C. S. 1973. Resilience and stability of ecological systems. *Annual Review of Ecology and Systematics* 4:1–23.

Holling, C. S. 1986. The resilience of terrestrial ecosystems; local surprise and global change. Pages 292–317 in W.C. Clark and R.E. Munn, editors. *Sustainable Development of the Biosphere*. Cambridge University Press, Cambridge, United Kingdom.

Holling, C. S., and G. K. Meffe. 1996. Command and control and the pathology of natural resource management. *Conservation Biology* 10:328–337.

Holloran, M. J. 2005. Greater sage-grouse (*Centrocercus urophasianus*) population response to gas field development in western Wyoming. Dissertation, University of Wyoming, Laramie.

Holloran, M. J., R. C. Kaiser, and W. A. Hubert. 2010. Yearling greater sage-grouse response to energy development in Wyoming. *Journal of Wildlife Management* 74:65–72.

Holloran, M. R. J., and S. H. Anderson. 2005. Spatial distribution of greater sage-grouse nests in relatively contiguous sagebrush habitats. *Condor* 107:742–752.

Horak, G. C., E. C. Vlachos, and E. W. Cline. 1983. Methodological guidance for assessing cumulative impacts on fish and wildlife. Report to U.S. Fish and Wildlife Service. Dynamic Corporation, Fort Collins, Colorado.

Houle, M., D. Fortin, C. Dussault, R. Courtois, and J. P. Ouellet. 2010. Cumulative effects of forestry on habitat use by gray wolf (*Canis lupus*) in the boreal forest. *Landscape Ecology* 25:419–433.

Housman, D. C., H. H. Powers, A. D. Collins, and J. Belnap. 2006. Carbon and nitrogen fixation differ between successional stages of biological soil crusts in the Colorado Plateau and Chihuahuan Desert. *Journal of Arid Environments* 66: 620–634.

Howell, R. G. 1982. The urban coyote problem in Los Angeles County. *Proceedings of the Vertebrate Pest Conference* 10:21–23.

Hubbard, M. W., B. J. Danielson, and R. A. Schmitz. 2000. Factors influencing the location of deer-vehicle accidents in Iowa. *Journal of Wildlife Management* 64:707–713.

Hudson, R. J. 2002. An evaluation of ALCES, a landscape cumulative effects simulator for us in integrated resource management in Alberta. Unpublished Report, CyberCervus International, New Sarepta, Alberta, Canada.

Huhtalo, H., and O. Järvinen. 1977. Quantitative composition of the urban bird community in Tornio, northern Finland. *Bird Study* 24:179–185.

Humphries, P., and K. O. Winemiller. 2009. Historical impacts on river fauna, shifting baselines, and challenges for restoration. *Bioscience* 59:673–684.

Ingelson, A. W. Holden, and M. Bravante. 2009. Philippine environmental impact assessment, mining and genuine development. *Law Environment and Development Journal* 5:1–7.

Integrated Land Management Bureau. 2006. A new direction for land use planning in BC. Integrated Land Management Bureau, Ministry of Agriculture and Lands, Victoria, British Columbia, Canada.

Inter-organizational Commission on Guidelines and Principles. 1994. Guidelines and principles for social impact assessment. *Impact Assessment* 12:107–152.

Ito, T. Y., N. Miura, B. Lhagvasuren, D. Enkhbileg, S. Takatsuki, A. Tsunekawa, and Z. Jiang. 2005. Preliminary evidence of a barrier effect of a railroad on the migration of Mongolian gazelles. *Conservation Biology* 19:245–248.

Jacks, S., S. Sharon, R. E. Kinnunen, D. K. Britton, D. Jensen, and S. S. Smith. 2009. Controlling the spread of invasive species while sampling. Pages 217–222 in S. A. Bonar, W. A. Hubert, and D. W. Willis, editors. *Standard Methods for Sampling North American Freshwater Fishes*. American Fisheries Society, Bethesda, Maryland.

Jackson, T. 2002. Consensus processes in land use planning in British Columbia: The nature of success. *Progress in Planning* 57:1–90.

Jackson, T., and J. Curry. 2002. Regional development and land use planning in rural British Columbia: Peace in the woods? *Regional Studies* 36:439–443.

Jakimäki, J., and J. Suhonen. 1993. Effects of urbanization on the breeding bird species richness in Finland: A biogeographic comparison. *Ornis Fennica* 70:71–77.

Jansen, K. P., A. P. Summers, and P. R. Delis. 2001. Spadefoot toads (*Scaphiopus holbrookii holbrookii*) in an urban landscape: Effects of nonnatural substrates on burrowing in adults and juveniles. *Journal of Herpetology* 35:141–145.

Jasoni, R. L., S. D. Smith, and J. A. Arnone III. 2005. Net ecosystem CO_2 exchange in Mojave Desert shrublands during the eighth year of exposure to elevated CO_2. *Global Change Biology* 11:749–756.

Jefferson, T. 1781–1782. Notes on the state of Virginia.

Johnson, C. J., and M. S. Boyce. 2004. A quantitative approach for regional environmental assessment: Application of a habitat-based population viability analysis to wildlife of the Canadian central Arctic. Canadian Environmental Assessment Agency Research and Development Monograph Series, <http://www.ceaa-acee.gc.ca/015/0002/index_e.htm>. Accessed November 15, 2009.

Johnson, C. J., and M. H. St. Laurent. 2010. A unifying framework for understanding the impacts of human developments for wildlife. In D. Naugle, editor. *Energy Development and Wildlife Conservation in Western North America*. Island Press, Washington, D.C., in press.

Johnson, C. J., and M. P. Gillingham. 2004. Mapping uncertainty: Sensitivity of wildlife habitat ratings to variation in expert opinion. *Journal of Applied Ecology* 41:1032–1041.

Johnson, C. J., and M. P. Gillingham. 2005. An evaluation of mapped species distribution models used for conservation planning. *Environmental Conservation* 32:1–12.

Johnson, C. J., and M. P. Gillingham. 2008. Sensitivity of species distribution models to error, bias, and model design: An application to resource selection functions for woodland caribou. *Ecological Modelling* 213:143–155.

Johnson, C. J., M. S. Boyce, R. L. Case, H. D. Cluff, R. J. Gau, A. Gunn, and R. Mulders. 2005. Cumulative effects of human developments on arctic wildlife. *Wildlife Monographs* 160:1–36.

Johnson, C. J., S. E. Nielsen, E. H. Merrill, T. L. McDonald, and M. S. Boyce. 2006. Resource selection functions based on use-availability data: Theoretical motivation and evaluation methods. *Journal of Wildlife Management* 70:347–357.

Johnson, D. H., M. J. Holloran, J. W. Connelly, S. H. Hanser, C. L. Amundson, and S. T. Knick. 2010. Influences of environmental and anthropogenic features on greater sage-grouse populations, 1997–2007. *Studies in Avian Biology*: in press.

Joly, K., C. Nellemann, and I. Vistnes. 2006. A reevaluation of caribou distribution near an oilfield road on Alaska's North Slope. *Wildlife Society Bulletin* 34:866–869.

Jones, S. G., D. H. Gordon, G. M. Phillips, and B. R. D. Richardson. 2005. Avian community response to a golf-course landscape unit gradient and restoring nature in every day life. *Wildlife Society Bulletin* 33:422–434.

Joshua Tree National Park. 2002. Air quality. ParkNet: Nature & Science—Environmental Factors. U.S. Department of the Interior, Washington, D.C.

Jude, D. J., and J. Leach. 1999. Great lakes fisheries. Pages 623–664 in C. C. Kohler and W. A. Hubert, editors. *Inland Fisheries Management in North America*. American Fisheries Society, Bethesda, Maryland.

Kaiser, R. C. 2006. Recruitment by greater sage-grouse in association with natural gas development in western Wyoming. Thesis, University of Wyoming, Laramie.

Keane, R. M., and M. J. Crawley. 2002. Exotic plant invasions and the enemy release hypothesis. *Trends in Ecology and Evolution* 17:164–170.

Kellert, S. R. 1985. Public perceptions of predators, particularly the wolf and coyote. *Biological Conservation* 31:167–189.

Kellert, S. 2004. Ordinary nature: The value of exploring and restoring nature in everyday life. Pages 9–19 in W. W. Shaw, L. K. Harris, and L. VanDruff, editors. *Fourth International Symposium on Urban Wildlife Conservation*, Tucson, Arizona.

Kellert, S. R., and T. W. Clark. 1991. The theory and application of a wildlife policy framework. Pages 17–38 in W. R. Mangun, editor. *Public Policy Issues in Wildlife Management*. Greenwood Press, New York.

Kennett, S. A. 1999. Towards a new paradigm for cumulative effects management. Canadian Institute of Resources Law Occasional Paper Eight. University of Calgary, Calgary, Alberta, Canada.

Keynes, J. M. 1936. *The General Theory of Employment, Interest and Money*. Harcourt and Brace, New York.

Kiesecker, J. M., H. Copeland, A. Pocewicz, N. Nibbelink, B. McKenney, J. Dahlke, M. Holloran, and D. Stroud. 2009. A framework for implementing biodiversity offsets: Selecting sites and determining scale. *Bioscience* 59:77–84.

Kilpatrick, H. J., and W. D. Walter. 1997. Urban deer management: A community vote. *Wildlife Society Bulletin* 25:388–391.

Kilpatrick, H. J., and W. D. Walter. 1999. A controlled archery hunt in a residential community: Cost, effectiveness, and deer recovery rates. *Wildlife Society Bulletin* 27:115–123.

Kilpatrick, H. J. and S. M. Spohr. 2000a. Movements of female white-tailed deer in a suburban landscape: A management perspective. *Wildlife Society Bulletin* 28:1038–1045.

Kilpatrick, H. J., and S. M. Spohr. 2000b. Spatial and temporal use of a suburban landscape by female white-tailed deer. *Wildlife Society Bulletin* 28:1023–1029.

Kime, D. E. 1998. *Endocrine Disruption in Fish*. Kluwer Academic Publishers, Norwell, Massachusetts.

Kingdon, J. 1993. *Self-Made Man: Human Evolution from Eden to Extinction?* John Wiley & Sons, New York.

Kingsland, W. E. 2005. *The Evolution of American Ecology, 1890–2000*. The John Hopkins University Press, Baltimore, Maryland.

Kjoss, V. A., and J. A. Litvaitis. 2000. Community structure of snakes in a human-dominated landscape. *Biological Conservation* 98:285–292.

Klemmedson, J. O. 1967. Big game winter range—A diminishing resource. *Transactions of North American Wildlife and Natural Resources Conference* 32:259–269.

Knick, S. T., D. S. Dobkin, J. T. Rotenberry, M. A. Schroeder, W. M. Vander Haegen, and C. Van Riper. 2003. Teetering on the edge or too late? Conservation and research issues for avifauna of sagebrush habitats. *Condor* 105:611–634.

Knick, S. T., and S. E. Hanser. 2010. Connecting pattern and process in greater sage-grouse populations and sagebrush landscapes. *Studies in Avian Biology*: in press.

Knuth, B. A., W. F. Siemer, M. D. Duda, S. J. Bissel, D. J. Decker. 2001. Wildlife management in urban environments. Pages 195–219 in D. J. Decker, T. L. Brown, and W. F. Siemer, editors. *Human Dimensions of Wildlife Management in North America*. The Wildlife Society, Bethesda, Maryland.

Kohler, C. C., and W. A. Hubert, editors. 1999. *Inland Fisheries Management in North America*. American Fisheries Society, Bethesda, Maryland.

Kolstad, C. D. 2010. *Environmental Economics*. Oxford University Press, New York.

Krausman, P. R. 1993. The exit of the last wild mountain sheep. Pages 242–250 in G. P. Nabhan, editor. *Counting Sheep: Diversity in Nature Writing*. University of Arizona Press, Tucson.

Krausman, P. R. 1997. Human disturbance and the disappearance of bighorn sheep. *Caprinea* July: 3–4.

Krausman, P. R. 2002. *Introduction to Wildlife Management the Basics*. Prentice Hall, Upper Saddle River, New Jersey.

Krausman, P. R., G. Long, and L. Taragno. 1996. Desert bighorn sheep and fire, Santa Catalina Mountains, Arizona. Pages 162–168 in P. F. Folliott, L. Debano, M. B. Baker, Jr., G. J. Gottfried, G. Soils-Garza, C. B. Edminster, D. G. Neary, L. S. Allen, and R. H. Hamre, technical coordinators. *Effects of Fire on the Madrean Province Ecosystem*. U.S. Forest Service General Technical Report RM-289.

Krausman, P. R., M. I. Grinder, P. S. Gipson, G. L. Zuercher, G. C. Stewart. 2006. Molecular identification of coyote feces in an urban environment. *Southwestern Naturalist* 51:122–126.

Krausman, P. R., W. C. Dunn, L. K. Harris, W. W. Shaw, and W. B. Boyce. 2001. Can mountain sheep and humans co-exist? Pages 224–227 in R. Field, R. J. Warren, H. O'Korma, and P. R. Sievert, editors. Wildlife, land and people: Priorities for the 21st century. *Proceedings of the Second International Wildlife Management Congress*. The Wildlife Society, Bethesda, Maryland.

Kremen, C., N. M. Williams, M. A. Aizen, B. Gemmill-Harren, G. LeBuhn, R. Minckley, L. Packer, S. G. Potts, T. Roulston, I. Steffan-Dewenter, D. P. Vazquez, R. Winfree, L. Adams, E. E. Crone, S. S. Greenlead, T. H. Keitt, A. M. Klein, J. Regetz, and

T. H. Ricketts. 2007. Pollination and other ecosystem services produced by mobile organisms: A conceptual framework for the effects of land-use change. *Ecology Letters* 10:299–314.

Kruse, J. A., R. G. White, H. E. Epstein, B. Archie, M. Berman, S. R. Braund, F. S. Chapin, J. S. Charlie, C. J. Daniel, J. Eamer, N. Flanders, B. Griffith, S. Haley, L. Huskey, B. Joseph, D. R. Klein, G. P. Kofinas, S. M. Martin, S. M. Murphy, W. Nebesky, C. Nicolson, D. E. Russell, J. Tetlichi, A. Tussing, M. D. Walker, and O. R. Young. 2004. Modeling sustainability of Arctic communities: An interdisciplinary collaboration of researchers and local knowledge holders. *Ecosystems* 7:815–828.

Kucera, T. E., and C. McCarthy. 1988. Habitat fragmentation and mule deer migration corridors: a need for evaluation. *Transactions of the Western Section of the Wildlife Society* 24:61–67.

Kuhnlein, H. V., and O. Receveur. 1996. Dietary change and traditional food systems of indigenous peoples. *Annual Review of Nutrition* 16:417–442.

Kurz, W. A., C. C. Dymond, G. Stinson, G. J. Rampley, E. T. Neilson, A. L. Carroll, T. Ebata, and L. Safranyik. 2008. Mountain pine beetle and forest carbon feedback to climate change. *Nature* 452:987–990.

Laliberte, A. S., and W. J. Ripple. 2004. Range contractions of North American carnivores and ungulates. *BioScience* 54:123–138.

Lancaster, R. K., and W. E. Rees. 1979. Bird communities and the structure of urban habitats. *Canadian Journal of Zoology* 57:2358–2368.

Lane, P.A., and R.R. Wallace. 1988. A user's guide to cumulative effects assessment in Canada. Canadian Environmental Assessment Research Council. Ottawa, Ontario, Canada.

Langeland, K. A. 1996. *Hydrilla verticillata* (L.F.) Royle (Hydrocharitaceae), "The perfect aquatic weed." *Castanea* 61:293–304.

Laundre, J. W., and B. L. Keller. 1981. Home range use by coyotes in Idaho. *Animal Behavior* 29:449–461.

Lee, M. E., and R. Miller. 2003. Managing elk in the wildland-urban interface: Attitudes of Flagstaff, Arizona residents. *Wildlife Society Bulletin* 31:185–191.

Legat, A., G. Chocolate, M. Chocolate, and S. A. Zoe. 2001. Caribou migration and the state of their habitat. Final Report, Yellowknife, Northwest Territories, Canada.

Lehtinen, R. M., S. M. Galatowitsch, and J. R. Tester. 1999. Consequences of habitat loss and fragmentation for wetland amphibian assemblages. *Wetlands* 19:1–12.

Leigh, J., and M. J. Chamberlain. 2008. Effects of aversive conditioning on behavior of nuisance Louisiana black bears. *Human-Wildlife Conflicts* 2:175–182.

Lenart, E. A. 2007. Units 26B and 26C caribou. Pages 284–308 in P. Harper, editor. *Caribou Management Report of Survey and Inventory Activities July 1, 2004–June 30, 2006.* Alaska Department of Fish and Game. Project 3.0. Juneau, Alaska.

Lenth, B. A., R. L. Knight, and W. C. Gilgert. 2006. Conservation value of clustered housing developments. *Conservation Biology* 20:1445–1456.

Leonard, J. R. 1979. *The Fish Car Era of the National Fish Hatchery System.* Washington, D.C.

Leontief, W. 1949. Structural matrices of national economies. *Econometrica* 17:273–282.

Leopold, A. 1933. *Game Management.* Charles Scribner's Sons, New York.

Leopold, A. 1949. *A Sand County Almanac with Essays on Conservation from Round River.* Oxford University Press, New York.

Lepczyk, C. A., A. G. Mertig, and J. G. Liu. 2004. Landowners and cat predation across rural-to-urban landscapes. *Biological Conservation* 115:191–201.

Leu, M., S. E. Hanser, and S. T. Knick. 2008. The human footprint in the west: A large-scale analysis of anthropogenic impacts. *Ecological Applications* 18:1119–1139.

Lewin, W. C., R. Arlinghaus, and T. Mehner. 2006. Documented and potential biological impacts of recreational fishing: Insights for management and conservation. *Reviews in Fisheries Science* 14:305–367.

Li, H. W., and P. B. Moyle. 1999. Management of introduced fishes. Pages 345–374 in C. C. Kohler and W. A. Hubert, editors. *Inland Fisheries Management in North America*. American Fisheries Society, Bethesda, Maryland.

Lieberman, J. I. 2007. Letter to the U.S. Department of Homeland Security Secretary Michael Chertoff in regards to environmental waiver issued in support of the border infrastructure construction project near the San Pedro Riparian National Conservation Area. Senate Committee on Homeland Security and Governmental Affairs. Washington, D.C.

Lindstrom, M. J., and H. Bartling. 2003. *Suburban Sprawl: Culture, Theory, and Politics*. Rowman and Littlefield Publishers, Lanham, Maryland.

Linstone, H. A., and M. Turoff. 1975. *The Delphi Methods: Techniques and Applications*. Addison-Wesley Publishing, Reading, Massachusetts.

Loker, C. A., D. J. Decker, and S. J. Schwager. 1999. Social acceptability of wildlife management actions in suburban areas: Three cases from New York. *Wildlife Society Bulletin* 27:152–159.

Loomis, J. B., and R. G. Walsh. 1997. Recreation economic decisions. Venture, State College, Pennsylvania.

Lucchetti, G., and R. Fuerstenberg. 1992. Management of coho salmon habitat in urbanizing landscapes of King County, Washington. Pages 308–317 in L. Berg and P. Delaney, editors. *Proceedings of the 1992 Coho Workshop*, Nanaimo, British Columbia Canada, North Pacific Chapter of the American Fisheries Society.

Ludwig, D., B. Walker, and C. S. Holling. 1997. Sustainability, stability, and resilience. *Conservation Ecology* 1:7.

Lund, R. C. 1997. A cooperative, community-based approach for the management of suburban deer populations. *Wildlife Society Bulletin* 25:488–490.

Luniak, M. 2004. Synurbization—Adaptation of animal wildlife to urban development. Pages 50–55 in W. W. Shaw, L. K. Harris, and L. VanDruff, editors. *Fourth International Symposium on Urban Wildlife Conservation*, Tucson, Arizona.

Luo, H., W. C. Oechel, S. J. Hastings, R. Zulueta, Y. Qian, and H. Kwon. 2007. Mature semiarid chaparral ecosystems can be a significant sink for atmospheric carbon dioxide. *Global Change Biology* 13:386–396.

Lyon, A. G., and S. H. Anderson. 2003. Potential gas development impacts on sage grouse nest initiation and movement. *Wildlife Society Bulletin* 31:486–491.

Lyons, A. J. 2005. Activity patterns of urban American black bears in the San Gabriel Mountains of southern California. *Ursus* 16:255–262.

Lyons, J. R., and D. L. Leedy. 1984. The status of urban wildlife programs. *Transactions of the North American Wildlife and Natural Resources Conference* 49:233–251.

MacArthur, R. H., and E. O. Wilson. 1967. *The Theory of Island Biogeography*. Princeton University Press, Princeton, New Jersey.

MacDonald, L. 2000. Evaluating and managing cumulative effects: Process and constraints. *Environmental Management* 26:299–315.

Mace, R. D., J. S. Waller, T. L. Manley, L. J. Lyon, and H. Zuuring. 1996. Relationships among grizzly bears, roads, and habitat in the Swan Mountains, Montana. *Journal of Applied Ecology* 33:1395–1404.

Mace, R. D., J. S. Waller, T. L. Manley, K. Ake, W. T. Wittenger. 1998. Landscape evaluation of grizzly bear habitat in western Montana. *Conservation Biology* 13:367–377.

Mackie, R. J., and D. F. Pac. 1980. Deer and subdivisions in the Bridger Mountains, Montana. *Proceedings of the Western Association of Fish and Wildlife Agencies* 60:517–526.

MacLachlan, L. 1996. NWT diamonds project—Report of the Environmental Assessment Panel. Canadian Environmental Assessment Agency, Ottawa, Ontario, Canada.

Macnab, J. 1985. Carrying capacity and other slippery shibboleths. *Wildlife Society Bulletin* 13:403–410.

Mahan, C. G., and T. J. O'Connell. 2005. Small mammal use of suburban and urban parks in central Pennsylvania. *Northeastern Naturalist* 12:307–314.

Mahmoud M., Y. Liu, H. Hartmann, S. Stewart, T. Wagener, D. Semmens, R. Stewart, H. Gupta, D. Dominguez, F. Dominguez, D. Hulse, R. Letcher, B. Rashleigh, C. Smith, R. Street, J. Ticehurst, M. Twery, H. Van Delden, R. Waldick, and D. White. 2009. A formal framework for scenario development in support of environmental decision-making. *Environmental Modeling and Software* 24:798–808.

Major, J. T., and J. A. Sherburne. 1987. Interspecific relationships of coyotes, red foxes, and bobcats in western Maine. *Journal of Wildlife Management* 51:606–616.

Major, W. W. III, J. M. Grassley, K. E. Ryding, C. E. Grue, T. N. Pearsons, D. A. Tipton, and A. E. Stephenson. 2005. Abundance and consumption of fish by California gulls and ring-billed gulls at water and fish management structures within the Yakima River, Washington. *Waterbirds* 28:366–377.

Makings, E. 2005. Flora of the San Pedro Riparian National Conservation Area, Cochise, County, Arizona. Page 36 in Connecting mountain islands and desert seas: Biodiversity and management of the Madrean Archipelago II, Gottfried, Gerald J.; Gebow, Brooke S.; Eskew, Lane G.; and Edminster, Carleton B., compilers. U.S. Forest Service, Rocky Mountain Research Station, Fort Collins, Colorado.

Mangin, S. 2001. The 100th meridian initiative: A strategic approach to prevent the westward spread of zebra mussels and other aquatic nuisance species. U.S. Fish and Wildlife Service, Washington, D.C.

Mannan, R. W., W. W. Shaw, W. A. Estes, M. Alaner, and C. W. Boal. 2004. A preliminary assessment of the attitudes of people toward Cooper hawks nesting in an urban environment. Pages 87–92 in W. W. Shaw, L. K. Harris, and L. VanDruff, editors. *Fourth International Symposium on Urban Wildlife Conservation*, Tucson, Arizona.

Manor, R., and D. Saltz. 2003. Impact of human nuisance disturbance on vigilance and group size of a social ungulate. *Ecological Applications* 13:1830–1834.

Marchand, M. N., and J. A. Litvaitis. 2004. Effects of habitat features and landscape composition on the population structure of a common aquatic turtle in a region undergoing rapid development. *Conservation Biology* 18:758–767.

Martin, P. S. 1967. Prehistoric overkill. Pages 75–120 in P. S. Martin and H. E. Wright, editors. *Pleistocene Extinctions: The Search for a Cause*. Yale University Press, New Haven, Connecticut.

Marzluff, J. M., and K. Ewing. 2001. Restoration of fragmented landscapes for the conservation of birds: A general framework and specific recommendations for urbanizing landscapes. *Restoration Ecology* 9:280–292.

Mascarenhas, M., and R. Scarce. 2004. "The intention was good": Legitimacy, consensus-based decision making, and the case of forest planning in British Columbia. *Society and Natural Resources* 17:17–38.

Mathews, N. E. 1989. Social structure, genetic structure and anti-predator behavior of white-tailed deer in the central Adirondacks. Dissertation, State University of New York, Syracuse.

Mattson, D. J., and T. Merrill. 2002. Extirpations of grizzly bears in the contiguous United States, 1850–2000. *Conservation Biology* 16:1123–1136.

Mazaika, R., P. R. Krausman, and R. C. Etchberger. 1992. Forage availability for mountain sheep in Pusch Ridge Wilderness, Arizona. *Southwestern Naturalist* 37:372–378.

McAninch, J. B., and J. M. Parker. 1991. Urban deer management programs: A facilitated approach. *Transactions of the North American Wildlife and Natural Resources Conference* 56:428–436.

McArthur, K. L. 1981. Factors contributing to effectiveness of black bear transplants. *Journal of Wildlife Management* 45:102–110.

McCarthy, T. M., and R. J. Seavoy. 1994. Reducing nonsport losses attributable to food conditioning: Human and bear behavior and modification in an urban environment. *International Conference on Bear Research and Management* 9:75–84.

McClennen, N., R. R. Wigglesworth, and S. H. Anderson. 2001. The effect of suburban and agricultural development on the activity patterns of coyotes (*Canis latrans*). *American Midland Naturalist* 146:27–36.

McClure, M. F., N. S. Smith, and W. W. Shaw. 1995. Diets of coyotes near the boundary of Saguaro National Monument and Tucson, Arizona. *The Southwestern Naturalist* 40:101–125.

McClure, M. F., J. A. Bissonette, and M. R. Conover. 2005. Migratory strategies, fawn recruitment, and winter habitat use by urban and rural mule deer. *European Journal of Wildlife Research* 51:170–177.

McCold, L. N., and J. W. Saulsbury. 1996. Including past and present impacts in cumulative impact assessments. *Environmental Management* 20:767–776.

McCorquodale, S. M. 1999. Movements, survival, and mortality of black-tailed deer in the Klickitat Basin of Washington. *Journal of Wildlife Management* 63:861–871.

McCoy, J. E., D. G. Hewitt, and F. C. Bryant. 2005. Dispersal by yearling male white-tailed deer and implications for management. *Journal of Wildlife Management* 69:366–376.

McCullough, D. R. 1982. Behavior, bears, and humans. *Wildlife Society Bulletin* 10:27–33.

McDaniels, T. L., H. Dowlatabadi, and S. Stevens. 2005. Multiple scales and regulatory gaps in environmental change: The case of salmon aquaculture. *Global Environmental Change-Human and Policy Dimensions* 15:9–21.

McDonnell, M. J., S. T. A. Pickett, and R. V. Pouyat. 1993. The application of the ecological gradient paradigm to the study of urban effects. Pages 175–189 in M. J. McDonnell and S. T. A. Pickett, editors. *Humans as Components of Ecosystems*. Springer-Verlag, New York.

McKinney, M. L. 2002. Urbanization, biodiversity, and conservation. *BioScience* 52:883–890.

McKinney, M. L. 2006. Urbanization as a major cause of biotic homogenization. *Biological Conservation* 127:247–260.

McNay, R. S., and J. M. Voller. 1995. Mortality causes and survival estimates for adult female Columbian black-tailed deer. *Journal of Wildlife Management* 59:138–146.

Meffe, G. K. 1984. Effects of abiotic disturbance on coexistence of predator–prey fish species. *Ecology* 65:1525–1534.

Melles, S., S. Glenn, and K. Martin. 2003. Urban bird diversity and landscape complexity: Species-environment associations along a multi-scale habitat gradient. *Ecology and Society* 7:5.

Milder, J. C., J. P. Lassoie, and B. L. Bedford. 2008. Conserving biodiversity and ecosystem function through limited development: An empirical evaluation. *Conservation Biology* 22:70–79.

Miller, J. R., and R. J. Hobbs. 2002. Conservation where people live and work. *Conservation Biology* 16:330–337.

Miller, K. V., and J. M. Wentworth. 2000. Carrying capacity. Pages 140–155 in S. Demarais and P. R. Krausman, editors. *Ecology and Management of Large Mammals in North America*. Prentice Hall, Upper Saddle River, New Jersey.

Mills, E. L., J. H. Leach, J. T. Carlton, and C. L. Secor. 1993. Exotic species in the Great Lakes: A history of biotic crises and anthropogenic introductions. *Journal of Great Lakes Research* 19:1–54.

Mills, E. L., J. H. Leach, J. T. Carlton, and C. L. Secor. 1994. Exotic species and the integrity of the Great Lakes. *BioScience* 44:666–676.

Millsap, B. A., and C. Bear. 2000. Density and reproduction of burrowing owls along an urban development gradient. *Journal of Wildlife Management* 64:33–41.

Miltner, R. J., D. White, and C. Yoder. 2004. The biotic integrity of streams in urban and suburbanizing landscapes. *Landscape and Urban Planning* 69:87–100.

Minckley, W.L., and J. E. Deacon, editors. 1990. Battle against extinction. *Native Fish Management in the American West.* University of Arizona Press, Tucson.

Minnesota IMPLAN Group. 2009. IMPLAN Professional 2.0. Hudson, WI. http://www.implan.com/. Accessed August 3, 2010.

Minton, S. A. 1968. The fate of amphibians and reptiles in a suburban area. *Journal of Herpetology* 2:113–116.

Morrison, S. A., and D. T. Bolger. 2002. Lack of an urban edge effect on reproduction in a fragmentation-sensitive sparrow. *Ecological Applications* 12:398–411.

Morton, L. D. 1988. Winter ecology of the eastern coyote (*Canis latrans*) in Fundy National Park. Thesis, University of New Brunswick, Fredericton, Canada.

Moshiri, G. A., editor. 1993. *Constructed Wetlands for Water Quality Improvement.* Lewis Publishers, Boca Raton, Florida.

Moyle, P. B., and T. Light. 1996. Fish invasions in California: Do abiotic factors determine success? *Ecology* 77: 1666–1670.

Muir, J. 1912. *The Yosemite.* The Century Company, New York.

Munns, W. R. 2006. Assessing risks to wildlife populations from multiple stressors: Overview of the problem and research needs. *Ecology and Society* 11: 23 (online: http://www.ecologyandsociety.org).

Murphy, M. A., J. S. Evans, and A. Storfer. 2010. Quantifying Bufo boreas connectivity in Yellowstone National Park with landscape genetics. *Ecology* 91:252–261.

Murphy, S. M., D. E. Russell, and R. G. White. 2000. Modeling energetic and demographic consequences of caribou interactions with oil development in the Arctic. *Rangifer Special Issue* 12:107–109.

National Environmental Policy Act, 1969, United States, 42 U.S.C.A.:4321–4370.

National Petroleum Council. 2007. Facing the hard truths about energy: A comprehensive view to 2030 of global oil and natural gas. Washington, D.C.

Natural Resources Defense Council. 1996. Breath-taking: Premature mortality due to particulate air pollution in 239 American cities. <http://www.nrdc.org/air/pollution/bt/btinx.asp>. Accessed February 17, 2006.

Naugle, D. E., K. E. Doherty, B. L. Walker, M. J. Holloran, and H. E. Copeland. 2010. Energy development and sage-grouse. *Studies in Avian Biology*: in press.

Nellemann, C., P. Jordhøy, O-G Støen, and O. Strand. 2000. Cumulative impacts of tourist resorts on wild reindeer (*Rangifer tarandus tarandus*) during winter. *Arctic* 53:9–17.

Nellemann, C., P. Jordhøy, I. Vistnes, O. Strand, and A. Newton. 2003. Progressive impacts of piecemeal development. *Biological Conservation* 113:307–317.

Nelson, M. E., and L. D. Mech. 1992. Dispersal of female white-tailed deer. *Journal of Mammalogy* 73:891–894.

Nerbonne, B. A., and B. Vondracek. 2001. Effects of local land use on physical habitat, benthic macroinvertebrates, and fish in the Whitewater River, Minnesota, USA. *Environmental Management* 28:87–99.

Nicolson, C. R., A. M. Starfield, G. P. Kofinas, and J. A. Kruse. 2002. Ten heuristics for interdisciplinary modeling projects. *Ecosystems* 5:376–384.

Nielsen, C. K., W. F. Porter, and H. B. Underwood. 1997. An adaptive management approach to controlling suburban deer. *Wildlife Society Bulletin* 25:470–477.

Nielson, C. K., R. G. Anderson, and M. D. Grund. 2003. Landscape influences on deer-vehicle accident areas in an urban environment. *Journal of Wildlife Management* 67:46–51.

Nielsen, S. E., S. Herrero, M. S. Boyce, R. D. Mace, B. Benn, M. L. Gibeau, and S. Jevons. 2004. Modelling the spatial distribution of human-caused grizzly bear mortalities in the Central Rockies ecosystem of Canada. *Biological Conservation* 120:101–113.

Nitschke, C. R. 2008. The cumulative effects of resource development on biodiversity and ecological integrity in the Peace-Moberly region of northeast British Columbia, Canada. *Biodiversity and Conservation* 17:1715–1740.

Nixon, C. M., L. P. Hansen, P. A. Brewer, and J. E. Chelsvig. 1991. Ecology of white-tailed deer in an intensively farmed region of Illinois. *Wildlife Monographs* 118:1–77.

Noble, B. F. 2002. The Canadian experience with SEA and sustainability. *Environmental Impact Assessment Review* 22:3–16.

Noble, B. F. 2004a. A state-of-practice survey of policy, plan, and program assessment in Canadian provinces. *Environmental Impact Assessment Review* 24:351–361.

Noble, B. F. 2004b. Integrating Strategic Environmental Assessment with industry planning: A case study of the Pasquai-Porcupine forest management plan, Saskatchewan, Canada. *Environmental Management* 33:401–411.

Noble, B. F. 2009. Promise and dismay: The state of strategic environmental assessment systems and practices in Canada. *Environmental Impact Assessment Review* 29:66–75.

Noble, R. L., and T. W. Jones. 1999. Managing fisheries with regulations. Pages 455–480 in C. C. Kohler and W. A. Hubert, editors. *Inland Fisheries Management in North America*. American Fisheries Society, Bethesda, Maryland.

Noel, L. E., K. R. Parker, and M. A. Cronin. 2006. Response to Joly et al. (2006), A reevaluation of caribou distribution near an oilfield road on Alaska's North Slope. *Wildlife Society Bulletin* 34:870–873.

North Yukon Planning Commission. 2009. North Yukon regional land use plan—Nichih Gwanał' in—looking forward. Yukon and Vuntut Gwitchin Governments, Whitehorse, Yukon, Canada.

Northwest Territories Cumulative Effects Assessment and Management Framework (NWT CEAMF) Steering Committee. 2007. A Blueprint for Implementing the Cumulative Effects Assessment and Management Strategy and Framework in the NWT and Its Regions. Yellowknife, Northwest Territories, Canada.

Noss, R. F. 2004. Can urban areas have ecological integrity? Pages 3–8 in W. W. Shaw, L. K. Harris, and L. VanDruff, editors. *Fourth International Symposium on Urban Wildlife Conservation*, Tucson, Arizona.

O'Connor, J. E., L. L. Ely, E. E. Wohl, L. E. Stevens, T. S. Melis, V. S. Kale, and V. R. Baker. 1994. A 4500-year record of large floods on the Colorado River in the Grand Canyon, Arizona. *The Journal of Geology* 102:1–9.

Odell, E. A., and R. L. Knight. 2001. Songbird and medium-sized mammal communities associated with exurban development in Pitkin County, Colorado. *Conservation Biology* 15:1143–1150.

Office of the Deputy Prime Minister. 2005. A practical guide to the strategic environmental assessment direction. Department of the Environment, London, United Kingdom.

Office of the Secretary, Department of Homeland Security. 2008. Determination pursuant to section 102 of the Illegal Immigration Reform and Immigrant Responsibility Act of 1996, as amended. Washington, D.C.

Ogutu-Ohwayo, R. 1990. The decline of the native fishes in lakes Victoria and Kyoga (East Africa) and the impact of introduced species, especially the Nile perch, *Lates niloticus*, and the Nile tilapia, *Oreochromis niloticus*. *Environmental Biology of Fishes* 27, 81–96.

Olden, J. D., N. L. Poff, and M. L. McKinney. 2006. Forecasting faunal and floral homogenization associated with human population geography in North America. *Biological Conservation* 127:261–271.

Olson, D. E., and P. K. Cunningham. 1989. Sport-fisheries trends shown by an annual Minnesota fishing contest over a 58-year period. *North American Journal of Fisheries Management* 9:287–297.

Ormerod, P. 1997. *The Death of Economics*. John Wiley & Sons, New York.

Osborne, L. L., and D. A. Kovacic. 1993. Riparian vegetated buffer strips in water-quality restoration and stream management. *Freshwater Biology* 29:243–258.

Oscoz, J., P. M. Leunda, R. Miranda, C. Garcia-Fresca, F. Campos, and M. Carmen Escala. 2005. River channelization effects on fish population structure in the Larraun river (Northern Spain). *Hydrobiologia* 543:191–198.

Outdoor Resources Review Group. 2009. Great outdoors America. <http://www.orrgroup.org/documents/July2009_Great-Outdoors-America-report.pdf>. Accessed April 29, 2010.

Oyler-McCance, S. J., and T. W. Quinn. 2010. Molecular insights into the biology of greater sage-grouse. *Studies in Avian Biology*: in press.

Paci, C., A. Tobin, and P. Robb. 2002. Reconsidering the Canadian Environmental Impact Assessment Act: A place for traditional environmental knowledge. *Environmental Impact Assessment Review* 22:111–127.

Palomino, D., and L. M. Carrascal. 2007. Threshold distances to nearby cities and roads influence the bird community of a mosaic landscape. *Biological Conservation* 140:100–109.

Parlee, B., M. Manseau, and Łutsël K`é Dene First Nation. 2005. Using traditional knowledge to adapt to ecological change: Denesoline monitoring of caribou movements. *Arctic* 58:26–37.

Parry, B. 1990. Cumulative habitat loss: Cracks in the environmental review process. *Natural Areas Journal* 10:76–83.

Patterson, B. R., S. Bondrup-Nielsen, and F. Messier. 1999. Activity patterns and daily movements of the eastern coyote, *Canis latrans*, in Nova Scotia. *Canadian-Field Naturalist* 113:251–257.

Pearce, M. J. 2003. Interactive groundwater/surface-water regulation in Arizona. *Southwest Hydrology*, 2:16–17.

Pearl, C. A., M. J. Adams, N. Leuthold, and R. B. Bury. 2005. Amphibian occurrence and aquatic invaders in a changing landscape: Implications for wetland mitigation in the Willamette Valley, Oregon. *Wetlands* 25:76–88.

Peine, J. D. 2001. Nuisance bears in communities: Strategies to reduce conflicts. *Human Dimensions of Wildlife* 6:223–237.

Perdicoúlis, A., M. Hanusch, H. D. Kasperidus, and U. Weiland. 2007. The handling of causality in SEA guidance. *Environmental Impact Assessment Review* 27:176–187.

Peterson, G. D., G. S. Cumming, and S. R. Carpenter. 2003. Scenario planning: A tool for conservation in an uncertain world. *Conservation Biology* 17:358–366.

Peterson, M. N., R. R. Lopez, P. A. Frank, B. A. Porter, and N. J. Silvy. 2004. Key deer fawn response to urbanization: Is sustainable development possible? *Wildlife Society Bulletin* 32:493–499.

Phillips, J., E. Nol, D. Burke, and W. Dunford. 2005. Impacts of housing developments on wood thrush nesting success in hardwood forest fragments. *Condor* 107:97–106.

Piccolo, B. P., K. M. Hollis, R. E. Warner, T. R. Van Deelen, D. R. Etter, and C. Anchor. 2001. Variation of white-tailed deer home ranges in fragmented urban habitats around Chicago, Illinois. Pages 349–356 in M. C. Brittingham, J. Kays, and R. McPeake, editors. *The Ninth Wildlife Damage Management Conference Proceedings*, Oct. 5–8, 2000. State College, Pennsylvania.

Pidgeon, A. M., V. C. Radeloff, C. H. Flather, C. A. Lepczyk, M. K. Clayton, T. J. Hawbaker, and R. B. Hammer. 2007. Associations of forest bird species richness with housing and landscape patterns across the USA. *Ecological Applications* 17:1989–2010.

Pillsbury, F. C., and J. R. Miller. 2008. Habitat and landscape characteristics underlying anuran community structure along an urban-rural gradient. *Ecological Applications* 18:1107–1118.

Pima Association of Governments. 2009. Population facts. <http://www.pagnet.org/RegionalData/Population/tabid/104/Default.aspx>. Accessed April 29, 2010.

Pimentel, D., L. Lach, R. Zuniga, and D. Morrison. 2000. Environmental and economic costs of nonindigenous species in the United States. *BioScience* 50:53–65.

Piper, J. 2001. Barriers to implementation of cumulative effects assessment. *Journal Environmental Assessment Policy and Management* 3:465–481.

Polfus, J. L. 2010. Assessing cumulative human impacts on northwood land caribou with traditional ecological knowledge and resource selection functions. Thesis, University of Montana, Missoula.

Pollock, M. M., T. J. Beechie, M. Liermann, and R. E. Bigley. 2009. Stream temperature relationships to forest harvest in western Washington. *Journal of the American Water Resources Association* 45:141–156.

Polson, J. E. 1983. Application of aversion techniques for the reduction of losses to beehives by black bears in northeastern Saskatchewan. Department of Supply and Services, Ottawa, Canada. SRC Publication No. C-805-13-E-83.

Porter, W. F., N. E. Mathews, H. B. Underwood, R. W. Sage, Jr., and D. F. Behrend. 1991. Social organization in deer: Implications for localized management. *Environmental Management* 15:809–814.

Porter, W. F., H. B. Underwood, and J. L. Woodard. 2004. Movement behavior, dispersal, and the potential for localized management of deer in a suburban environment. *Journal of Wildlife Management* 68:247–256.

Prange, S., S. D. Gehrt, and E. P. Wiggers. 2003. Demographic factors contributing to high raccoon densities in urban landscapes. *Journal of Wildlife Management* 67:324–333.

Prange, S., and S. D. Gehrt. 2004. Changes in mesopredator-community structure in response to urbanization. *Canadian Journal of Zoology-Revue Canadienne de Zoologie* 82:1804–1817.

Price, S. J., M. E. Dorcas, A. L. Gallant, R. W. Klaver, and J. D. Willson. 2006. Three decades of urbanization: Estimating the impact of land-cover change on stream salamander populations. *Biological Conservation* 133:436–441.

Putman, R. J. 1997. Deer and road traffic accidents: Options for management. *Journal of Environmental Management* 51:43–57.

Quinn, T. 1991. Distribution and habitat associations of coyotes in Seattle, Washington. *Wildlife Conservation in Metropolitan Environments Symposium Series* 2:48–51.

Quinn, T. 1995. Using public sighting information to investigate coyote use of an urban habitat. *Journal of Wildlife Management* 59:238–245.

Quinn, T. 1997a. Coyote (*Canis latrans*) food habits in three urban habitat types of western Washington. *Northwest Science* 71:1–5.

Quinn, T. 1997b. Coyote (*Canis latrans*) habitat selection in urban areas of western Washington via analysis of routine movements. *Northwest Science* 71:289–297.

Quiñonez-Piñon, R, A. Mendoza-Duran, and C. Valeo. 2007. Design of an environmental monitoring program using NDVI and cumulative effects assessment. *International Journal of Remote Sensing* 28:1643–1664.

Randa, L. A., and J. A. Yunger. 2006. Carnivore occurrence along an urban–rural gradient: A landscape-level analysis. *Journal of Mammalogy* 87:1154–1164.

Rapport, D. J., H. A. Regier, and T. C. Hutchinson. 1985. Ecosystem behavior under stress. *American Naturalist* 125:617–640.

Rasker, R., and A. Hackman. 1996. Economic development and the conservation of large carnivores. *Conservation Biology* 10:991–1002.

Reed, D. F. 1981. Conflicts with civilization. Pages 509–535 in O. C. Wallmo, editor. *Mule and Black-Tailed Deer of North America*. University of Nebraska Press, Lincoln.

Reed, D. F., and T. N. Woodard. 1981. Effectiveness of highway lighting in reducing deer-vehicle collisions. *Journal of Wildlife Management* 45:721–726.

Reimers, E., S. Eftestol, and J. E. Coleman. 2003. Behavior responses of wild reindeer to direct provocation by a snowmobile or skier. *Journal of Wildlife Management* 67:747–754.

Reisner, M. 1986. Cadillac desert. *The American West and Its Disappearing Water*. Penguin Books, New York.

Renton, S. D., and J. M. Bailey. 2002. Policy development and the environment. *Impact Assessment and Project Appraisal* 18:245–251.

Retief, F., C. Jones, and S. Jay. 2008. The emperor's new clothes—Reflections on strategic environmental assessment practice in South Africa. *Environmental Impact Assessment Review* 28:504–514.

Riley, S. P. D., G. T. Busteed, L. B. Kats, T. L. Vandergon, L. F. S. Lee, R. G. Dagit, J. L. Kerby, R. N. Fisher, and R. M. Sauvajot. 2005. Effects of urbanization on the distribution and abundance of amphibians and invasive species in southern California streams. *Conservation Biology* 19:1894–1907.

Riley, S. P., R. M. Sauvajot, T. K. Fuller, E. C. York, D. A. Kamradt, C. Bromley, and R. K. Wayne. 2003. Effects of urbanization and habitat fragmentation on bobcats and coyotes in southern California. *Conservation Biology* 17:566–576.

Riley, W. D., D. L. Maxwell, M. G. Pawson, and M. J. Ives. 2009. The effects of low summer flow on wild salmon (*Salmo salar*), trout (*Salmo trutta*) and grayling (*Thymallus thymallus*) in a small stream. *Freshwater Biology* 54:2581–2599.

Roberts, G. 1996. Why individual vigilance declines as group size increases. *Animal Behavior* 51:1077–1086.

Rodriguez-Freire, M., and R. Crecente-Maseda. 2008. Directional connectivity of wolf (*Canis lupus*) populations in northwest Spain and anthropogenic effects on dispersal patterns. *Environmental Modeling and Assessment* 13:35–51.

Rogers, L. L. 1970. Black bear of Minnesota. *Naturalist* 21:42–47.

Rogers, L. L. 1986. Effects of translocation distance on frequency of return of adult black bears. *Wildlife Society Bulletin* 14:76–80.

Rogers, L. L., D. W. Kuehn, A. W. Erickson, E. M. Harger, L. J. Verme, and J. J. Ozoga. 1976. Characteristics and management of black bears that feed in garbage dumps, campgrounds, or residential areas. *Third International Conference on Bears* 15:169–175.

Romin, L. A., and J. A. Bissonette. 1996. Deer-vehicle collisions: Status of state monitoring activities and mitigation efforts. *Wildlife Society Bulletin* 24:276–283.

Romme, W. H. 1997. Creating pseudo-rural landscapes in the mountain west. Pages 141–161 in J. I. Nassauer, editor. *Placing Nature: Culture and Landscape Ecology*. Island Press, Washington, D.C.

Rondeau, D., and J. M. Conrad. 2003. Managing urban deer. *American Journal of Agricultural Economics* 85:266–281.

Roni, P., T. J. Beechie, R. E. Bilby, F. E. Leonetti, M. M. Pollock, and G. R. Pess. 2002. A review of stream restoration techniques and a hierarchical strategy for prioritizing restoration in Pacific Northwest watersheds. *North American Journal of Fisheries Management* 22:1–20.

Rosemont Cooper. 2007. Mine plan of operations. <http://www.rosemontcopper.com/mpo_official.html>. Accessed April 29, 2010.

Ross, W. A. 1998. Cumulative effects assessment: Learning from Canadian case studies. *Impact Assessment and Project Appraisal* 16:267–276.

Rosen, P. C., and C. Funicelli. 2008. Conservation of urban amphibians in Tucson. Final report for Pima County Regional Flood Control District, Tucson, Arizona.

Rottenborn, S. C. 1999. Predicting the impacts of urbanization on riparian bird communities. *Biological Conservation* 88:289–299.

Rubbo, M. J., and J. M. Kiesecker. 2005. Amphibian breeding distribution in an urbanized landscape. *Conservation Biology* 19:504–511.

Rubin, E. S. W. M. Boyce, M. C. Jorgensen, S. G. Torres, C. L. Hayes, C. S. O'Brien, and D. A. Jessup. 1998. Distribution and abundance of bighorn sheep in the Peninsular Ranges, California. *Wildlife Society Bulletin* 26:539–551.

Rubin, E. S., C. J. Stermer, W. M. Boyce, and S. G. Torres. 2009. Assessment of habitat models for bighorn sheep in California's Peninsular Ranges. *Journal of Wildlife Management* 73: 859–869.

Rudd, H., J. Vala, and V. Schaefer. 2002. Importance of backyard habitat in a comprehensive biodiversity conservation strategy: A connectivity analysis of urban green spaces. *Restoration Ecology* 10:368–375.

Ruggerone, G. T. 1986. Consumption of migrating juvenile salmonids by gulls foraging below a Columbia River dam. *Transactions of the American Fisheries Society* 115:736–742.

Runge, M. C., J. F. Cochrane, S. J. Converse, J. A. Szymanski, D. R. Smith, J. E. Lyons, M. J. Eaton, A. Matz, P. Barrett, J. D. Nichols, M. J. Parkin, K. Motivans, and D. C. Brewer. 2009. *Introduction to Structured Decision Making*, fifth edition. U.S. Fish and Wildlife Service, National Conservation Training Center, Shepherdstown, West Virginia.

Rupp, T. S., M. Olson, L. G. Adams, B. W. Dale, K. Joly, J. Henkelman, W. B. Collins, and A. M. Starfield. 2006. Simulating the influences of various fire regimes on caribou winter habitat. *Ecological Applications* 16:1730–1743.

Russell, D. E., A. M. Martell, and W. A. C. Nixon. 1993. Range ecology of the porcupine caribou herd in Canada. *Rangifer Special Issue* No. 8:1–168.

Russell, D. E., R. G. White, and C. J. Daniel. 2005. Energetics of the Porcupine Caribou herd: A computer simulation model. Technical Report series No. 431. Canadian Wildlife Service. Ottawa, Ontario, Canada.

Russell, J. S. 2008. The case for including integrated land management within forest management plans: An opinion. *Forestry Chronicle* 84:369–374.

Rutberg, A. T. 1997. Lessons from the urban deer battlefront: A plea for tolerance. *Wildlife Society Bulletin* 25:520–523.

Saalfeld, S. T., and S. S. Ditchkoff. 2007. Survival of neonatal white-tailed deer in an exurban population. *Journal of Wildlife Management* 71:940–944.

Samarakoon, M., and J. S. Rowan. 2008. A critical review of environmental impact statements in Sri Lanka with particular reference to ecological impact assessment. *Environmental Management* 41:441–460.

Sasidharan, V., and B. Thapa. 2004. Ethnicity and variations in wildlife concerns: Exploring the socio-structural and socio-psychological basis of wildlife values. Pages 305–316 in W. W. Shaw, L. K. Harris, and L. VanDruff, editors. *Fourth International Symposium on Urban Wildlife Conservation*, Tucson, Arizona.

Sato, T., and Y. Harada. 2008. Loss of genetic variation and effective population size of Kirikuchi charr: Implications for the management of small, isolated salmonid populations. *Animal Conservation* 11:153–159.

Sato, T., and Y. Harada. 2010. Human-induced population fragmentation and management of small, isolated Kirikuchi charr populations. *Animal Conservation* 13:24–25.

Sauer, J. R., J. E. Fallon, and R. Johnson. 2003. Use of North American breeding bird survey data to estimate population change for bird conservation regions. *Journal of Wildlife Management* 67:372–389.

Savard, J. P. L., P. Clergeau, and G. Mennechez. 2000. Biodiversity concepts and urban ecosystems. *Landscape and Urban Planning* 48:131–142.

Schade, C. B., and S. A. Bonar. 2005. Distribution and abundance on nonnative fishes in streams of the western United States. *North American Journal of Fisheries Management* 25:1386–1394.

Schaefer, J. A. 2003. Long-term range recession and the persistence of caribou in the Taiga. *Conservation Biology* 17:1435–1439.

Schmidt, E., and C. E. Bock. 2005. Habitat associations and population trends of two hawks in an urbanizing grassland region in Colorado. *Landscape Ecology* 20:469–478.

Schmitt, C. J., C. A. Caldwell, B. Olsen, D. Serdar, and M. Coffey. 2002. Inhibition of erythrocyte δ-aminolevulinic acid dehydratase activity in fish from waters affected by lead smelters. *Environmental Monitoring and Assessment* 77:99–119.

Schneider, R. R., J. B. Stelfox, S. Boutin, and S. Wasel. 2003. Managing the cumulative impacts of land uses in the Western Canadian Sedimentary Basin: A modelling approach. *Conservation Ecology* 7:8.

Schroeder, M. A., C. L. Aldridge, A. D. Apa, J. R. Bohne, C. E. Braun, S. D. Bunnell, J. W. Connelly et al. 2004. Distribution of sage-grouse in North America. *Condor* 106:363–376.

Schroeder, M. A., J. R. Young, and C. E. Braun. 1999. Greater sage-grouse (*Centrocercus urophasianus*). A. Poole, and F. Gill, editors. *The Birds of North America* Number 425. The Birds of North America Incorporated, Philadelphia, Pennsylvania.

Schueler, T. R. 2000. The importance of imperviousness. *Watershed Protection Techniques* 1(3):100–111.

Schulze, W. D., and D. S. Brookshire. 1983. The economic benefit of preserving visibility in the national parklands of the Southwest. *Natural Resources Journal* 23:149–174.

Scottish Executive. 2005. SEA tool kit-offers guidance for the environmental assessment (Scotland) act. <http://www.scotland.gov.uk/Publications/2006/09/13104943/0>. Accessed October 10, 2009.

Seibert, H. C., and C. W. Hogen. 1947. Studies on a population of snakes in Illinois. *Copeia* 1:6–22.

Seip, D. R., C. J. Johnson, and G. S. Watts. 2007. Displacement of mountain caribou from winter habitat by snowmobiles. *Journal of Wildlife Management* 71:1539–1534.

Shargo, E. S. 1988. Home range, movements, and activity patterns of coyotes (*Canis latrans*) in Los Angeles suburbs. Dissertation, University of California, Los Angeles, California.

Sharp, R., and B. Walker. 2002. Particle Civics. Environmental Working Group, Washington, D.C., <http://www.ewg.org>. Accessed September 21, 2007.

Shaw, W. W., W. R. Mangug, and R. R. Lyons. 1985. Residential enjoyment of wildlife by Americans. *Leisure Science* 7:361–375.

Shaw, W. W., L. K. Harris, and L. VanDruff, editors. 2004. *Fourth International Symposium on Urban Wildlife Conservation*. Tucson, Arizona.

Shaw, W. W., R. McCaffry, and R. J. Steidl. 2009. Integrating wildlife conservation into land-use plans for rapidly growing cities. Pages 117–131 in A. X. Esparza and G. McPherson, editors. *The Planners Guide to Natural Resource Conservation: The Science of Land Development beyond the Metropolitan Fringe*. Springer Science + Business Media, New York.

Shea, K., and P. Chesson. 2002. Community ecology theory as a framework for biological invasions. *Trends in Ecology and Evolution* 17:164–170.

Shenstone, J. C. 1912. The flora of London building sites. *Journal of Botany* 50:117–124.

Shifley, S. R., F. R. Thompson, W. D. Dijak, and Z. F. Fan. 2007. Forecasting landscape-scale cumulative effects of forest fragmentation on vegetation and wildlife habitat: A case study of issues, limitations, and opportunities. *Forest Ecology and Management* 254:474–483.

Shively, K. J., A. W. Alldredge, and G. E. Phillips. 2005. Elk reproductive response to removal of calving season disturbance by humans. *Journal of Wildlife Management* 69:1073–1080.

Shuman, J. R. 1995. Environmental considerations for assessing dam removal alternatives for river restoration. *Regulated Rivers: Research and Management* 11:249–261.

Siemer, W. F., T. B. Lauber, L. C. Chase, D. J. Decker, R. J. McPeak, and C. A. Jacobson. 2004. Pages 228–237 in W. W. Shaw, L. K. Harris, and L. VanDruff, editors. *Fourth International Symposium on Urban Wildlife Conservation*. Tucson, Arizona.

Simberloff, D. 1998. Flagships, umbrellas, and keystones: Is single species management passé in the landscape era? *Biological Conservation* 83:247–257.

Sinclair, A. J, L. Sims, and H. Spaling. 2009. Community-based approaches to strategic environmental assessment: lessons from Costa Rica. *Environmental Impact Assessment Review* 29:147–156.

Skidds, D. E., F. C. Golet, P. W. C. Paton, and J. C. Mitchell. 2007. Habitat correlates of reproductive effort in wood frogs and spotted salamanders in an urbanizing watershed. *Journal of Herpetology* 41:439–450.

Smit, B., and H. Spaling. 1995. Methods for cumulative effects assessment. *Environmental Impact Assessment Review* 15:81–106.

Smith, C. M., and D. G. Wachob. 2006. Trends associated with residential development in riparian breeding bird habitat along the Snake River in Jackson Hole, WY: Implications for conservation planning. *Biological Conservation* 128:431–446.

Smith, D. O., M. Conner, and E. R. Loft. 1989. The distribution of winter mule deer use around homesites. *Transactions of the Western Section of the Wildlife Society* 25:77–80.

Smith, M. B., and J. S. Sharp. 2005. Growth, development, and farming in an Ohio exurban region. *Environment and Behavior* 37:565–579.

Smith, P. G. R. 2007. Characteristics of urban natural areas influencing winter bird use in southern Ontario, Canada. *Environmental Management* 39:338–352.

Soderman, T. A. 2006. Treatment of biodiversity issues in impact assessment of electricity power transmission lines: A Finnish case review. *Environmental Impact Assessment Review* 26:319–338.

Sonntag, N., R. Everitt, L. Rattie, D. Colnett, C. Wolf, J. Truett, A. Dorcey, and C. Holling. 1987. Cumulative effects assessment: A context for further research and development. Canadian Environmental Assessment Research Council, Vancouver, British Columbia.

Sorensen, T., P. D. McLoughlin, D. Hervieux, E. Dzus, J. Nolan, B. Wynes, and S. Boutin. 2007. Determining sustainable levels of cumulative effects for boreal caribou. *Journal of Wildlife Management* 72:900–906.

Sparks, D. W., C. M. Ritzi, J. E. Duchamp, and J. O. Whitaker. 2005. Foraging habitat of the Indiana bat (*Myotis sodalis*) at an urban-rural interface. *Journal of Mammalogy* 86:713–718.

Sparrowe, R. D., and P. F. Springer. 1970. Seasonal activity patterns of white-tailed deer in eastern South Dakota. *Journal of Wildlife Management* 34:420–431.

Spencer, R. D., R. A. Beausoleil, and D. A. Martorello. 2007. How agencies respond to human-black bear conflicts: A survey of wildlife agencies in North America. *Ursus* 18:217–229.

Spies, T. A., K. N. Johnson, K. M. Burnett, J. L. Ohmann, B. C. McComb, G. H. Reeves, P. Bettinger, J. D. Kline, and B. Garber-Yonts. 2007. Cumulative ecological and socioeconomic effects of forest policies in Coastal Oregon. *Ecological Applications* 17:5–17.

Stabler, L. B., C. A. Martin, and A. J. Brazel. 2005. Microclimates in a desert city were related to land use and vegetation index. *Urban Forestry and Urban Greening* 3:137–147.

Stakhiv, E. Z. 1991. A cumulative impact analysis framework for the U.S. Army Corps of Engineers regulatory program. Institute for Water Resources, U.S. Army Corps of Engineers, Fort Belvoir, Virginia.

Stankowich, T. 2008. Ungulate flight responses to human disturbance: A review and meta-analysis. *Biological Conservation* 141:2159–2173.

Starfield, A. M. 1997. A pragmatic approach to modeling for wildlife management. *Journal of Wildlife Management* 61:261–270.

Steidl, R. J., W. W. Shaw, and P. Fromer. 2009. A science-based approach to regional conservation planning. Pages 217–233 in A. X. Esparza and G. McPhearson, editors. *The Planners Guide to Natural Resource Conservation: The Science of Land Development beyond the Metropolitan Fringe*. Springer Science + Business Media, New York.

Steinitz, C., R. Anderson, H. Arias, S. Bassett, M. Flaxman, T. Goode, T. Maddock III, D. Mouat, R. Peiser, and A. Shearer. 2005. Alternative futures for landscapes in the Upper San Pedro River Basin of Arizona and Sonora. Pages 93–100 in J. C. Ralph and T. D. Rich, editors, *Bird Conservation Implementation and Integration in the Americas: Proceedings of the Third International Partners in Flight Conference 2002* March 20–24; Asilomar, California. Volume 1 General Technical Report PSW-GTR-191, U.S. Forest Service, Pacific Southwest Research Station, Albany, California.

Stewart, A. M., D. Keith, and J. Scottie. 2004. Caribou crossings and cultural meanings: placing traditional knowledge and archaeology in context in an Inuit landscape. *Journal of Archaeological Method and Theory* 11:183–211.

Storfer, A., M. A. Murphy, J. S. Evans, C. S. Goldberg, S. Robinson, S. F. Spear, R. Dezzani, E. Delmelle, L. Vierling, and L. P. Waits. 2007. Putting the "landscape" in landscape genetics. *Heredity* 98:128–142.

Storm, D. J., C. K. Nielsen, E. M. Schauber, and A. Woolf. 2007. Space use and survival of white-tailed deer in an exurban landscape. *Journal of Wildlife Management* 71:1170–1176.

Stull, E.A., K.E. LaGory, and W.S. Vinikour. 1987. Methodologies for assessing the cumulative environmental effects of hydroelectric developmental fish and wildlife in the Columbian River Basin. Volume one: Recommendations. Final Report 1946 1-3 to Bonneville Power Administration, Portland, Oregon, USA.

Sullivan, T. L., A. F. Williams, T. A. Messmer, L. A. Hellinga, and S. Y. Kyrychenko. 2004. Effectiveness of temporary warning signs in reducing deer-vehicle collisions during mule deer migrations. *Wildlife Society Bulletin* 32:907–915.

Suring, L., K. Barber, C. Schwartz, T. Bailey, W. Shuster, M. Tetreau. 1998. Analysis of cumulative effects on brown bears on the Kenai Peninsula, south central Alaska. *Ursus* 10:107–117.

Sutton, P. C., and R. Costanza. 2002. Global estimates of market and non-market values derived from nighttime satellite imagery, land cover, and ecosystem service valuation. *Ecological Economics* 41:509–527.

Swenson, J. E., C. A. Simmons, and C. D. Eustace. 1987. Decrease of sage grouse *Centrocercus urophasianus* after ploughing of sagebrush steppe. *Biological Conservation* 41:125–132.

Swihart, R. K., P. M. Picone, A. J. DeNicola, and L. Cornicelli. 1995. Ecology of urban and suburban white-tailed deer. Pages 35–44 in J. B. McAninch, editor. *Proceedings of the 1993 Symposium on the North Central Section of the Wildlife Society*. Bethesda, Maryland.

Tabor, R. A., G. S. Brown, and V. T. Luiting. 2004. The effect of light intensity on sockeye salmon fry migratory behavior and predation by cottids in the Cedar River, Washington. *North American Journal of Fisheries Management* 24:128–145.

Tack, J. E. 2009. Sage-grouse and the human footprint: implications for conservation of small and declining populations. Thesis, University of Montana, Missoula.

Talberth, J., C. Cobb, and N. Slattery. 2007. The genuine progress indicator 2006: A tool for sustainable development. Redefining Progress, Oakland, California.

Ternent, M. A., and D. L. Garshelis. 1999. Taste-aversion conditioning to reduce nuisance activity by black bears in a Minnesota military reservation. *Wildlife Society Bulletin* 27:720–728.

Thatcher, T. L. 1990. Understanding interdependence on the natural environment: Some thoughts on cumulative impact assessment under the National Environmental Policy Act. Environmental Law 20:611–633.

The Nature Conservancy. 2010. San Pedro River, Arizona. <http://www.nature.org/initiatives/freshwater/work/sanpedroriver.html>. Accessed March 16, 2010.

Theobald, D. M., J. R. Miller, and N. T. Hobbs. 1997. Estimating the cumulative effects of development on wildlife habitat. *Landscape and Urban Planning* 39:25–36.

Therivel, R. and B. Ross. 2007. Cumulative effects assessment: Does scale matter? *Environmental Impact Assessment Review* 27:365–385.

Thiel, D., S. Jenni-Eiermann, V. Braunisch, R. Palme, and L. Jenni. 2008. Ski tourism affects habitat use and evokes a physiological stress response in Capercaillie *Tetrao urogallus*: A new methodological approach. *Journal of Applied Ecology* 45:845–853.

Thomas, B. E. 2006. Trends in streamflow in the San Pedro River, southeastern Arizona. U.S. Geological Survey Fact Sheet 2006-3004.

Thomas, D. C., S. J. Barry, and G. Alaie. 1996. Fire-caribou-winter range relationships in northern Canada. *Rangifer* 16:57–67.

Thompson, G. 2005. Arizona minutemen driven largely by sense of insecurity, victimization. *The New Standard*, April 15, 2005.

Thorington, K. K., and R. Bowman. 2003. Predation rate on artificial nests increases with human housing density in suburban habitats. *Ecography* 26:188–196.

Tierson, W. C., G. F. Mattfeld, R. W. Sage, and D. F. Behrend. 1985. Seasonal movements and home ranges of white-tailed deer in the Adirondacks. *Journal of Wildlife Management* 49:760–769.

Timm, R. M., R. O. Baker, J. R. Bennett, and C. C. Coolahan. 2004. Coyote attacks: An increasing suburban problem. *Proceedings Vertebrate Pest Conference* 21:47–57.

Torchin, M. E., K. D. Lafferty, A. P. Dobson, V. J. McKenzie, and A. M. Kuris. 2003. Introduced species and their missing parasites. *Nature* 421:628–630.

Trautman, M. B. 1981. *The Fishes of Ohio*. Second edition. Ohio State University Press, Columbus.

Truett, J. C., H. L. Short, and S. C. Williamson. 1994. Ecological impact assessment. Pages 607–622 in T. A. Bookhout, editor. *Research and Management Techniques for Wildlife and Habitats*. Fifth edition. The Wildlife Society, Bethesda, Maryland.

Turner, J. C., C. L. Douglas, C. R. Hallum, P. R. Krausman, and R. R. Ramey. 2004. Determination of critical habitat for the endangered Nelson's bighorn sheep in southern California. *Wildlife Society Bulletin* 32:427–448.

U.S. Census Bureau. 2000. <http://quickfacts.census.gov/qfd/states/04/04003.html>. Accessed January 25, 2010.

U.S. Census Bureau. 2002. Statistical abstract of the United States. Washington, D.C.

U.S. Congress, Office of Technology Assessment. 1993. Harmful non-indigenous species in the United States. OTA-F-565. Washington, D.C.

U.S. Department of Agriculture. 1979. The mitigation symposium: A national workshop on mitigating losses of fish and wildlife habitats. G. A. Swanson, technical coordinator. General technical report RM–65. Rocky Mountain Forest and Range Experiment Station, Fort Collins, Colorado.

U.S. Department of Defense. 1992. Final report to Congress. Conduct of the Persian Gulf War. Washington, D.C.

U.S. Department of Energy. 2005. Tucson Electric Power Company Sahuarita-Nogales transmission line final environmental impact statement and comment response documents. <http://nepa.energy.gov/finalEIS-0336.htm>. Accessed April 30, 2010.

U.S. Department of Fish and Wildlife Service. 2006. 2006 national survey of fishing, hunting, and wildlife-associated recreation. U.S. Department of the Interior, Fish and Wildlife Service, U.S. Department of Commerce, and U.S. Census Bureau, Washington, D.C.

U.S. Department of Homeland Security.2008. <http://www.cbp.gov/linkhandler/cgov/newsroom/highlights/border_sec_news/dhs_waiver/segment_waiver.ctt/segment_waiver.pdf>. Accessed March 2010.

U.S. Department of the Interior, Fish and Wildlife Service and U.S. Department of Commerce, U.S. Census Bureau. 2001. National survey of fishing, hunting and wildlife-associated recreation. Washington, D.C.

U.S. Fish and Wildlife Service. 2010. Southwest Region Ecological Services. Threatened and Endangered Species List for Cochise County, ArizonaPhoenix, Arizona. <http://www.fws.gov/southwest/es/arizona/Documents/CountyLists/Cochise.pdf>. Accessed March 2010.

U.S. Forest Service. 1974. *National Forest Landscape Management*, Agriculture handbook 462, Washington, D.C.

U.S. Forest Service. 1995. *Landscape Aesthetics: A Handbook for Scenery Management*, Agriculture Handbook 701, Washington, D.C.

U.S. Forest Service. 1998. *Coronado National Forest land and resource management plan.* Coronado National Forest, Tucson, Arizona.

U.S. Forest Service. 2005. Values, attitudes and beliefs towards National Forest System lands: the Coronado National Forest. http://www.fs.fed.us/r3/coronado/plan-revision/plan-revision-documents.shtml.Accessed April 29, 2010.

U.S. Forest Service. 2008. *National visitor use monitoring results.* Coronado National Forest, Tucson, Arizona. <http://www.fs.fed.us/recreation/programs/nvum/>. Accessed April 29, 2010.

U.S. Forest Service. 2010. Coronado National Forest home page. <http://www.fs.fed.us/r3/coronado/>. Accessed April 29, 2010.

U.S. Senate Committee on Homeland Security and Governmental Affairs. 2009. Testimony of Larry A. Dever, Sheriff, Cochise County, Arizona, 20 April (Case: Border Security: Moving beyond the Virtual Fence).

United Nations. 2006. 2005–2015 Water for Life, Web Services Section, Department of Public Information. <http://www.un.org/waterforlifedecade/scarcity.html>.

United Nations. 2010. Environmental-Economic accounting. <http://unstats.un.org/unsd/envaccounting/default.asp>. Accessed January 10, 2010.

University of Arizona. 2006. Climate change and wildfire impacts on Southwest forests and woodlands. Arizona Cooperative, Tucson.

Usher, P. J. 2000. Traditional ecological knowledge in environmental assessment and management. *Arctic* 53:183–193.

Valdez, R., and P. R. Krausman. 1999. Description, distribution, and abundance of mountain sheep in North America. Pages 3–22 in R. Valdez and P. R. Krausman, editors. *Mountain Sheep of North America.* University of Arizona Press, Tucson.

Van der Ree, R., and M. A. McCarthy. 2005. Inferring persistence of indigenous mammals in response to urbanization. *Animal Conservation* 8:309–319.

Vargas, E. E., D. F. Schreiner, G. Tembo, and D. W. Marcouiller. 1999. Computable general equilibrium modeling for regional analysis. In *The Web Book of Regional Science*, second edition (www.rri.wvu.edu/regscweb.htm), S. Loveridge, editor, Regional Research Institute, West Virginia University, Morgantown, West Virginia.

Vistnes, I., and C. Nellemann. 2008. The matter of spatial and temporal scales: A review of reindeer and caribou response to human activity. *Polar Biology* 31:399–407.

Vogel, W. O. 1983. The effects of housing developments and agriculture on the ecology of white-tailed deer and mule deer in the Gallatin Valley, Montana. Thesis, Montana State University, Bozeman.

Vogel, W. O. 1989. Response of deer to density and distribution of housing in Montana. *Wildlife Society Bulletin* 17:406–413.

Vondracek, B., K. L. Blann, C. B. Cox, J. F. Nerbonne, K. G. Mumford, B. A. Nerbonne, L. A. Sovell, and J. K. H. Zimmerman. 2005. Land use, spatial scale, and stream systems: Lessons from an agricultural region. *Environmental Management* 36:775–791.

Vors, L. S., J. A. Schaefer, B. A. Pond, A. R. Rodgers, and B. R. Patterson. 2007. Woodland Caribou extirpation and anthropogenic landscape disturbance in Ontario. *Journal of Wildlife Management* 71:1249–1256.

Vors, L. S., and M. S. Boyce. 2009. Global declines of caribou and reindeer. *Global Change Biology* 15:2626–2633.

Wagner, D., and J. B. Jones. 2006. The impact of harvester ants on decomposition, N mineralization, litter quality, and the availability of N to plants in the Mojave Desert. *Soil Biology and Biochemistry* 38:2593–2601.

Wait, S., and H. McNally. 2004. Selection of habitats by wintering elk in a rapidly subdividing area of La Plata County, Colorado. Pages 200–209 in W. W. Shaw, L. K. Harris, and L. VanDruff, editors. *Fourth International Symposium on Urban Wildlife Conservation*, Tucson, Arizona.

Walker, B. H., and D. Salt. 2006. *Resilience Thinking: Sustaining Ecosystems and People in a Changing World*. Island Press, Washington, D.C.

Walker, B. L., D. E. Naugle, and K. E. Doherty. 2007a. Greater sage-grouse population response to energy development and habitat loss. *Journal of Wildlife Management* 71:2644–2654.

Walker, B. L., D. E. Naugle, K. E. Doherty, and T. E. Cornish. 2004. Outbreak of West Nile virus in greater sage-grouse and guidelines for monitoring, handling, and submitting dead birds. *Wildlife Society Bulletin* 32:1000–1006.

Walker, B. L., D. E. Naugle, K. E. Doherty, and T. E. Cornish. 2007b. West Nile virus and greater sage-grouse: Estimating infection rate in a wild population. *Avian Diseases* 51:691–696.

Walsh, C. J., A. H. Roy, J. W. Feminella, P. D. Cottingham, P. M. Groffman, and R. P. Morgan II. 2005. The urban stream syndrome: Current knowledge and the search for a cure. *Journal of the North American Benthological Society* 24:706–723.

Wärnbäck, A., and T. Hilding-Rydevik. 2009. Cumulative effects in Swedish EIA practice—Difficulties and obstacles. *Environmental Impact Assessment Review* 29:107–115.

Wang, L., J. Lyons, and P. Kanehl. 2001. Impacts of urbanization on stream habitat and fish across multiple spatial scales. *Environmental Management* 28:255–266.

Wanielista, M. P., and Y. A. Yousef. 1993. *Stormwater Management*. John Wiley & Sons, New York.

Warsh, D. 2006. *Knowledge and the Wealth of Nations*. W. W. Norton, New York.

Water Education Foundation and University of Arizona Water Resources Research Center. 2007. Layperson's guide to Arizona Water. Tucson, Arizona.

Way, J. G., S. M. Cifun, D. L. Eatough, and E. G. Strauss. 2006. Rat poison kills a pack of eastern coyotes, *Canis latrans*, in an urban area. *Canadian Field-Naturalist* 120:478–480.

Weaver, J. L., R. E. F. Escano, and D. S. Winn. 1987. A framework for assessing cumulative effects on grizzly bears. *Transactions of the North American Wildlife and Natural Resources Conference* 52:364–376.

Weaver, J. L., P. C. Paquet, and L. F. Ruggiero. 1996. Resilience and conservation of large carnivores in the Rocky Mountains. *Conservation Biology* 10:964–976.

Weber, D. A. 2004. Winter raptor use of prairie dog towns in the Denver, Colorado vicinity. Pages 195–199 in W. W. Shaw, L. K. Harris, and L. VanDruff, editors. *Fourth International Symposium on Urban Wildlife Conservation*, Tucson, Arizona.

Weiler, S. A., J. B. Loomis, R. B. Richardson, and S. A. Schwiff. 2003. Driving regional economic models with a statistical model: Possibilities for hypothesis tests on economic impacts. *Review of Regional Studies* 32(1): 97–111.

Weiss, M. H., and R. Figura. 2003. Provisional typology of highway economic development projects. *Transportation Research Record: Journal of the Transportation Research Board* 1839:115–119.

West Central Alberta Caribou Landscape Planning Team. 2008. West central Alberta caribou landscape plan. Unpublished report submitted to Alberta Caribou Committee Governance Board, Edmonton, Alberta, Canada.

Whittaker, R. H. 1967. Gradient analysis of vegetation. *Biological Reviews* 49:207–264.

Williams, R., D. Lusseau, and P. S. Hammond. 2009. The role of social aggregations and protected areas in killer whale conservation: The mixed blessing of critical habitat. *Biological Conservation* 142:709–719.

Winemiller, K. O., F. H. Lopez, D. C. Taphorn, L. G. Nico, and A. Barbarino-Duque. 2008. Fish assemblages of the Casiquiare River, a corridor and zoogeographic filter for dispersal between the Orinoco and Amazon basins. *Journal of Biogeography* 35:1551–1563.

Wolch, J., and U. Lassiter. 2004. Pages 255–263 in W. W. Shaw, L. K. Harris, and L. VanDruff, editors. *Fourth International Symposium on Urban Wildlife Conservation*, Tucson, Arizona.

Wolfe, S. A., B. Griffiths, and C. A. G. Wolfe. 2000. Response of reindeer and caribou to human activities. *Polar Research* 19:63–73.

Wood, C., and M. Dejeddour. 1992. Strategic environmental assessment: EA of policies, plans and programmes. *Impact Assessment Bulletin* 10:3–22.

Wood, W. E., and S. M. Yezerinac. 2006. Song sparrow (*Melospiza melodia*) song varies with urban noise. *Auk* 123:650–659.

Woodruff, R. A., and B. L. Keller. 1982. Dispersal, daily activity, and home range of coyotes in southeastern Idaho. *Northwest Science* 56:199–207.

Woodruff, T. J., J. Grillo, and K. C. Schoendorf. 1997. The relationship between selected causes of post-neonatal infant mortality and particulate air pollution in the United States. *Environmental Health Perspectives* 105(6): 608–613.

Wu, J., R. Adams, and W. Boggess. 2000. Cumulative effects and optimal targeting of conservation efforts: Steelhead trout habitat enhancement in Oregon. *American Journal of Agricultural Economics* 82:400–413.

Wydoski, R. S., and R. W. Wiley. 1999. Management of undesirable fish species. Pages 403–430 in C. C. Kohler and W. A. Hubert, editors. *Inland Fisheries Management in North America*. American Fisheries Society, Bethesda, Maryland.

Yamasaki, S. H., R. Duchesneau, F. Doyon, J. S. Russell, and T. Gooding. 2008. Making the case for cumulative impacts assessment: Modelling the potential impacts of climate change, harvesting, oil and gas, and fire. *Forestry Chronicle* 84:349–368.

Yoder, J. M., E. A. Marschall, and D. A. Swanson. 2004. The cost of dispersal: Predation as a function of movement in ruffed grouse. *Behavioral Ecology* 15:469–476.

Zalatan, R., A. Gunn, and G. H. R. Henry. 2006. Long-term abundance patterns of barren-ground caribou using trampling scars on roots of *Picea mariana* in the Northwest Territories, Canada. *Arctic, Antarctic and Alpine Research* 38:624–630.

Zhu, D., and J. Ru. 2008. Strategic environmental assessment in China: Motivations, politics, and effectiveness. *Journal of Environmental Management* 88:615–626.

Zipperer, W. C., J. G. Wu, R. V. Pouyat, and S. T. A. Pickett. 2000. The application of ecological principles to urban and urbanizing landscapes. *Ecological Applications* 10:685–688.

Index

F

G

H

M

N

O

P

FSC
www.fsc.org
MIX
Papier | Fördert
gute Waldnutzung
FSC® C083411

Kolibri 360 GmbH
Ferdinand-Jühlke-Straße 7
99095 Erfurt, Deutschland
produktsicherheit@kolibri360.de